CHEMISTRY RESEARCH AND APPLICATIONS

ORGANIC RADICAL REACTIONS IN WATER AND ALTERNATIVE MEDIA

Chemistry Research and Applications

Additional books in this series can be found on Nova's website
under the Series tab.

Additional E-books in this series can be found on Nova's website
under the E-books tab.

CHEMISTRY RESEARCH AND APPLICATIONS

ORGANIC RADICAL REACTIONS IN WATER AND ALTERNATIVE MEDIA

J. ALBERTO POSTIGO
EDITOR

Nova Science Publishers, Inc.
New York

For permission to use material from this book please contact us:
Telephone 631-231-7269; Fax 631-231-8175
Web Site: http://www.novapublishers.com

Library of Congress Cataloging-in-Publication Data

Organic radical reactions in water and alternative media / [edited by] J. Alberto Postigo.
p. cm.
Includes index.
ISBN 978-1-61209-648-3 (hardcover)
1. Free radical reactions. 2. Chemical bonds. 3. Free radicals (Chemistry) I. Postigo, J. Alberto.
QD471.O695 2011
547'.2--dc22

2011002580

Published by Nova Science Publishers, Inc. □ + New York

DEDICATION

To my parents José Edmundo and Maria Re who always encouraged me and supported me unconditionally

CONTENTS

ACKNOWLEDGMENTS

I wish to thank William J. Leigh for being my first mentor as supervisor of my Ph.D. thesis, in the Department of Chemistry, Faculty of Science, at McMaster University, Hamilton, Canada, who introduced me to the field of reactive intermediates.

I also wish to thank Roberto A. Rossi, from the Departamento de Química Orgánica, Facultad de Ciencias Químicas, Universidad Nacional de Córdoba, Argentina for bringing up his enthusiasm and illustrating me on the subject of radicals and radical ions in Chemistry. When I arrived in his laboratory, little didI know about electron transfer reactions. His advice and expertise helped me in the taking up of future projects, and in the understanding of basic principles.

I also wish to thank Chryssostomos Chatgilialoglu, for having introduced me to the subject of Radicals in Water, and all the training and experience I received in his laboratory, at the Istituto per la Sintesi Organica e la Fotoreattivita, ISOF, from the ConsiglioNazionalledelleRicerche, CNR, in Bologna, Italy. I greatly benefitted from his vast experience on silyl radicals reactivity.

I wish to thank especially to all contributors of this book, for allowing me to present to the reader a focus on Radical Chemistry in Water and non-Conventional media with an up-to-date glance of this vast and active area of research: Tamara Perchyonok, Ioannis Lykakis, and Stanislav A. Grabovskiy.

Finally I do wish to thank all my collaborators in these last years:Norma SbarbatiNudelman, ChryssostomosChatgilialoglu, Carla Ferreri, Santiago Vaillard, Sergey Kopsov, Sebastián Barata Vallejo, Julia Calandra, and Roberto A. Rossi.

INTRODUCTION

This book is intended for the practical synthetic organic chemist whose interest is devoted to studying radical reactions of synthetic utility in water and non-conventional media.

Gomberg's epoch-making discovery of the existence of the stable triphenyl methyl radical in 1900 marked the genesis of a new domain, radical chemistry, in the vast realm of

Organic Chemistry. Almost four decades later, significant contributions to the development of this area were made by Hey and Waters and Kharasch, who carried out elaborate studies on the mechanisms of radical reactions.

In spite of a clear understanding of the mechanistic background, radical reactions found little application in synthesis, largely due to the erroneous notion that they are prone to give intractable mixtures. A dramatic change in this outlook which triggered an upsurge of interest in this approach, particularly over the past three decades, can be attributed to the conceptualization and demonstration by Stork that the controlled generation and addition of vinyl radicals to olefins constitutes a unique and powerful method for complex carbocyclic construction. It is noteworthy that the investigations by Julia, Beckwith, Ingold, and Giese have contributed to a deeper understanding of the structure and reactivity of radicals. A number of others, most notably Curran, Chatgilialoglu, Crich. and Pattenden, have made significant contributions to the application of radical methodology in organic synthesis. Today, radical methodology has evolved as a prominent tool in the arsenal of the synthetic organic chemist.

During the last twenty years, an increase attention to reactions carried out in friendly, economic media has led radical chemists to focus on water, whose research in organic radical chemistry had allowed them to accumulate enough knowledge of radical reactivity in organic solvents.

Water is an ideal system and support for radical reactions, since the H-OH bond strength is significantly-high enough not to react with carbon-centered and other radicals, under standard conditions. Only recently, by the assistance of certain transition metals such as titanium, has the H-OH bond from water been able to cleave homolytically by common organic radicals. Thus efficient hydrogen-atom transfer reactions have been reported from water-Ti(III) complexes and alkyl radicals (Newcomb and Oltra).

Except for these interesting examples reported of hydrogen donation from water, it is considered under ordinary conditions to be inert to organic carbon radicals. However inert water is, large solvents effects are noticeable in radical reactions carried out in water when compared to organic solvents.

The logical progression of research in this area fostered radical methodologies including surfactants, water-soluble radical initiators and water-soluble chain-carriers that enabled the use of water or aqueous solvent systems, thus reducing the environmental impact. The present paradigm relies on the expectation to replace all organic radical reactions performed in organic solvents by water, or water-content media. Reference material that has been dealt with in review chapters is not treated, but properly cited and referenced.

The book is divided into several chapters, thus covering different aspects of Synthetic Radical Chemistry in Water and non-conventional media. Emphasis is made throughout the book of synthetic methods, not covering other important aspects of radical chemistry in water, such as Computational studies aimed at clarifying the powerful effect of water as a selective solvating media for radical chemistry, and reasons for the higher selectivity and reaction yields when water is employed as solvent. Aspects covering studies of rate constants of radical reactions in water are briefly summarized but not treated in depth, although proper citation is given. Notably, prestigious researchers in this field, Ingold, Giese, Curran, Chatgilialoglu, Newcomb, have advanced our understanding of rates of radical reactions in organic solvents through the use of radical-clock reactions, laser flash photolysis and radiolysis methods, thus allowing us to extrapolate data for some of these reactions in non-conventional media.

The chapter dedicated to Carbon and Sulfur Centered Radicals in water, Chapter 1, intends to cover recent literature, up to 2010, on these types of radicals, with a synthetic goal, thus helping the Radical Synthetic Chemist to grab the fundamental aspects devoted to the synthesis of a wide array of organic compounds through radical methods in water.This chapter excludes the syntheses of organic substrates through the aid of organometallic-centered radicals, which is the subject of successive chapters of this book. The array of substrates synthesized through this method includes radical cyclization reactions, atom-transfer radical cyclization, aryl radical addition to double bonds, carbon-carbon radical bond formation reactions, radical addition to carbon-nitrogen double bonds, and carboaminohydroxylation of alkenes, and the analogues of Knovonoegel reactions, synthesis of functionalized enol ethers, carboazidation and azidation, and ynol formation. The end of Chapter I, tries to rationalize the rate acceleration effects observed when radical reactions are performed in water, as compared to organic solvents. For doing so, some examples are presented to the reader, where the transition states for radical reactions show enhanced stability in water as compared to the transition states of the same reactions in other media, a contender for rate increase.

Chapter II is concerned with the chemistry of silicon-centered radicals in water. The attention devoted to silicon reactions in wateris not casual, given the need to replace the known reactivity of tin radicals in water as fast reductive agents. Thus, a series of interesting transformations employing silyl radicals were recently uncovered. With the aid of "polarity reversal methods" similar rates of radical reactivity were achieved with silyl radicals as compared with tin radicals. These transformations (with silyl radicals) encompass organic halide reductions, deoxygenation of organic alcohols, azide-to-amine reduction reactions, hydrosilylationof carbon-carbon and carbon-heteroatom multiplebonds, silyl radical-induced cyclization reactions, and more recently, consecutive radical addition reactions of perfluoroalkyl radicals onto multiple bonds, mediated by silyl radicals in water. In doing so, the radical triggering events for production of silicon-centered radicals are discussed.

As for radical synthetic reactions carried out in water with the aid of organometallic-centered radicals (Chapter III), emphasis is again made on the types of organic transformations achieved by these metallic radicals.

Metals have been used in organic synthesis for many decades, employing organic solvents. Since the last two decades, however, chemists have gradually realized that in dealing with transition metal radicals, water and aqueous media can fairly well replace organic solvents. Furthermore, reaction yields employing metal-centered radicals are improved when the reactions are carried out in water than in organic solvents, enhancing the scope of metal-mediated radical reactions in water.

The radical transformations employing metals encompass Reformatzky –type reactions, alkylation and allylations of both carbonyl compounds and imine derivatives, radical conjugate addition reactions, metal-mediated Mannich-type reactions and cyclization reactions, pinacol coupling reactions, metal-mediated reduction and oxidation reactions performed in water, and miscellaneous reactions. Radical triggering events can be spontaneous single electron transfer processes in water, thermal-induced decomposition of azo compounds, or the mediation of oxygen as a radical initiator. Particularly, the low reduction potential of various elements (i.e.: indium, gallium, zinc, Group II, and several low-valent elements) that act as excellent electron donors has awaken a genuine interest in using these metals for several types of carbon-carbon bond formation reactions. This has triggered an active research in the area in the last ten years.

The array of metal-centered radicals employed for these latter transformations encompasses main-group elements (Groups XIII, XIV), and several transition metals.

In Chapter IV, radical reactions performed in alternative media are discussed. These alternative media, supercritical carbon dioxide fluids and radical reactions in ionic liquids broaden the scope of radical reactions for the synthetic chemist. Also, radical fluorous technologies are introduced as a means of facilitating rapid product separations and easiness in extraction and purification techniques, thus permitting the synthetic chemist to bypass difficult separation and purification protocols resorting to multiphase chemistry.

Chapter V deals with combinatorial polymers and enzyme mimics and other types of catalysis. These state of the art approaches shows a new facet of radical chemistry and assets to the general scope achieved by these intermediates, both for laboratory organic chemists and for biochemists as well.

ACRONYMS AND ABBREVIATIONS

ABAP	azobis-(amidinopropane)dihydrochloride
ABCVA	4,4′-azobis(4-cyanovaleric acid)
ACCN	1,1′-azobis(carbonitrilecyclohexane)
AcOH	acetic acid
ACTR	addition-cyclization-trap reaction
AIBN	azo(*iso*butyro)nitrile
Ar	aromatic
atm	atmosphere
ATRA	atom transfer radical addition
ATRP	atom transfer radical polymerization
BNP	2-bromo-2-nitropropane
BR	bilirubin
BTF	benzotrifluoride
t-BuOOH	*tert*-butylhydroperoxide
t-BuO•	*tert*-buthoxide radical
CAN	cerium ammonium nitrate
CD	cyclodextrin
CLIP	cross-linked imprinted proteins
concd	concentrated
CTAB	cetyltrimethyl ammonium bromide
ΔGcat	Catalyzed free energy
ΔG_{uncat}	Uncatalyzed free energy
ΔG_{catETS}	Gibbs energy for the transition state
DBM	dibenzoylmcthanc
DEPO	diethylphosphine oxide
DMF	dimethylformamide
DMPA	2,2-dimethoxy-2-phenyl acetophenone
DMSO	dimethyl sulfoxide
dr	diastereoselectivity
Ea	activation energy
EPHP	1-ethylpiperidine hypophosphite
eq	equation
equiv	equivalent

Ered	reduction potential
ESR	electron spin resonance
Et	ethyl
Et_3B	triethylborane
EtOH	ethanol
EtSiH	triethylsilane
Et_2O	diethyl ether
$Eu(Otf)_3$	europium triflate
Ev	electron volt
EWG	electron-withdrawinggroup
GPX	glutathione peroxidase
h	hour
HAT	halogen atom transfer
HOO·	hydroperoxyl radical
HPLC	high performance liquid chromatography
hrs	hours
IR	infrared
KIE	kinetic isotope effect
LFP	laser flash photolysis
LUMO	lowest unoccupied molecular orbital
M	monomer
Me	methyl
MeCN	acetonitrile
MeOH	methanol
$(Me_3Si)_3SiH$	tris(trimethylsilyl)silane
Min	minute
MIP	molecular imprinted polymer
MOM	methoxy methyl ether
nBu	n-butyl
NBS	N-bromosuccinimide
NHPI	N-hydroxyphthalimide
NMR	nuclear magnetic resonance
OBz	benzyloxy group
OTBDMS	tert-butyl dimethyl silyloxy group
PAD	poly(aminomethylstyrene-co-divinylbenzene)
P_c	critical pressure
PEI	poly(ethylenimine)
PINO	phthalimide N-oxyl
PPTS	
PV	phase vanishing
R_fX	perfluoroalkyl halides
R_fI	perfluoroalkyl iodide
R_fBr	perfluoroalkyl bromide
SCF	supercritical fluids
$scCO_2$	supercritical carbon dioxide
SDS	sodium dodecyl sulfate

SET	single electron transfer
S_Hi	intramolecular substitution mechanism
S_N2	bimolecular nucleophilic substitution mechanism
SOMO	singly occupied molecular orbital
r.t.	room temperature
RCA	radical conjugate addition
RLi	organolithium reagent
RMgX	Grignard reagent
T	template molecule
TBHP	tert-butyl hydroperoxide
T_c	critical temperature
TEMPO	tetramethylpiperidinyl-1-oxy
THF	tetrahydrofuran
TLC	thin layer chromatography
$(TMS)_3SiH$	tris(trimethylsilyl)silane
TSA	transition state analogue
v/v	volumen in volume
VA-044	(1,2-bis(2-(4,5-dihydro-1H-imidazol-2-yl)propan-2-yl)diazenehydrochloride
VCP	vinylidenecyclopropane
Vic	vicinal
vs.	versus

In: Organic Radical Reactions in Water and Alternative Media ISBN 978-1-61209-648-3
Editor: J. Alberto Postigo, pp. 1-43

Chapter 1

CARBON AND SULFUR CENTERED RADICALS IN WATER

J. Alberto Postigo[*]
Department of Chemistry-Faculty of Science,University of Belgrano,
Villanueva 1324 CP 1426-Buenos Aires, Argentina

ABSTRACT

This chapter is dedicated to carbon- and sulfur-centered radicals in water. Chapter I, intends to cover recent literature, up to 2010, on these types of radicals, with a synthetic goal, thus helping the Radical Synthetic Chemist to grab the fundamental aspects devoted to the synthesis of a wide array of organic compounds through radical methods in water.This chapter excludes the syntheses of organic substrates through the aid of organometallic / metal-centered radicals, which is the subject of successive chapters of this book. The array of substrates synthesized through this method includes radical cyclization reactions, atom-transfer radical cyclization, aryl radical addition to double bonds, carbon-carbon radical bond formation reactions, radical addition to carbon-nitrogen double bonds, and carboaminohydroxylation of alkenes, and the analogues of Knovonoegel reactions, synthesis of functionalized enol ethers, carboazidation and azidation, and ynol formation. Section 9 of this Chapter intends to throw some light on the rates of radical reactions in water, exploring acceleration effects, and stabilization of intermediates.

I. INTRODUCTION

More general compound reaction mechanisms lead to systems of differential equations of different orders. They can sometimes be treated by applying a quasi-stationary or a quasi-equilibrium approximation. Often, this may even work for simple chain reactions. Chain

[*]Corresponding Author:Dr. Al Postigo. Fax: (+)54 11 4511-4700. E-mail: alberto.postigo@ub.edu.ar

reactions generally consist of four types of reaction steps: In the *chain initiation*steps, reactive species (radicals) are produced from stable species (reactants or catalysts). They react with stable species to form other reactive species in the *chain propagation.* Reactive species recovered in the chain propagation steps are called *chain carriers* (see Chapter II). Propagation steps where one reactive species is replaced by another less-reactive species are sometimes termed *inhibiting. Chain branching*occurs if more than one reactive species are formed. Finally, the chain is *terminated* by reactions of reactive species, which yield stable species, for example through recombination in the gas phase or at the surface of the reaction vessel. In most chain reactions, radicals are the dominant reactive species involved. These species are considered to haveunpaired electrons and some in a singly occupied orbital.

Although radicals are usually viewed as neutral, relatively nonpolar species, polar effects in radical chemistry are well known and continue to be exploited in organic synthesis.For example, nucleophilic radicals, or radicals with a singly occupied molecular orbital (SOMO) of relatively high energy, prefer to react with electron-deficient alkenes. In this case, the electron-withdrawing group attached to the alkene lowers the energy of the lowest unoccupied molecular orbital (LUMO), thereby permitting greater overlap between the SOMO of the radical and the LUMO of the alkene. This principle is illustrated by the work of Della, in which the 5-carbomethoxy-5-hexenyl radical preferred cyclization via the 6-endo mode rather than the typically favored 5-exo pathway.

The chemistry of free radicals has undergone a massive renaissance over the past thirty years. The real burst in synthetic applications arose from the use of trialkyltin and triaryltin hydrides as radical-chain carriers. However, the toxicity of the tin reagents, coupled with the difficulty in separation of their byproducts from the desired reaction products, meant that they were never acceptable to the pharmaceutical industry. Although improved methods of separation and operation have been developed, there is still a reluctance to use toxic tin reagents. This is unfortunate, since the properties of free radicals-*e.g.* lack of solvation (useful in assembling congested quaternary carbons) and predictable kinetics regardless of the reaction solvent (useful for predicting the relative speed of desired *vs.* side-reactions)-give them a unique advantage in certain synthetic maneuvers.

In radical chain processes, the initial radicals are generated by some initiation. In organic solvents, the most popular initiator is 2′,2′-azobis*iso*butyronitrile (AIBN), with a half-life of 1 h at 81 °C, generating the incipient radicals that commence the radical chain reaction. Other azo-compounds are used from time to time as well as the thermal decomposition of di-*tert*butyl peroxide depending on the reaction conditions. Triethylborane (Et_3B) in the presence of very small amounts of oxygen is an excellent initiator for low temperature reactions (down to – 78 °C). Also air-initiated reactions have recently been reported in aqueous and neat mixtures. However, in water, the radical initiation varies, and several studies have been undertaken to assess the best methodology. In the last decade, Et_3B/ dioxygen has also been used as radical initiator in water.

Tris(trimethylsilyl)silane, $(Me_3Si)_3SiH$, as a pure material or in organic solution reacts spontaneously and slowly at ambient temperature with molecular oxygen from air, to form the siloxane. The mechanism of this unusual process has been studied in some detail (Chapter II). Absolute rate constants for the spontaneous reaction of $(Me_3Si)_3SiH$ with molecular oxygen have been determined to be 3.5×10^{-5} M^{-1} s^{-1} at 70 °C, and theoretical studies elucidated the reaction coordinates. The couple $(Me_3Si)_3SiH$ / dioxygen, has lately been used asradical triggering event in water.In organic solvents (benzene), Curran et al. reported the $(Me_3Si)_3SiH$ / dioxygen-mediated radical initiation reaction in the absence of an azo compound or light. Thus, it was shown that $(Me_3Si)_3SiH$ mediates in the radical addition reaction of aryl iodides to benzene, and the rearomatization is achieved through oxygen. Silyl radicals, generated from reaction of $(Me_3Si)_3SiH$ with oxygen, abstract the iodine atom from iodobenzene generating aryl radicals that suffer intramolecular addition to benzene (solvent) to form the cyclohexadienyl radical. Oxygen-induced rearomatization affords products along with hydroperoxyl radicals (HOO•). This radical abstracts hydrogen from the silane yielding $(Me_3Si)_3Si$• radicals completing the chain reaction. Also, recently, $(Me_3Si)_3SiH$ / dioxygen has been applied for the radical initiation reactions in water (Chapter II).

Following the progressive exploration of radical reactions in non-conventional media, water has been adopted as the solvent of choice as advantages such as rate acceleration, stereoselectivity, and regioselectivity have been observed when radical reactions are performed in this latter medium as compared to organic solvents.

1. Radical Carbon-carbon Bond Formation in Water

1.1. Radical Atom Transfer IntermolecularCarbon-carbon Bond Formation in Water

Halogen atom-transfer (HAT) has been extensively studied and widely used in organic synthesis. The addition of a carbon-halogen bond across a double bond was pioneered by Kharasch (Scheme 1) and provides new carbon-carbon and carbon-halogen bonds in a single operation. The choice of the halogen that transfers in the reaction is crucial for the success of atom-transfer additions.

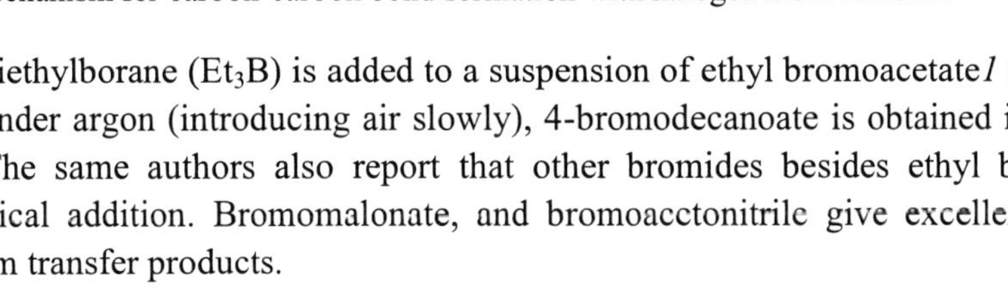

R• + R′ ⟶ R R′ •

RX + R R′ • ⟶ R• + R R′ X

Scheme 1. Mechanism for carbon-carbon bond formation with halogen atom transfer.

When triethylborane (Et_3B) is added to a suspension of ethyl bromoacetate*1* and 1-octene *2* in water under argon (introducing air slowly), 4-bromodecanoate is obtained in 80% yield (eq 1).[1] The same authors also report that other bromides besides ethyl bromoacetate undergo radical addition. Bromomalonate, and bromoacctonitrilc give excellent results of bromine atom transfer products.

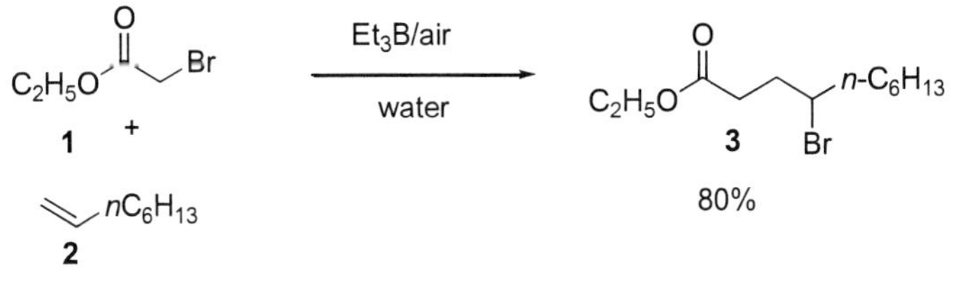

C_2H_5O O Br **1** + nC_6H_{13} **2** —Et_3B/air, water⟶ C_2H_5O O n-C_6H_{13} Br **3** 80% (1)

A tandem radical addition-oxidation sequence which converts alkenylsilanes into ketones has been elegantly described by Oshima et al.[2]

When 2-silyl-1-alkenes are made to react with an excess amount of Et_3B (2 equiv) in air, methyl ketonesare afforded as major products as shown in Scheme 2 below. Water was found to be an excellent solvent. The addition of ammonium chloride is important as without it, the yields drop substantially.

When the methyl-substituted alkenylsilane*4* is allowed to react with benzyl-2-iodoacetate under the reaction conditions described above, compound *5* is obtained in 80% yield (Scheme 2). Analogously, from phenyl-substituted alkenylsilane*6* and benzyl-2-iodoacetate, compound *7* is obtained in 77% yield. When butyl-2-iodoacetate and 2-iodo-1-phenylethanone react with *6*, compounds*8* and *9* are obtained in 74% and 84% yields, respectively. When iodoperfluoro-*n*hexane is allowed to react with *6*, compound *10* is obtained in 81% isolated yield. The reaction of methylformate-substituted alkenylsilane*11* with benzyl-2-iodoacetate affords product *12* in 61% yield (Scheme 2).

The authors[2] propose a plausible reaction mechanism such as that depicted in Scheme 3.

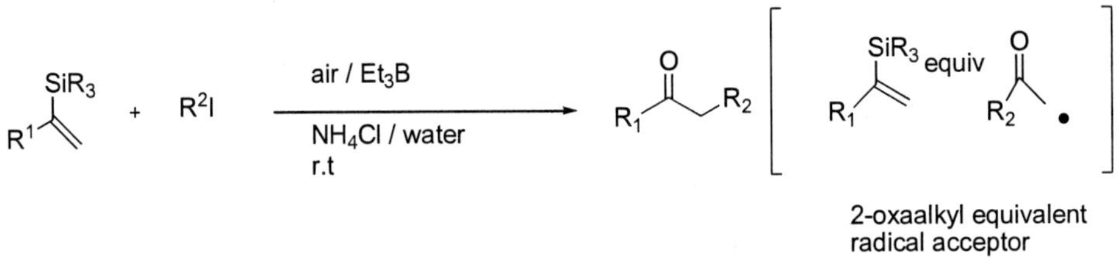

air / Et_3B; NH_4Cl / water; r.t

2-oxaalkyl equivalent radical acceptor

4 R^1 = Me, R^2 = CH_2COOBn — **5, 80%**
6 R^1 = Ph, R^2 = CH_2COOBn — **7, 77%**
6 R^1 = Ph, R^2 = $CH_2COOnBu$ — **8, 74%**
6 R^1 = Ph, R^2 = CH_2COPh — **9, 84%**
6 R^1 = Ph, R^2 = C_6F_{13} — **10, 81%**
11 R^1 = CO_2Me, R^2 = CH_2COOBn — **12, 61%**

Scheme 2. Synthesis of ketones from silylalkenes and alkyl iodides in water.

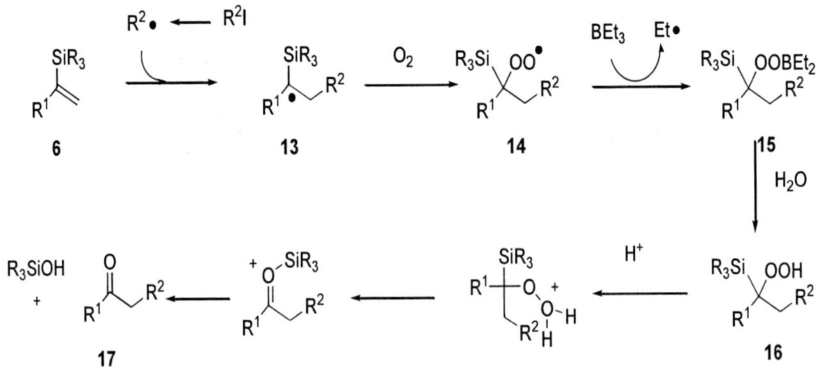

Scheme 3. Reaction mechanism of Et3B-induced synthesis of ketones from alkenylsilanes and iodoalkanes in water.

Addition of a radical arising from benzyl-2-iodoacetate, for instance, to alkenylsilane*6*, affords an α-silyl radical *13*, which then reacts with dioxygen, to afford peroxyl radical *14*. The reaction of radical *14* with Et_3B furnishes peroxyborane*15*. Hydroperoxy*16*, derived from hydrolysis of *15*, is eventually converted to carbonyl compound *17*, through migration of the silyl group to the internal oxygen atom. The ethyl radical which results from reaction *14* → *15*, regenerates an alkyl radical from RI (Scheme 3).

Radical Carbon-carbon Bond Formation from Imines in Water

The carbon-nitrogen double bond of imine derivatives has emerged as a radical acceptor, and thus numerous synthetically useful carbon-carbon bond-forming reactions are available (see also Chapter III).

Naito et al.[3] utilized oximes, oxime ethers, hydrazones, and nitrones as radical acceptors for construction of new carbon-carbon bonds.

Thus, when the glyoxylicoxime ether *18* is added to *iso*propyl iodide in a methanol / water mixture and Et_3B as initiator, a quantitative conversion to *19* is obtained (eq 2).

MeO_2C–CH=NOBn (i-PrI, Et_3B; H_2O:MeOH 1:1) → MeO_2C–CH(i-Pr)–NHOBn (2)

18 **19**

Conventional condensation of glyoxylic acid hydrate *20* with benzyloxyamine hydrochloride proceeds smoothly in water. Subsequently, alkyl iodide (RI) and Et_3B are added to the reaction vessel to afford excellent yields of α-amino acid derivatives*21a-d* (eq 3) after the purification. It should be noted that the one-pot reactions in water are much more effective compared to the reactions carried out in organic solvents such as toluene and methylene chloride.[3]

HO_2C–CH(OH)–OH (1.-$BnONH_2$.HCl; 2.-RI; 3.-Et_3B; H_2O) → HO_2C–CH(R)–NHOBn (3)

20 **21a-d**

a:R=i-Pr (97%)
b:R=c-hexyl (99%)
c:R=s-Bu (95%)
d:R=c-pentyl (97%)

Other primary and secondary alkyl radical additions have also been attempted successfully on glyoxylicoxime ethers. Scheme 4 depicts the scope of glyoxylicoxime ethers as excellent radical acceptors in water / methanol mixtures.[4]

When glyoxylicoxime ether *22* reacts in a methanol / water mixture with dihydro-3-iodofuran-2(*3H*)-one *23*, in the presence of Et_3B, product *24* is obtained in 82% yield, with a diastereomeric ratio equal to 2:1 (Scheme 4). Iodoethylacetate, under the same reaction conditions affords product *25* in 76% yield, while benzyl iodide, affords product *26* in 62% yield.[4]

Ketimines have also been used as alkyl radical acceptors in water to construct new carbon-carbon bonds.[5] Especially ketimines having *N*-heteroatom substituents stabilizing the intermediate aminyl radical or the *N*-alkyl group, in the intermolecular radical reaction.

The aqueous-medium alkyl-radical addition to hydrazone*27* proceeds efficiently in the presence of alkyl iodide and Et_3B / dioxygen as initiator (eq 4) to afford *28*. The aqueous-medium reaction of isatinhydrazone*29* affords the product *30* (eq 5).

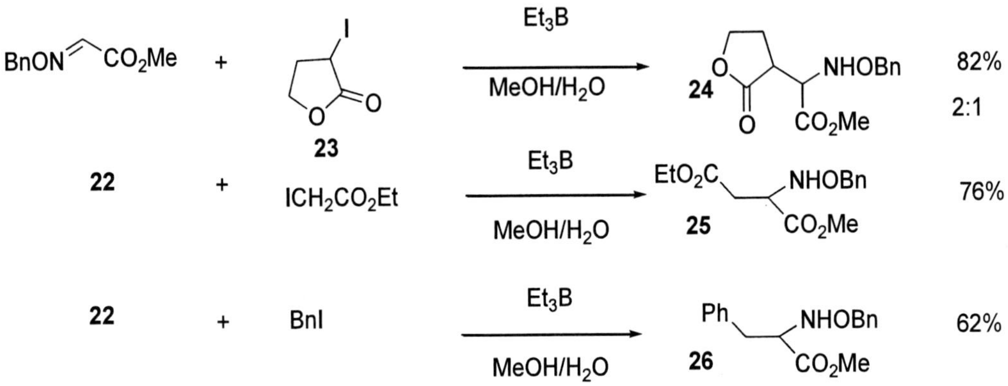

BnON=CHCO$_2$Me (**22**) + **23** → (Et_3B, MeOH/H_2O) → **24**, 82%, 2:1

22 + ICH_2CO_2Et → (Et_3B, MeOH/H_2O) → **25**, 76%

22 + BnI → (Et_3B, MeOH/H_2O) → **26**, 62%

Scheme 4. Reaction of glyoxylicoxime ethers with iodocompounds in water.

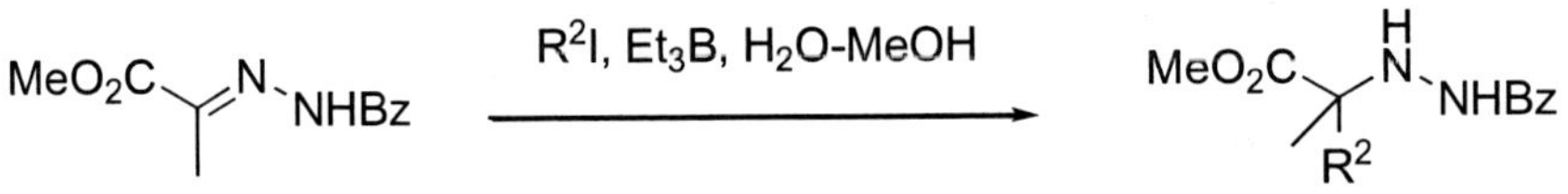

27 → (R^2I, Et_3B, H_2O-MeOH) → **28** (4)

28a:R^2=*i*-Pr, 88%
28b:R^2=*c*-hexyl, 91%
28c:R^2=*t*-Bu, 51%

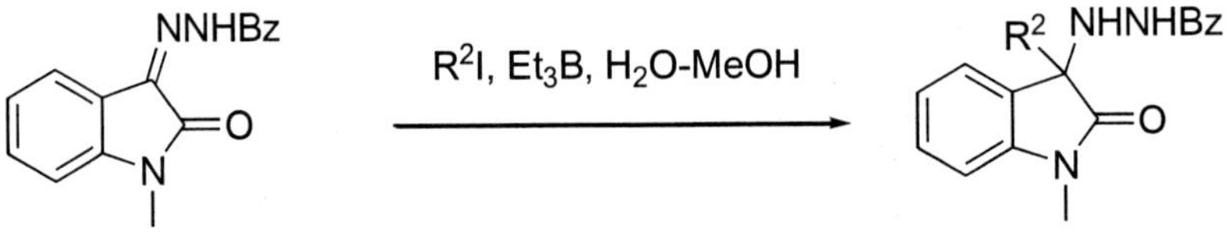

29 → (R^2I, Et_3B, H_2O-MeOH) → **30** (5)

29a:R^2=*i*-Pr
29b:R^2=*c*-hexyl
29c:R^2=*t*-Bu

30a, 73%
30b, 72%
30c, 40%

1.3. KnoevenagelReaction in Aqueous Media

Carbon-carbon bond formation through a free radical pathway has led to a variety of useful applications in organic synthesis (*vide supra*). Thus, a series of functionalized reactions of ylidenemalonitriles prepared *in-situ* by reaction of carbonyl compounds and malononitrile mediated by Et_3B in the presence of ammonium acetate in aqueous media (biphasic with

Et_2O)have been attempted to render compounds *34* in 80-96% yields with an array of carbonyl compounds *31a-c* (eq 6).[6]

R^1COR^2 + $CH_2(CN)_2$ —(H_2O-NH_4OAc)→ [$R^1R^2C=C(CN)_2$] —(Et_3B, Et_2O-H_2O)→ $R^1R^2C(Et)CH(CN)_2$ (34)
—(RI (**35**) / Et_3B, Et_2O-H_2O)→ $R^1R^2C(R)CH(CN)_2$ (36,37) (6)

31 **32** **33** **36,37**

31,33,34	**35**	**36**	**37**
a:R^1=Ph, R^2=H	**a:R=*i*-Pr**	**a:R^1=Ph, R^2=H, R=*i*-Pr**	**a:R^1=Ph, R^2=H, R=*t*-Bu**
b:R^1=C_4H_9, R^2=H	**b:R=*t*-Bu**	**b:R^1=C_4H_9, R^2=H, R=*i*-Pr**	**b:R^1=C_4H_9, R^2=H, R=*t*-Bu**
c: R^1+R^2=-$(CH_2)_5$-		**c:R^1+R^2=-$(CH_2)_5$-, R=*i*-Pr**	**c:R^1+R^2=-$(CH_2)_5$-, R=*t*-Bu**

Similarly, high yields of *36* and *37* were obtained when *33a*was reacted with Et_3B (3 equiv.) and *iso*propyl iodide or *tert*butyl iodide under similar reaction conditions. In addition to malononitrile*32,* dimethylmalonate was employed.

1.4. Carboaminohydroxylation of Olefins in Water

Ingold and Lusztyk pointed out that aqueous solvents are a critical element for controlled reactions of aryl radicals.[7]

When aryldiazonium salts*38*, tetramethylpiperidineoxyl radical (TEMPO, a persistent radical) *39*, and various olefins*40* are made to react in DMSO-water, adduct *41* is afforded (Scheme 5).In all cases, the yields of the reactions demonstrate that the method is synthetically useful.[8]

Interestingly, the reactions including ethyl acrylate (*40a*) (Scheme 5, column 2) do not generally give the best yields, although they are favored in two ways: first by the fast addition of the aryl radical to the olefin and second by the fact that carbodiazenylation(radical trapping by an aryl diazonium salt) is not a competing process. The carbodiazenylation accounts for the slightly decreased yield of *41h*, as it is favored by both the reactive diazonium salt *38c* and the nucleophilicity of the intermediate alkyl and alkoxy radicals arising from the aryl radical addition to allyl acetate *40b*.

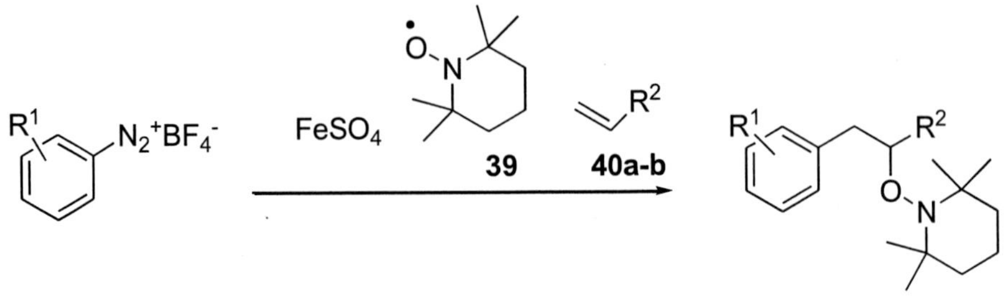

diazonium salt **38**

38a R^1= *p*-OMe
38b R^1=*p*-F
38c R^1=*o*-CO_2Me

olefins **40**

	40a R^2=CO_2Et	**40b** R^2=OAc
38a	**41a (68%)**	**41b (60%)**
38b	**41c (62%)**	**41f (62%)**
38c	**41g (63%)**	**41h (48%)**

Scheme 5.Carboaminohydroxylation of olefins in water.

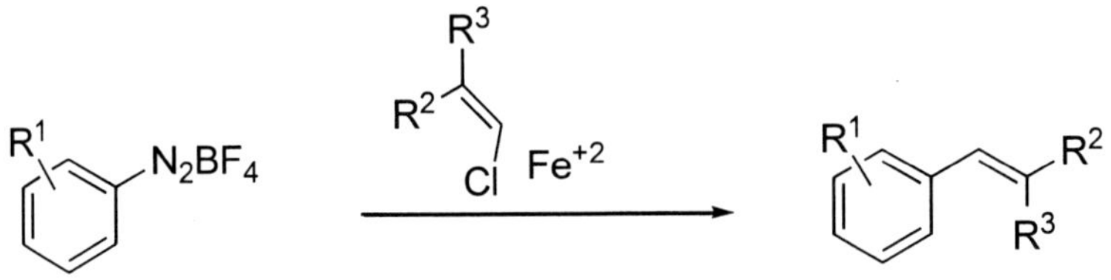

R^1= 4-CH_3O, R^2 = H, R^3 = Cl	53%
R^1= 4-CH_3O, R^2 = Cl, R^3 = Cl	61%
R^1= 4-F, R^2 = Cl, R^3 = Cl	46%
R^1 =4-CH_3O, R^2 = H, R^3 =4-MeOPh	61%

Scheme 6. Radical reactions of diazonium salts with vinyl chlorides in water.

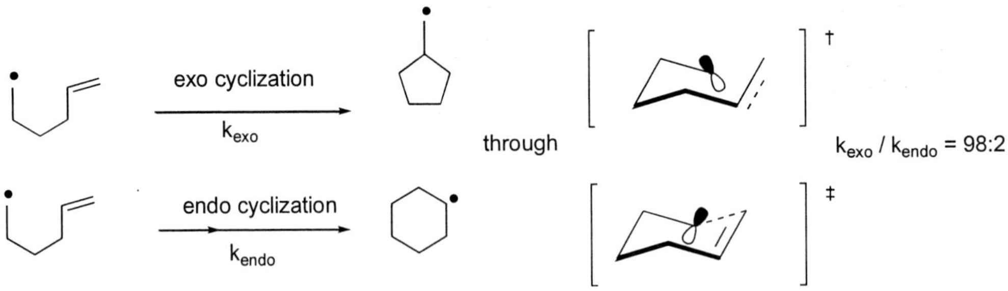

Figure 1.*Exo versus endo* cyclization of 5-hexyl radicals.

Carboaminohydroxylation of vinylcarbonate proceeded with moderate diastereoslectivity, but with methylcyclohexene-1-carboxylate only one product isomer was detected. The attempted addition of phenylacetylene failed and furnished the unstable benzoine derivative.[7]

Allylation and vinylation of aryl radicals can also be accomplished through the use of diazonium salts.[9] A key element to achieve selectivity in reaction of aryl radicals appears to be the use of water as a solvent, or at least co-solvent. Heinrich et al. achieved arylation reactions in good yields by adjusting the concentration of diazonium salt, as these act as efficient nitrogen-centered radical scavengers.[9] The olefinic substrates employed were chlorides and bromides.

A general reaction is described in Scheme 6.

2. Cyclization Reactions in Water

Although many reagents do exist for radical generation and trapping, establishing a single prevailing mechanism is not possible. However, once a radical is generated, it can react with multiple bonds in an intramolecular fashion to yield cyclized radical intermediates. The two ends of the multiple bond constitute two possible sites of reaction. If the radical in the resulting intermediate ends up outside of the ring, the attack is termed "*exo*"; if it ends up inside the newly formed ring, the attack is called "*endo*." In many cases, exo cyclization is favored over endo cyclization (macrocyclizations constitute the major exception to this rule). 5-Hexenyl radicals are the most synthetically useful intermediates for radical cyclizations, because cyclization is extremely rapid and *endo* selective. Although the *exo* radical is less thermodynamically stable than the *endo* radical, the more rapid *exo* cyclization is rationalized by better orbital overlap in the chair-like *exo*transition state (Figure 1, *vide infra*).

Substituents that affect the stability of these transition states can have a profound effect on the site selectivity of the reaction. Carbonyl substituents at the 2-position, for instance, encourage 6-*endo* ring closure. Alkyl substituents at positions 2, 3, 4, or 6 enhance selectivity for 5-*exo* closure.

Cyclization of the homologous 6-heptenyl radical is still selective, but is much slower—as a result, competitive side reactions are an important problem when these intermediates are involved. Additionally, 1,5-shifts can yield stabilized allylic radicals at comparable rates in these systems. In 6-hexenyl radical substrates, polarization of the reactive double bond with electron-withdrawing functional groups is often necessary to achieve high yields. Stabilizing the *initially formed* radical with electron-withdrawing groups provides access to more stable 6-*endo* cyclization products preferentially.

Cyclization reactions of vinyl, aryl, and acyl radicals are also known. Under conditions of kinetic control, 5-*exo* cyclization takes place preferentially. However, small concentrations of a radical scavenger establish thermodynamic control and provide access to 6-*endo* products—not via 6-*endo* cyclization, but by 5-*exo* cyclization followed by 3-*endo* closure and rearrangement. Aryl radicals exhibit similar reactivity (Figure 2).

Cyclization can involve heteroatom-containing multiple bonds such as nitriles, oximes, and carbonyls. Attack at the carbon atom of the multiple bond is almost always observed. In the latter case attack is reversible; however alkoxy radicals can be trapped using a stannane trapping agent.

The use of metal hydrides (tin, silicon, and mercury hydrides) is common in radical cyclization reactions; the primary limitation of this method is the possibility of reduction of the initially formed radical by H-M. Fragmentation methods avoid this problem by incorporating the chain-transfer reagent into the substrate itself—the active chain-carrying radical is not released until after cyclization has taken place. The products of fragmentation methods retain a double bond as a result, and extra synthetic steps are usually required to incorporate the chain-carrying group. In this Chapter we shall deal with carbon-centered radical cyclizations in water. Radical cyclizations in waterinvolving metal-centered radicals or where other non-carbon radicals participate, will be dealtwith in Chapters II and III.

Atom-transfer methods rely on the movement of an atom from the acyclic starting material to the cyclic radical to generate the product. These methods use catalytic amounts of weak reagents, preventing problems associated with the presence of strong reducing agents (such as tin hydride). Hydrogen- and halogen-transfer processes are known; the latter tend to be more synthetically useful (Figure 3).

Oxidative and reductive cyclization methods also exist. These procedures require fairly electrophilic and nucleophilic radicals, respectively, to proceed effectively. Cyclic radicals are either oxidized or reduced and quenched with either external or internal nucleophiles or electrophiles, respectively.

The use of radicals in organic chemistry has increased substantially within the last three decades.Several methods that lead to the synthesis of carbon-centered radicals have been developed.Among the radical initiators employed, Et_3B -induced radical reactions have attracted the attention of chemists. Firstly, the reactions with this initiator can be conducted at low temperatures, such as –78 °C, in the presence of trace amounts of oxygen.Secondly, various solvents including alcohol and water could be used because of the stability of Et_3B in aqueous media. Among the reactions performed with this radical initiator in water, cyclizations were shown to be more efficient than in organic solvents.

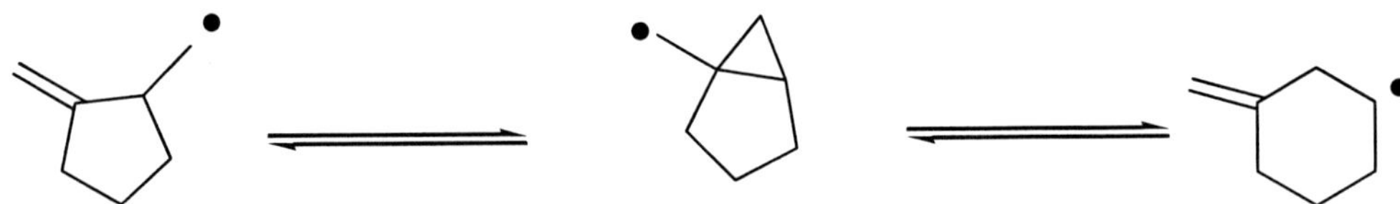

Figure 2. 5-*exo versus* 6-*endo* cyclization.

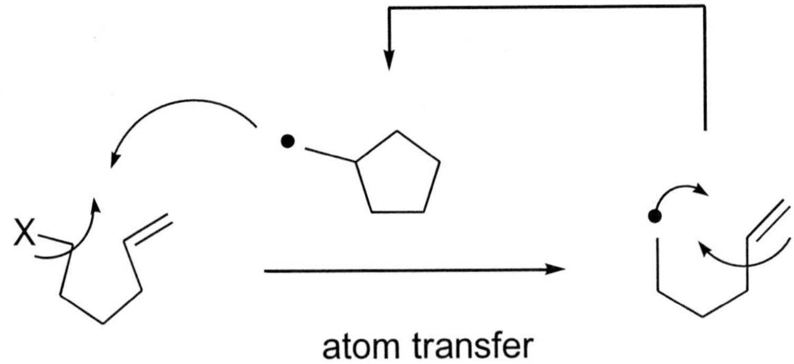

Figure 3. Atom transfer radical cyclization (ATRC).

2.1. Atom Transfer Radical Cyclization in Water

Oshima et al.[10] reported that theEt_3B-mediated radical cyclization reactions are much more efficient in water than in benzene or hexane.

Indeed, treatment of allyliodoacetate*42a* (eq 7)with Et_3Bin benzene or hexane gives no lactone *43a*.In contrast, in water,*42a* cyclized to give*43a* efficiently (67-78%).This powerful solvent effect operates also in crotyliodoacetate and 2-pentenyliodoacetate.

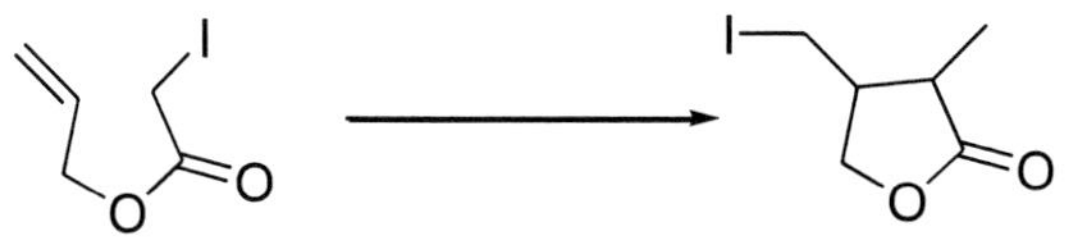

42a → **43a** (7)

3-Butenyl iodoacetate*44* gave δ-lactone *45* (eq 8) which is generated through 6-*exo* cyclization in 42% yield upon treatment with Et_3B in water.

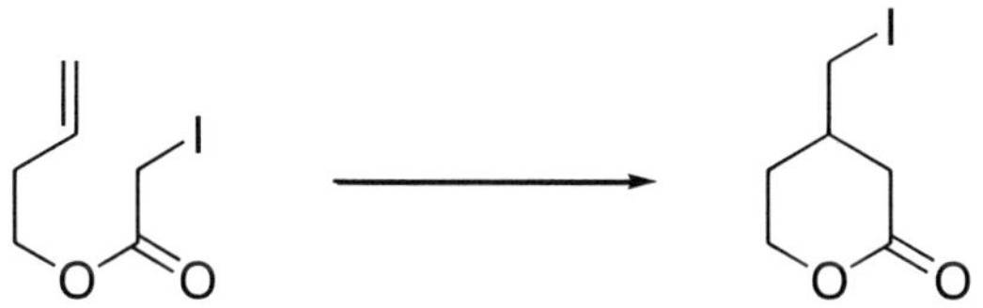

44 → **45** (8)

Treatment of α-iodoester*46* with Et_3B in water provided 9-membered lactone *47* in 68% yield (eq 9).

As a matter of fact, larger member rings are facilitated in water than in organic solvents, as is observed in Scheme 7.

The authors speculate that water can effectively decrease the barrier to rotation between the major *Z*-rotamer*Z-1* and minor *E*-rotamer*E-1* which can cyclize (Scheme 8), whereas in benzene, the *Z* and *E* conformers of the radical do not interconvert (see also section *9.*- of this Chapter).

48 → **49**

	benzene, %49	water, %49
48a n=2	78	84
48b n=3	43	86
48c n=4	83	98

Scheme 7. Radical formation of large-membered rings in different solvents.

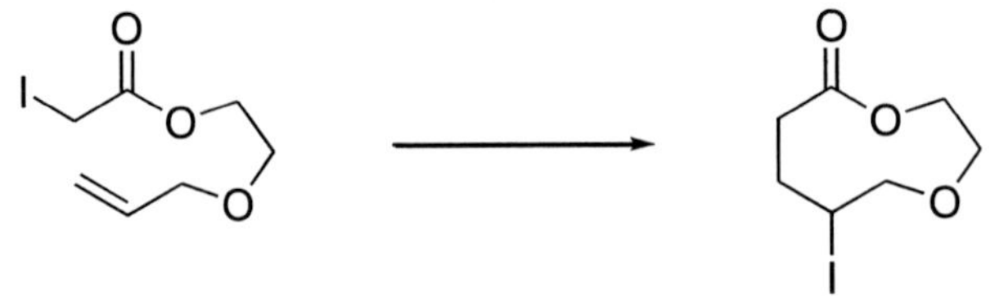

46 → 47 (9)

The cyclization of *N*-allyliodoamides induced by Et_3B proceeds smoothly in water to afford γ-lactams in good yields (Scheme 9).[11]

The low yield of the reaction from substrate*55* might be ascribed to the presence of a disfavored conformation for cyclization. Examination of ^1H NMR spectrum of *55* in D_2O proved that two methyl signals (in a ratio 42:58) appeared at δ = 2.82 and 3.02 ppm, respectively. The presence of two methyl signals is caused by restrictedrotation around the C(O)-N bond attributed to the resonance form of C(O−)=N+(Scheme 10A).

Radical chemistry has relied on the use of tributyltin hydride (TBTH) as mediator. While extremely useful, this compound is toxic and its by-products are difficult to remove from reaction mixtures. For these reasons, and because radical reactivities are complementary to other reactivities, the quest for alternative mediators has been extremely active.

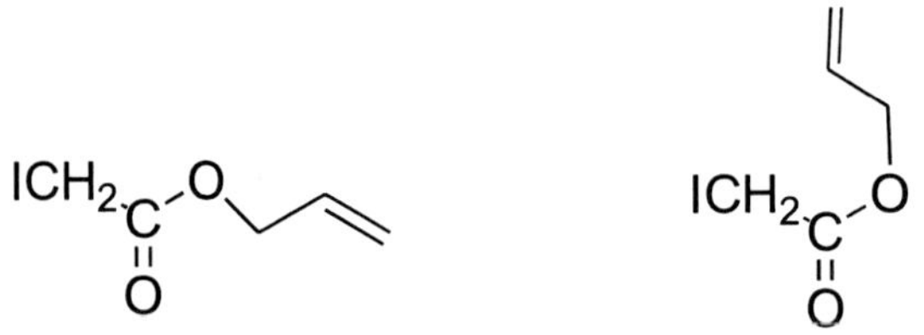

***Z*-1** ***E*-1**

Scheme 8. Rotamers from iodovinyl acetate.

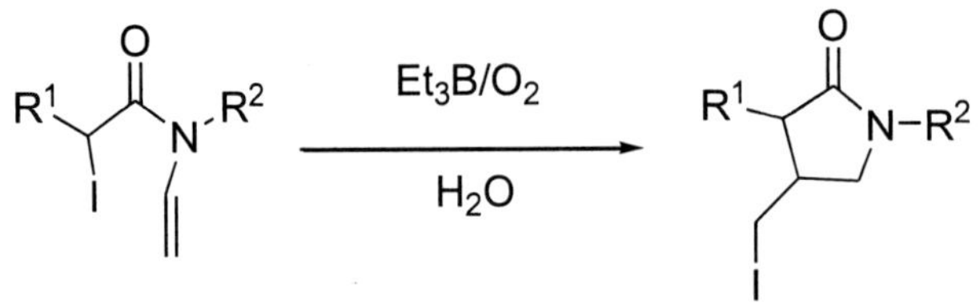

Et_3B/O_2, H_2O

50 R^1=H, R^2=SO_2CH_3	77%
51 R^1=H, R^2=SO_2CF_3	61%
52 R^1=H, R^2=cyclohexyl	60%
53 R^1=CH_3, R^2=SO_2CH_3	78%
54 R^1=CH_3, R^2=CH_3	75%
55 R^1=H, R^2=CH_3	22%

Scheme 9. Production of pyrrolidin-2-ones in water.

The best salt proved to be the N-ethylpiperidinium salt (EPHP). Dialkylphosphonates were also examined, but they require initiation by peroxides and can lead to undesired by-products.

Rapidly spawned several new contributions. In particular, Jang—who contributed to the initial work—showed that the sodium salt of hypophosphorous acid could reduce water soluble organohalides in water.

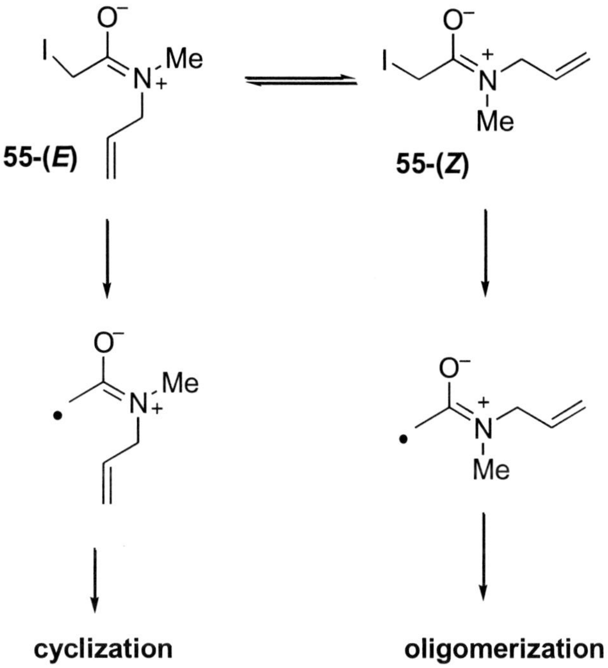

55-(*E*) 55-(*Z*)

cyclization oligomerization

Scheme 10A. Resonance forms from *N*-allyl amides.

He also introduced dibutylphosphine and diphenylphosphine oxides as new reducing agents. Because they are not ionic, they are less hygroscopic than EPHP and thus could be used with water-sensitive substrates. In any case, deoxygenation of hindered substrates was possible. Comparison of the yields to those obtained via Barton's method shows that the three mediators are complementary.

Once these two main families of P-based mediators had been introduced, rapid progress arose. Murphy and Stoodley simultaneously reported that EPHP could trigger formation of carbon–carbon bonds either through a 6-*exo-trig* cyclization of an aryl radical obtained from an iodide (Murphy).[14b]

Oshima introduced deuteratedhypophosphorous acid potassium salts to achieve radical deuteration. Deuteration of hydrophobic substrates was possible, albeit the incorporation of deuterium was not optimal because of hydrogen atom abstraction from either the solvent or the various additives used. As water is not prone to transfer a deuterium atom, less hydrophobic substrates led to deuteration with total incorporation.

Kita built on the previous studies to report EPHP-mediated cyclization of hydrophobic substrates in water (*vide infra*) This breakthrough was made possible by running the reaction in the presence of a water-soluble initiator (2,2′-azobis[2-(imidazolin-2-yl)propane , VA-061) and a surfactant (cetyltrimethylammonium bromide, CTAB). The authors explain this outstanding result by a micellar effect generated by CTAB. The organic ammonium probably contributes to the incorporation of the hypophosphoric acid in the micelles. By trapping hypophosphorous acid with atertiary amine, Jang introduced a surfactant-type chain carrier and reported good yields for deoxygenations of alcohols in water, without additive.

Eventually, Murphy introduced a water-soluble phosphine oxide which permits higher isolated yields than the corresponding reaction using EPHP, with no additional additive.

Upon using diethylphosphine oxide (DEPO), one can carry out sophisticated tin-free tandem radical reactions. Because DEPO is more lipophilic than hypophosphorous acid yet still water-soluble, it can facilitate the interaction between the water-soluble mediator and initiator and the lipophilic substrates without requiring a phase-transfer agent. Moreover, its pKa is 6, thus ensuring that this almost neutral excess reagent can be extracted into base during workup.

The use of phosphorous acid as mediator in radical cyclization reactions in water can be illustrated by the synthesis of indoles from thioanilides, Scheme 10B.[14d]

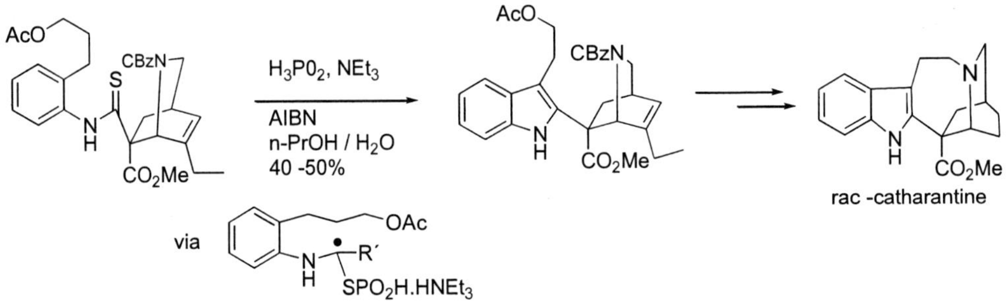

AcO CBzN S N H CO_2Me H_3PO_2, NEt_3 AIBN n-PrOH / H_2O 40 -50% AcO CBzN N H CO_2Me N N H CO_2Me rac -catharantine

via OAc N H R′ $SPO_2H.HNEt_3$

Scheme 10B.Synthesis of *rac*.Catharantine in water.

The use of 1-ethylpiperidine hypophosphite (EPHP), and the azo compound 2,2′-azobis[2-(imidazolin-2-yl)propane (VA-061) as initiator, together with a surfactant (cetyltrimethylammonium bromide, CTAB) provide the bestyields of cyclized products (eq 10).[12]

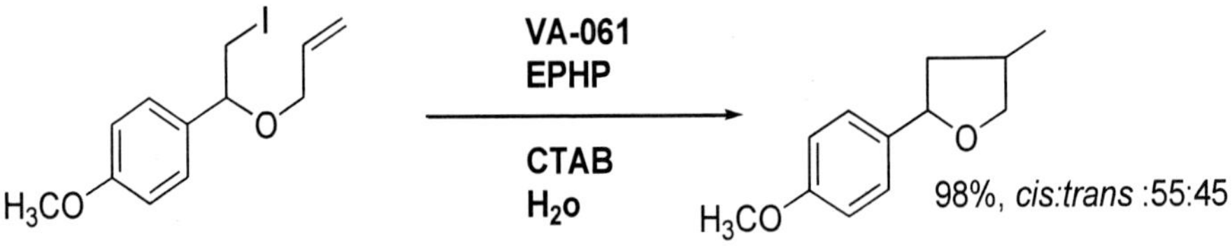

I O H_3CO **VA-061** **EPHP** **CTAB** **H_2o** O H_3CO 98%, *cis:trans* :55:45 (10)

Oshima et al. also described[13] that the addition of PhSH to alkene or alkyne proceeds smoothly to give the corresponding adducts (eq 11-12). Also, the atom transfer cyclization of diallyl-2-iodoacetamide affords lactams in excellent yields (eq 13), and the addition of 2-iodoacetamide to alkene followed by ionic cyclization gives lactone in excellent yields (eq 14).

Thus, heating a mixture of *N,N*-diallylacetamide*58,* benzenethiol, and azo initiator *56* (eq 11) in water at 60 °C, provided *N*-acetylpirrolidine*59* in 96% yield (eq 11).Treatment of diallylic ether *60* with benzenethiol in the presence of the azo compound *57* (eq 12) at 75 °C, afforded tetrahydrofuran derivative *61* (eq 12) in 75% yield.

NAc + PhSH → (H_2O, 80 C; 56) PhS / NAc (11)

58 **56** **59**, 96%

O + PhSH → (H_2O, 75 C; 57) PhS / O / HO (12)

HO **60** **57** **61**, 75%

Stirring a mixture of *N,N*-diallyl-2-iodoaetamide *62* and azo initiators *56* or *57* at 75 °C in water yielded γ-lactam *63* (80-99% yield, eq 13).

56 or 57 (13)

H_2O
R=H
R=CH_3

62 **63**

Intermolecular radical addition reactions give also good yields of lactones. The addition of 2-iodoacetamide to alkenol in water affords γ-substituted γ-lactones (*64*) in high yields (eq 14).

57 / H_2O (14)

64

X=NH$_2$	**R^2=$(CH_2)_n$OH n=2-4**	**85 - 95%**
X=OH	**R^2=$(CH_2)_n$OH n=1,2,4**	**76 - 95%**

The authors[13a]suggested a mechanism such as that proposed in Scheme 11A.

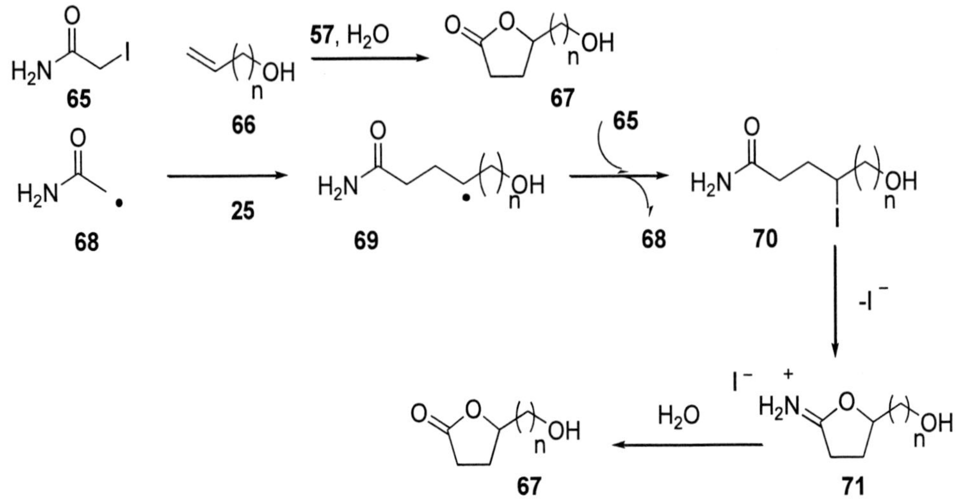

Scheme 11A. Mechanism proposed for the formation of lactones from iodoacetamides and alkenols in water.

Radical *68* (derived from thermal treatment of iodide *65* in the presence of azo compound *57*), adds readily to the alkenyl terminal carbon of alkenol*66* to render radical *69*. The iodine atom transfer between the radical *69* and iodide *65* affords a hydroxy-4-iodoamide *70*, and regenerates radical *68*. Compound *70* cyclizes to γ-lactone *67* via *71* under the reaction conditions due to the well-known ionic lactonization of 4-iodoalkanamide (Scheme 11A).

Interestingly enough, the same authors argue that the solvent effect of water is exerted in the initiation step, activating the carbon-iodine bond in *65*.

Also, oxime etherscan undergo tandem radical cyclization.[3] Treatment of oxime ether *72* with Et_3B and isopropyl iodide in water at 80 °C affords the cyclized product *73* in 63% yield via two carbon-carbon bond-forming steps (eq 15).

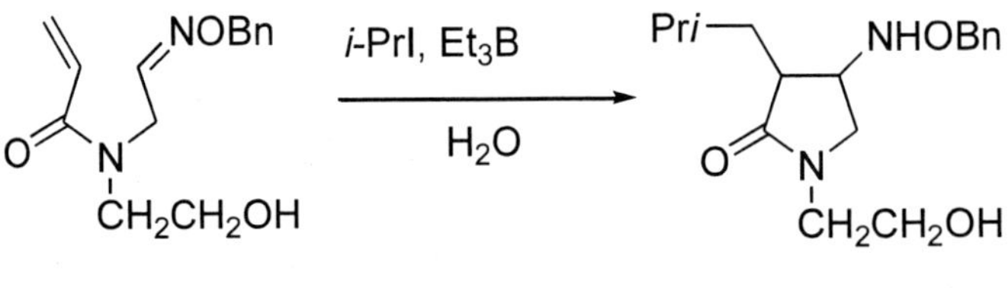

(15)

Landais and coworkers reported on the radical addition-*5-exo-trig* cyclization on ketoximes, Scheme 11B.[13b]

For instance, when the radical reaction was carried out on *73b*, and water was used as a solvent, this led to the unique formation of diester*73c*. Similar results were obtained when additives such as AcOH (1-5 equiv) were employed. Strikingly, the reaction can even be conducted in 1 N HCl or in concd H_2SO_4 and still leads to the exclusive formation of *73c* (albeit in lower chemical efficiency). Electrophiles such as $(Boc)_2O$ and trifluoroacetic anhydride were also tested to trap the putative amidoborane*iv* (Scheme 11C), but *73c* was invariably produced as a unique product. These experiments clearly rule out the formation of the C-N bond and the incorporation of the second ester fragment through an ionic pathway.[13b]The reaction mechanism is depicted in Scheme 11C.Precedents from the literature effectively reveal that alkoxyaminyl radicals are persistent due to steric shielding.[13c]In the case depicted inScheme 11C, intermediate radical *73d* may fit into this category and therefore possess a long enough lifetime to recombine with an excess of radical precursor ($EtOCOCH_2$)arising from iodoester. Several experiments were designed to reduce the alkoxyaminyl radical intermediate *73d* under free-radical conditions.

For instance, addition of ethyl xanthate in isopropanol as both a solvent and a reducing agent, according to Zard's procedure,[13d] led to *73c* in low yield (8%) along with recovered starting material (78%). No trace of the desired reduced product was observed. The use of Roberts's polarity-reversal catalysis[13e] (Ph_3SiH, $HSCH_2CO_2Et$ (cat.), $BrCH_2CO2Et$, see also Chapter II) led in turn to a complex mixture of products in which the desired product was absent. Finally, substitution of Et_3B for indium in a MeOH--H_2O mixture led to recovered allylsilane*73b* and ethyl acetate resulting from complete reduction of ethyliodoacetate. In parallel, the authors envisaged being able to trap the alkoxyaminyl radical (*73d*) in an intramolecular fashion through the introduction of an olefinic appendage. Recent studies effectively showed that a 5-*exo-trig* cyclization onto a ketoxime, followed by a second 5-*exo-trig* addition of an alkoxyaminyl radical onto an unsaturated Michael acceptor, was feasible, leading to a good yield of the desired bicyclic system. Radical cyclizations in water are successfully accomplished also by silyl radical mediators such as tris(trimethylsilyl)silane, $(Me_3Si)_3SiH$, as is depicted in Scheme 12A (see also Chapter II).[14]

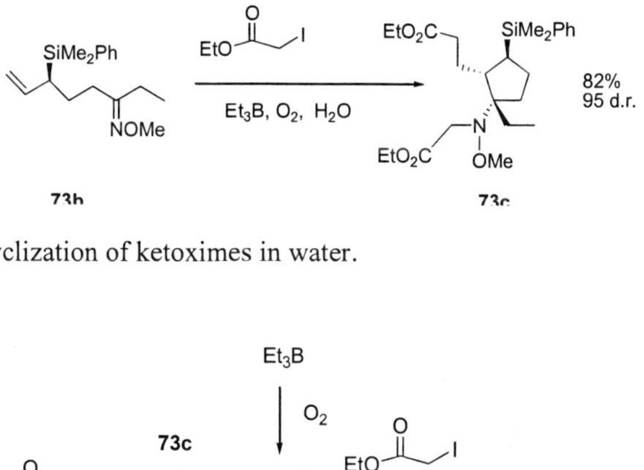

Scheme 11B.Radical cyclization of ketoximes in water.

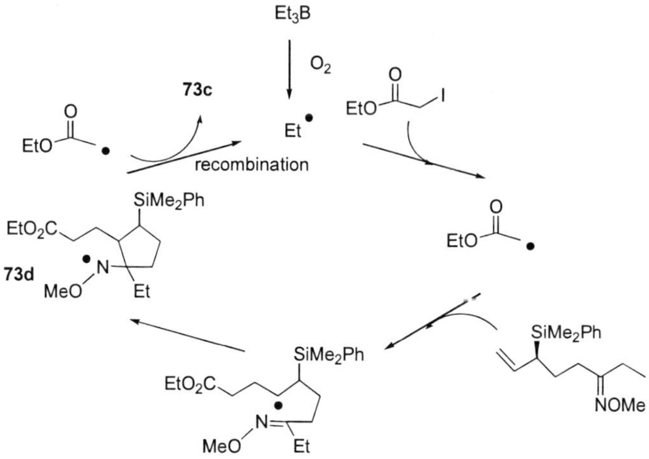

Scheme 11C. Mechanism for the formation of product *73c* in water.

In another report by the same authors,[14b], 6-bromo-3,3,4,4,5,5,6,6-octafluoro-1-hexene (12 mM) *75b* was subjectedto reaction (24 h) with $(Me_3Si)_3SiH$ (8 mM) and dioxygen in water (5 mL), and obtained the *exo-trig* cyclization product 1,1,2,2,3,3,4,4-octafluoro-5-methylcyclopentane *75c* (Scheme 12B) in 76% yield (isolated, see also Chapter II).

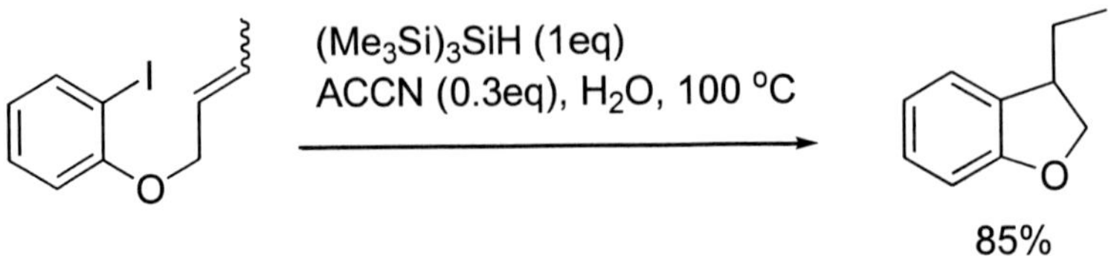

$(Me_3Si)_3SiH$ (1eq)
ACCN (0.3eq), H_2O, 100 °C

85%

74 **75**

Scheme 12A. $(Me_3Si)_3SiH$-mediated radical cyclization of 1-allyloxy-2-iodobenzene in water.

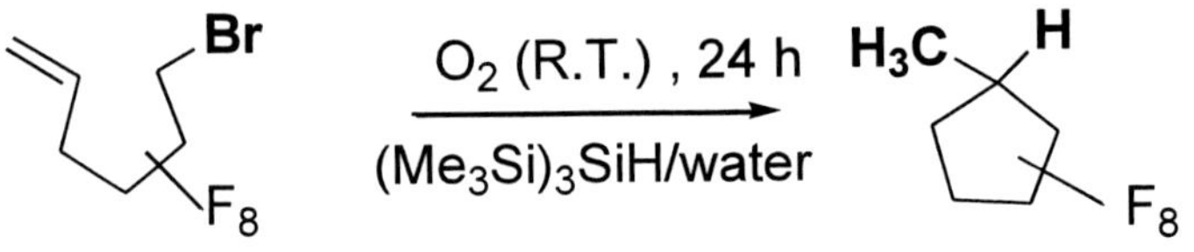

Br F_8 O_2 (R.T.) , 24 h $(Me_3Si)_3SiH$/water H_3C H F_8

75b **75c**

Scheme 12B. $(Me_3Si)_3SiH$-mediated radical Cyclization of 6-bromo-3,3,4,4,5,5,6,6-octafluoro-1-hexene in water.

Though the measurement of the rate constant for cyclization in the heterogeneous water system is difficult to be obtained, the cyclohexane cyclized product has not been observed in water under the reaction conditions reported. No uncyclized-reduced product is either observed.[14b]

Analogously, cyclization of 5-bromo-1,1,2,3,3,4,4,5,5-nonafluoro-pent-1-ene (12 mM) *75d* in water triggered by $(Me_3Si)_3SiH$ (8 mM) / dioxygen leads to nonafluorocyclopentane, the *exo-trig* cyclization product in 68% yield (isolated). No reduced product could be isolated from the reaction mixture. The reaction carried out in benzene-d_6 does not lead to cyclization product (Scheme 12C).

When 3,3,4,4-tetrafluoro-1,5-hexadiene (40 mM) *75e* is allowed to react (24 h) in water with$(Me_3Si)_3SiH$ (5 mM) / dioxygen and C_2F_5I (10 mM), product *75f* is obtained in 61% yield, based on C_2F_5I (Scheme 12D).[14b]

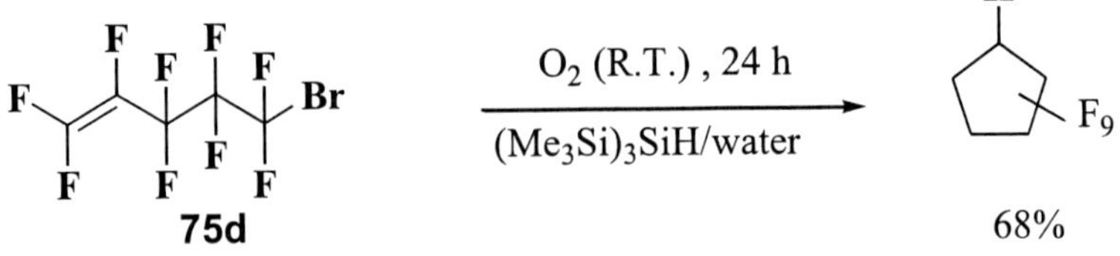

F F F F F Br F F F F O_2 (R.T.) , 24 h $(Me_3Si)_3SiH$/water H F_9

75d 68%

Scheme 12C.$(Me_3Si)_3SiH$-mediated radical cyclization of 5-bromo-1,1,2,3,3,4,4,5,5-nonafluoro-pent-1-ene in water.

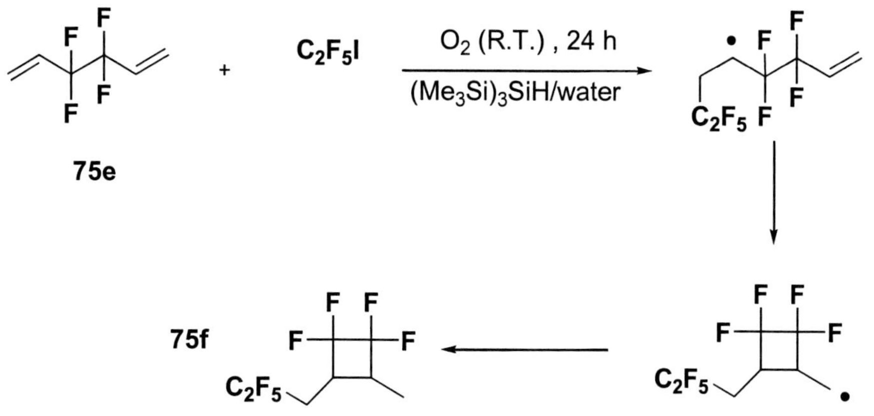

Scheme 12D.$(Me_3Si)_3SiH$-mediated radical cyclization of 3,3,4,4-tetrafluoro-1,5-hexadiene in water.

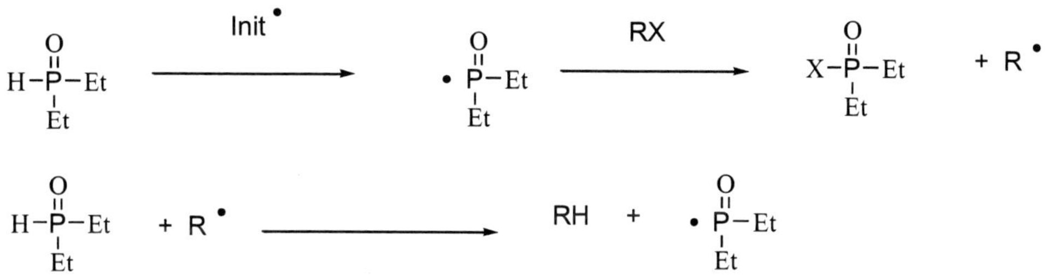

Figure 4. Mediation of DEPO inradical reactions in water.

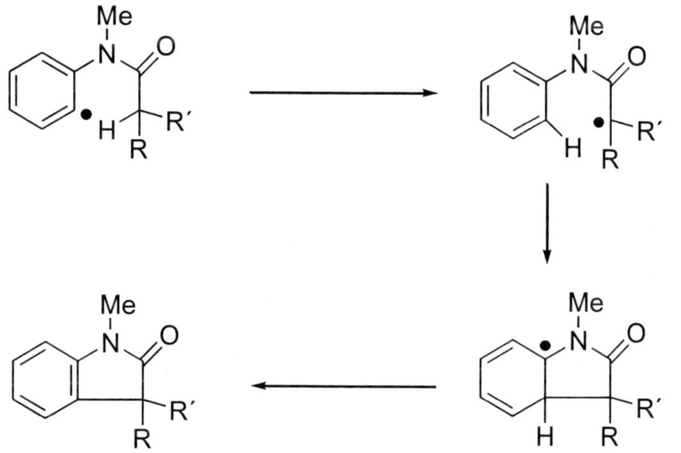

Figure 5.Reaction mechanism for indolone formation from aryl radicals, with ulterior rearomatization.

Murphy and collaborators[14c] have used diethylphosphine oxide (DEPO) as a radical mediator in the preparation of indolones in water, in good yields, through a sequence of aryl radical formation ($R^{\bullet}$), hydrogen atom abstraction (R-H) (Figure 4), cyclization and rearomatization.

Several indolones could be synthesized with this methodology, and the general reaction mechanism is depicted in Figure 5.

The initiator that gave the best reaction yields is shown in Figure 6.

Arylation of lactams in water was obtained by the use of DEPO and V-501 as initiator in water (Figure 7). [14e,f]

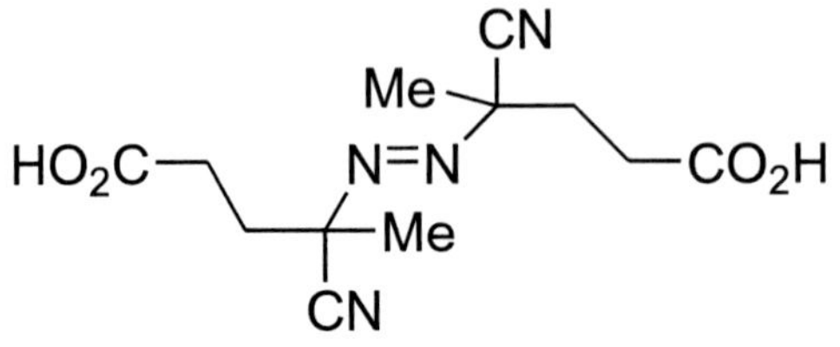

CN
Me
HO_2C N=N CO_2H
Me
CN

V-501

Figure 6. Structure of initiator V-501.

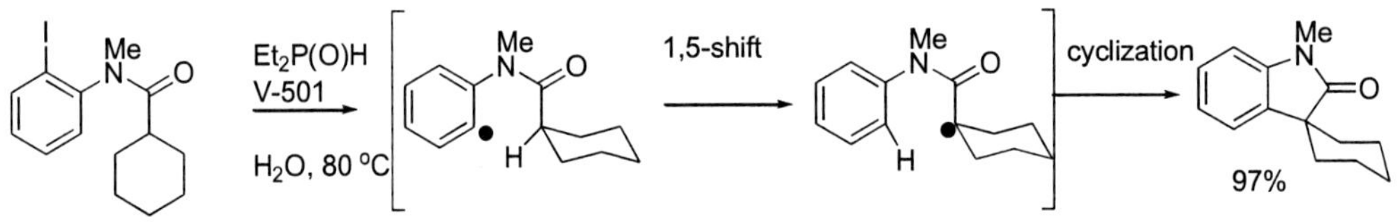

I Me N O
$Et_2P(O)H$
V-501
H_2O, 80 °C
Me N O H
1,5-shift
Me N O H
cyclization
Me N O
97%

Figure 7A.DEPO-mediated arylation of of lactams.

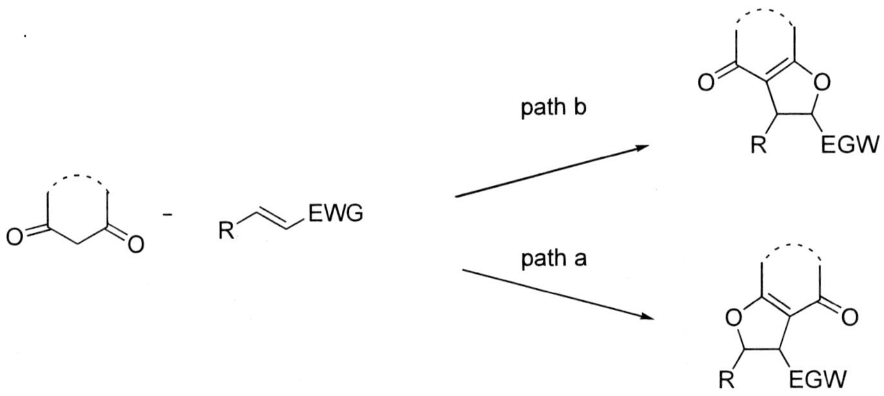

O O - R EWG
path b
O O R EGW
path a
O O R EGW
75g

Figure 7B. Radical synthesis of dihydrofurans from 1,3-dicarbonyl compounds.

2.2.Synthesis of Fused Dihydrofuran Derivatives

The radical cyclic addition of cyclic 1,3-dicarbonyl compounds to appropriate olefins provides a versatile method for the synthesis of fused dihydrofuran derivatives.[14g] When electron-poor alkenes were employed as the substrates, the radical reaction pathway resulted in the regioselective generation of product *75g* with the carbon of 1,3-dicarbonyl compounds added at the α-position of electron-poor alkenes (path a, Figure 7b). Wang and co-workers reported a $Mn(OAc)_3$- mediated reversed regioselective radical cyclic addition (path b, Figure 7b). However, according to the plausible reaction pathway, only 1-(pyridin-2-yl)enones were suitable substrates (see also Chapter III for radical cyclizations mediated by metals).[14h]

More recently, Fan and coworkers[14i] reported on the radical cyclization using a Michael adduct, and found that the best conditions entailed the use of PhIO, Bu_4NI in water, as shown in Scheme 12E.

The scope of this reaction was then investigated under optimized conditions (those from Scheme 12E), and the results were very promising. The Michael adducts of chalcone with 5,5-dimethylcyclohexane- 1,3-dione (dimedone), 6-methyl-*3H*-pyran- 2,4-dione, chroman-2,4-dione (4-hydroxycoumarin), and 6-fluorochroman-2,4-dione (4-hydroxy-6-fluorocoumarin) were effective substrates, and their reactions gave rise to the corresponding fused dihydrofurans in good to excellent yields (up to 90% yield).[14i] This is an efficient method developed for the construction of functionalized fused dihydrofurans via an aqueous PhIO / Bu_4NI-mediated stereoselective oxidative cyclization of Michael adducts of cyclic 1,3-dicarbonyl compounds with chalcones. The potential of this reaction system can be evaluated by its simple procedure, mild conditions, and adaptability to a wide variety of substrates.

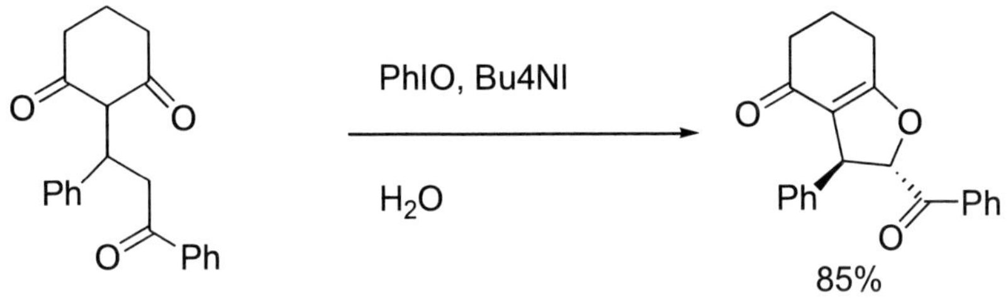

PhIO, Bu4NI

H_2O

85%

Scheme 12E.Radical synthesis of fused dihydrofurans in water.

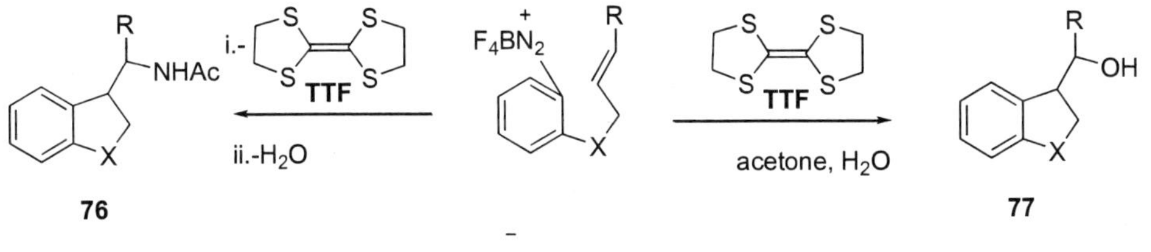

i.- TTF

ii.-H_2O

TTF

acetone, H_2O

76

77

Scheme 13. Reaction of diazonium salts with tetrathiafulvalene (**TTF**) in water.

3.Radical-Polar CrossoverReactionswith a Water-Soluble TetrathiafulvaleneDerivative

The radical-polar crossover reaction[15] in Scheme 13 affords a means of performing radical-based carbon-carbon bond formation. Thus, tetrathiafulvalene (TTF, Scheme 13) donates an electron to the diazonium group, which upon loss of dinitrogen affords an aryl radical which cyclizes, and the resulting radical couples with TTF+ to form a sulfonium salt, which undergoes solvolysis to afford alcohol *77* or an amide *76*, affording a polar termination to the radical reaction (Scheme 13).

More complex amines*78* and *79* can be converted into the corresponding diazoniumtetrafluoroborates through the use of a water-soluble tetrathiafulvalene-derivative (eq 16), which then successfully undergo radical cyclization, radical-polar crossover and intramolecular termination in 1:1 acetone: water, demonstrating thescope of the reaction.[16]

NH2 OH X → O X (16)

78, X=NMs, **41%**

79, X=O, **39%**

4.Radical Conjugate Addition

The reaction of trialkylboranes with 1,4-benzo-quinones to give in quantitative yield 2-alkylhydro-quinones was the first reaction of this type occurring without the assistance of a metal mediator. An ionic mechanism was originally proposed but rapidly refuted since the reaction is inhibited by radical scavengers such as galvinoxyl and iodine. Then, it was demonstrated that trialkylboranes, readily available via hydroboration, are excellent reagents for conjugate addition to vinyl ketones (Figure 8), acrolein, α-methylacrolein, α-bromoacrolein, 2-methylenecyclohexanone, and quinones (Figure 8). Traces of oxygen present in the reaction mixture are sufficient to initiate these reactions. Their free-radical nature was demonstrated by their complete inhibition in the presence of free-radical scavengers such as galvinoxyl (5 mol %).

The mechanism of these reactions could be illustrated in Figure 9.

Brown proposed a mechanism where the enolate radical resulting from the radical addition reacts with the trialkylborane to give a boron enolate and a new alkyl radical that can propagate the chain (Figure 9). The formation of the intermediate boron enolate was confirmed by ^{1}H NMR spectroscopy. The role of water present in the system is to hydrolyze the boron enolate and to prevent its degradation by undesired free-radical processes. This hydrolysis step is essential when alkynones and acrylonitrile are used as radical traps since the resulting allenes or keteneimines, respectively, react readily with radical species. Recently, Maillard and Walton have shown by ^{11}B NMR, ^{1}H NMR, and IR spectroscopy that triethylborane does complex methyl vinyl ketone, acrolein, and 3-methylbut-3-en-2-one. They proposed that the reaction of triethylborane with these traps involves complexation of the trap by the Lewis acidic borane prior to conjugate addition (Figure 10).

The reaction between trialkylboranes and enones has not found many synthetic applications. An exception is the preparation of prostaglandin precursors from *exo*-methylene cyclopentanone, generated *in situ* from a Mannich base. After dehydrogenation, a second conjugate addition of trioctylborane was used to introduce the ω-chain.

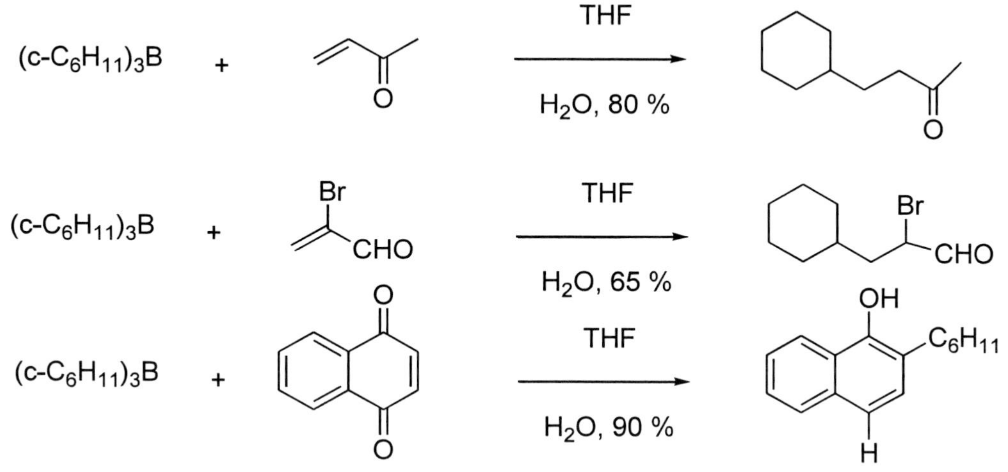

$(c\text{-}C_6H_{11})_3B$ + (methyl vinyl ketone) —THF / H_2O, 80 %→ (product)

$(c\text{-}C_6H_{11})_3B$ + (2-bromoacrolein, Br, CHO) —THF / H_2O, 65 %→ (product, Br, CHO)

$(c\text{-}C_6H_{11})_3B$ + (1,4-naphthoquinone, O, O) —THF / H_2O, 90 %→ (product, OH, C_6H_{11}, H)

Figure 8.Radical conjugate additions of cyclohexylboranes to different α,β-unsaturated compounds.

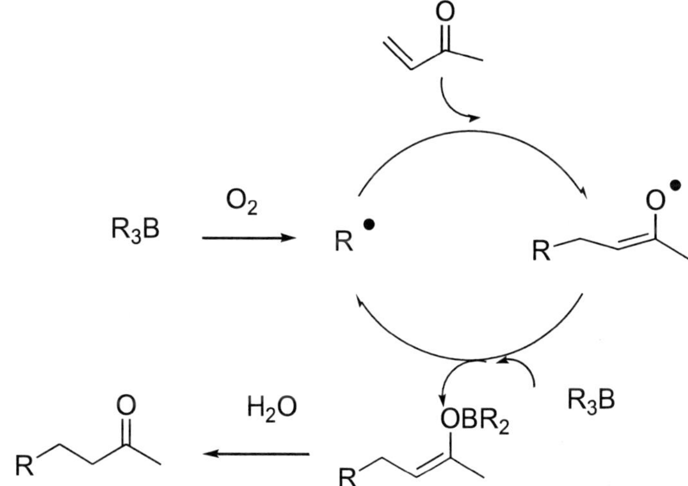

Figure 9. Mechanism for the radical conjugate addition of alkyl boranes to α,β-unsaturated compounds.

Synthetically, a serious drawback of the trialkylborane approach is that it requires a 1:1 trialkylborane/ radical trap ratio to obtain good yields. Therefore, the method is restricted to trialkylboranes obtained by hydroboration of easily available and cheap alkenes. To overcome this limitation α-alkyl-boracyclanes have been used. According to Brown and Negishi and co-workers, 3,3-dimethyl-borinane, prepared from BH_3 and 2,4-dimethyl-1,4- pentadiene, is the most efficient reagent. With this system, a selective cleavage of the boron-alkyl bond is possible for secondary and tertiary alkyl groups. This method, referred to later as the Brown-Negishi reaction, is not suitable for primary alkyl radicals (yield <35%) and for radical traps substituted at the β-position. With these traps, the addition of extra oxygen is necessary to run the chain reaction, and under these conditions the cleavage of the carbon-boron bond is no longer selective.

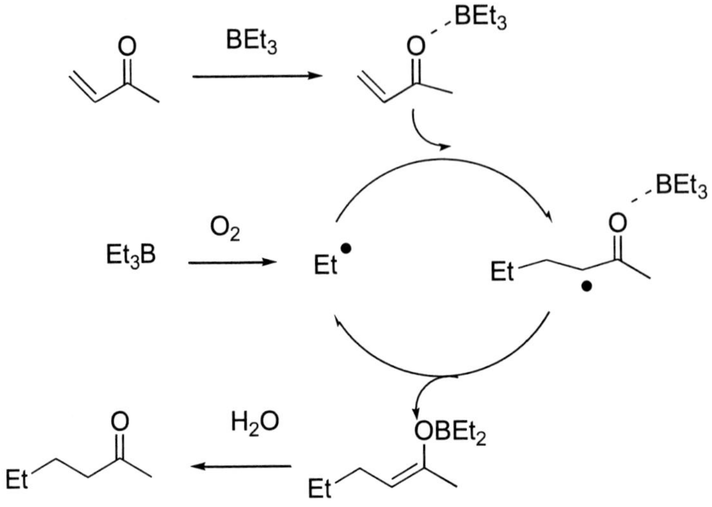

Figure 10. Complete proposed mechanism for the radical conjugate addition of alkyl boranes to α,β-unsaturated compounds.

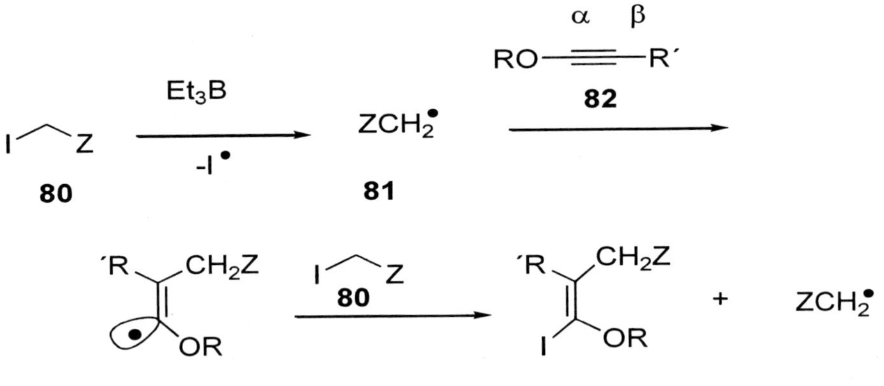

Scheme 14. Enol ether formation in water.

5. Construction of Carbon-oxygen Bonds by Radical Reactions in Water

5.1. Synthesis of FunctionalizedEnol Ethers and Allylic and AllenicAlcohols in Water

The use of ynol ethers in water is very scarse. However, the radical reaction between ethyl iodoacetate and an ynol ether has been explored successfully in water by Daoust et. al, and the method is currently regarded as the first to allow the production of enol ethers in water.[17]

The electrophilic radical *81*, formed by homolytic cleavage of activated methylenes*80* (Scheme 14, Z= EWG, electron withdrawing group) adds to the electron-rich β-carbon of ynol ether *82* leading to vinyl radical *83*, whose configuration is governed by stereoelectronic factors.The use of iodides, which are efficient halogen-atom transfer agents, ensures rapid trapping of vinyl radical *83* before its interconversion. The result is an *E* isomer *84* obtained with high stereoselectivity. The radical initiator is Et_3B in methanol (Scheme 14).

The reaction has to be carried out introducing the iodoacetate and Et_3B very slowly to the reaction mixture, in order to avoid homocoupling of $\bullet CH_2CO_2Et$ radicals.The reaction proceeds with high regio- and stereoselectivity.

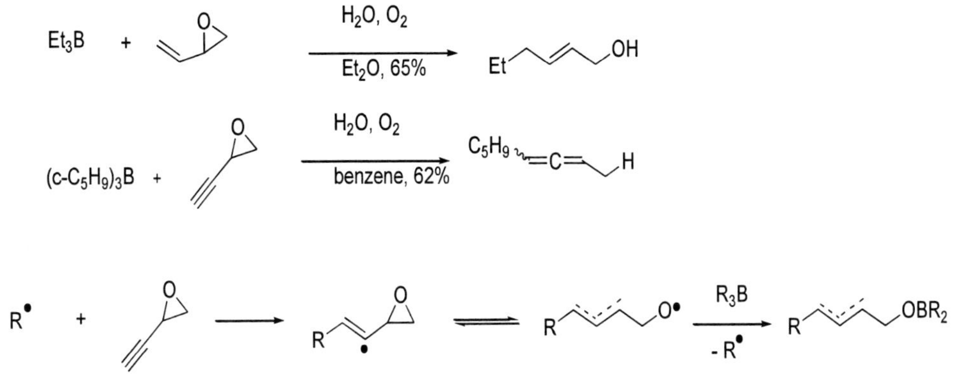

Figure 11. Reactions of trialkylboranes with ethenyl- and ethynyloxiranes in water.

Different ynol ethers (*82*) were used (R=Et, R′=H; R= menthyl, R¨=H; and R=menthyl, R¨= Me) with ethyliodoacetate (*80*, Z=COOEt) as the source of electrophilic radical, and the yields of substituted enol ethers*84* range from 70 to 80% in water.Iodoacetonitrile reacts modestly, while iodoacetamide adds poorly. The reaction does not occur with very bulky triple bonds. Ethyl bromoacetate, methyl bromide, and methyl iodide do not add to ynol ethers in water.

Brown, Suzuki and co-workers have shown that treatment of trialkylboranes with ethenyl and ethynyloxiranes in the presence of a catalytic amount of oxygen affords the corresponding allylic or allenic alcohols. The mechanism may involve the addition of alkyl radicals to the unsaturated system leading to 1-(oxiranyl)-alkyl and 1-(oxiranyl)-alkenyl radicals followed by rapid fragmentation to give alkoxyl radicals that finally complete the chain process by reacting with the trialkylborane (Figure 11).[17b]

5.2. Synthesis of CycloalkaneCarboxylicAcidsin Water

The aqueous carboxylation of a series of cycloalkanes has been undertaken by Pombeiro ct al.[18]

The carboxylation reactions of cycloalkanes are studied in stainless steel autoclave by allowing to react, at low temperature and in neutral H_2O / MeCNmedium, a cycloalkane with carbon monoxide, potassium peroxodisulfate and water, either in the absence (metal-free) or in the presence (metal-promoted) of a metal promoter (eq 17).

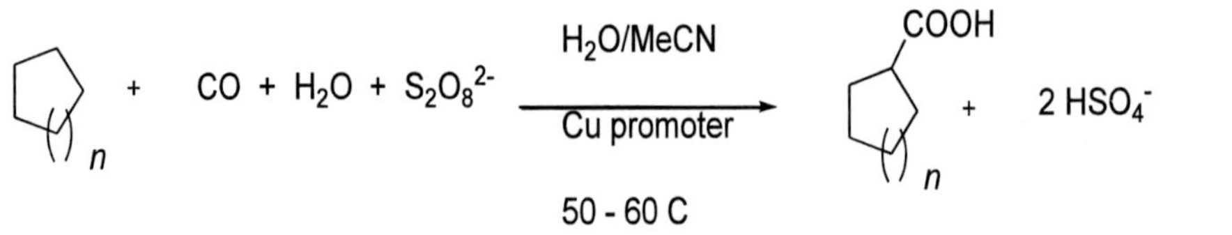

$$\text{cycloalkane} + CO + H_2O + S_2O_8^{2-} \xrightarrow[\text{Cu promoter, } 50-60\ C]{H_2O/MeCN} \text{cycloalkane-}COOH + 2\,HSO_4^- \qquad (17)$$

A crucial role in the unusual metal-free and mild transformation of an inert alkane is played by the active radical sulfate $SO_4^{\cdot-}$ (known as an efficient single electron-transfer oxidant).

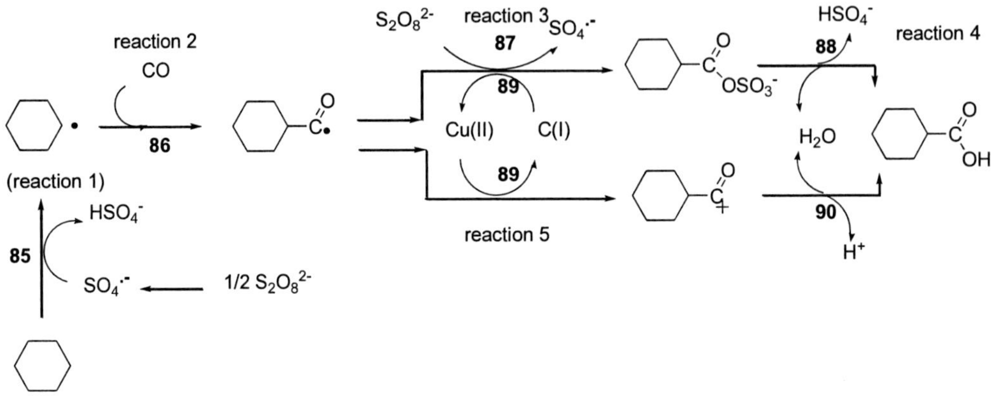

Scheme 15. Mechanism for the synthesis of cyclohexane carboxylic acid in water.

Important features of this reaction consist in the use of water as medium, the fact that the source of the OH functionality of the carboxylic acid moiety comes from water, the possibility of working in the absence of metals, and mild temperatures (considering the high inertness of C-H bonds in cycloalkanes). Also, high efficiency and selectivity have been invoked as salient aspects of this methodology.

Among those metal compounds that exhibit a promoting effect, various derivatives of copper and potassium dichromate revealed the highest activity, leading to yields of $C_6H_{11}COOH$ in the 70% range. The formation of cyclohexanone and cyclohexanol as by-products due to the partial oxidation of cyclohexane is also detected.The most effective promoting behavior is exhibited by the hydrosolubletetracopper(II) triethanolaminate derivative $[O\subset Cu_4\{N(CH_2CH_2O)_3\}_4(BOH)_4][BF_4]_2$.

The authors proposed a reaction mechanism as that depicted in Scheme 15.

The mechanism involves the formation of a free cyclohexyl radical, which is generated by hydrogen abstraction from cyclohexane (reaction 1, Scheme 15) by the sulfate radical.The latter is derived from thermolytic and copper-promoted decomposition of $K_2S_2O_8$. The involvement of cyclohexyl radical is confirmed when the reaction is carried out in the presence of the radical trap $CBrCl_3$, which results in the full suppression of cyclohexane carboxylic acid and the complete formationof bromocyclohexane.This radical path is also inhibited by the presence of oxygen, acting as a cyclohexyl trap to afford $C_6H_{11}OO\bullet$ peroxyl radical.Subsequent carbonylation of the cyclohexyl radical by carbon monoxide results in the acyl radical $C_6H_{11}CO\bullet$ (reaction 2, Scheme 15), that upon oxidation with further$S_2O_8^{2-}$ generates the acyl sulfate $C_6H_{11}C(O)OSO_3^-$ (reaction 3, Scheme 15). This is hydrolyzed by water (reaction 4) furnishing the cyclohexane carboxylic acid. In the copper-promoted process, an alternative route (reaction 5, Scheme 15) can occur, where the tetracopper complex ([O⊂Cu_4{N(CH_2CH_2O)$_3$}$_4$(BOH)$_4$][BF_4]$_2$) can behave as an oxidant of the acyl radical (reaction 5, Scheme 15).

6. Radical Carboazidationand Azidationin Water

As has been explained in section *1.a.*-(this Chapter), the radical atom transfer reactions can effectively be initiated by Et_3B in water. Since radical carboazidations are occurring via an initial transfer of iodine atom or xanthate group, it becomes reasonable that these reactions were also initiated by Et_3B in water.

The Et_3B -induced reaction of alkene *91* with iodoacetate ethyl ester and phenylsulfonyl azide in water yields azide *92* in 85% yield (eq 18).[19] Cycloalkene*93*, under the same reaction conditions, affords *94* in 90% yield (eq 19).

nC$_6$H$_{13}$ **91** —(EtOOCCH$_2$I; PhSO$_2$N$_3$, Et$_3$B/H$_2$O)→ **92, 85%** (18)

93 —(EtOOCCH$_2$I; PhSO$_2$N$_3$, Et$_3$B/H$_2$O)→ **94, 90%** (19)

The one-pot radical addition of the iodomalonate *95* to 1-octene followed by successive cyclization and azidation afforded the tertiary azide *96* in 72% yield as a 4:1 mixture of diastereomers (eq 20).

95 —(1-octene, PhSO$_2$N3; Et$_3$B, H$_2$O)→ **96** (20)

7. Radical Polymerization of Alkenes in Water

Free radical polymerization of alkenes has been carried out under aqueous conditions successfully for many years. Aqueous emulsion and suspension polymerization is carried out today on a large scale by free-radical routes. Polymer latexes can be obtained as a product, that is, stable aqueous dispersion of polymer particles. Such latexes possess a unique property profile and most studies on this subject are in patent literature. Atom transfer radical addition

(ATRA) of carbon tetrachloride and chloroform to unsaturated compounds including styrene and 1-octene was investigated using rutheniumindenylidene catalysts. The reaction was extended to atom transfer radical polymerization (ATRP) by changing the monomer halide ratio and can work in aqueous media.[20]

8. Sulfur-centered Radicals in Water

8.1. Thioesterification with Aldehydes

Kita et al.[21] reported on an effective intermolecular radical reaction of carbon-sulfur bond formation in a micellar system using the combination of a water-soluble radical initiator and surfactant in water (Scheme 16).

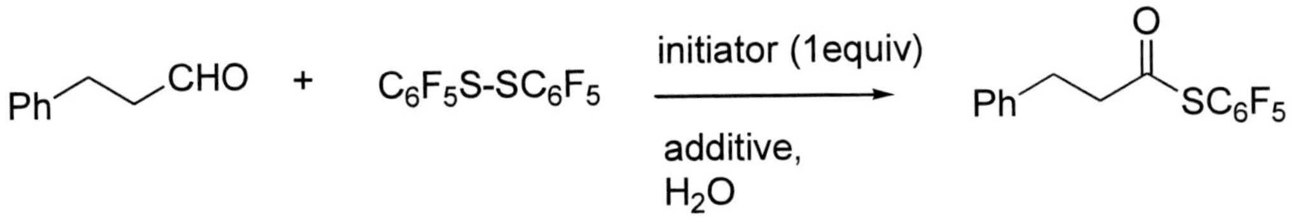

Ph CHO + $C_6F_5S\text{-}SC_6F_5$ —initiator (1equiv), additive, H_2O→ Ph O SC_6F_5

Scheme 16. Carbon-sulfur bond formation in micellar systems.

The initiator that gave the best reaction yields is VA-044 (1,2-bis(2-(4,5-dihydro-1*H*-imidazol-2-yl)propan-2-yl)diazene hydrochloride), and the additive used is the cationic surfactant CTAB.When galvinoxyl free radical was used as a radical scavenger, no thioesterification occurred. Hence, this reaction proceeds via a radical mechanism. Water was also shown to be the best solvent for the reaction.

Various disulfides were used in the above reaction, among which, pentafluorodiphenyl sulfide gave the corresponding thioesters in best yields. Among aldehydes, aliphatic aldehydes and aromatic aldehydes with electron releasing groups afforded the best thioester yields.

A plausible reaction mechanism is depicted in Scheme 17.

The disulfide *98* dissociates upon reaction with the initiator to yield thiyl radicals*A*.Secondly, the hydrogen from the aldehyde *97* is trapped by the thiyl radical *A* and the acyl radical *B* is formed. The acyl radical reacts with the disulfide *98* or thiyl radical *A*, and the thioester *99* is formed.

The authors also achieved the direct amidation of aldehydes using a one-pot synthetic methodology. Several aldehydes such as *p*-methoxybenzaldehyde, 2,4,6-trimethyl-benzaldehyde, 2,4,6-trimethoxybenzaldehyde, 3,4-dimethoxybenzaldehyde were used in reaction.[21b]

8.2. Hydrogen Abstraction by Thiyl Radicals in Water

Zhao et al.[22] had derived rate constants for H-abstraction by cysteine thiyl radicals at pH 10.5 from anionic glycine (3.2×10^5 $M^{-1}s^{-1}$) and alanine, respectively ($7{,}7 \times 10^5$ M^{-1} s^{-1}). The deprotonated amino group ensures optimum captodative stabilization of the $^{\alpha}C\bullet$ radical. However, deprotonated aliphatic amines are physiologically unrealistic.

Schöneik et al.[23] reported a H atom abstraction reaction from peptides by thiyl radicals, directly relevant for aminoacids within proteins. The substrates in their work cover a broad range of $^{\alpha}$C-H bond energies. For example, in *N*-formyl-Asp-NH_2, the calculated BDE($^{\alpha}$C-H) = 332KJ/mol, and in *N*-acetyl-Pro- NH_2, BDE($^{\alpha}$C-H) = 369KJ/mol (*trans*-Pro) and 358 KJ/mol (*cis*-Pro).

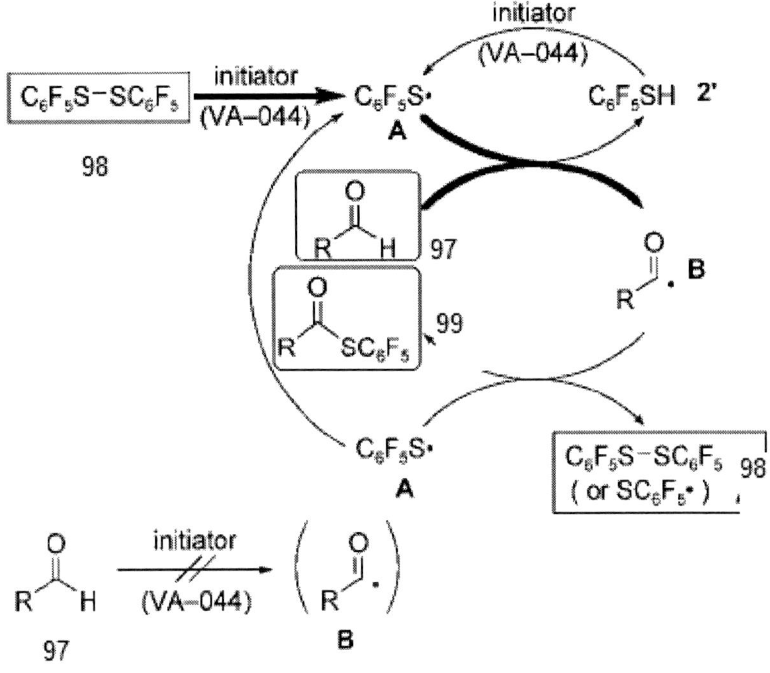

Scheme 17. Proposed reaction mechanism for a carbon-sulfur bond formation in water.

$PhSO_2CHXPO(OEt)_2$

HMDS/THF/-78 °C

100

101 X = H, X = F 102

	R	R[1]
a	C_6H_5	H
b	4-$(CH_3)C_6H_4$	H
c	4-$(CF_3)C_6H_4$	H
d	3,4-$(MeO)_2C_6H_3$	H
e	$C_6H_5CH_2CH_2$	H
f	o-C_6H_{11}	H
g	-$(CH_2)_5$-	
h	C_6H_5	CH_3

Scheme 18. Synthesis of *E*-vinyl and (*E/Z*)- (α-fluoro)vinyl sulfones.

8.3. Synthesis of (α-fluoro)vinyl Sulfides by Thiodesulfonylation of Vinyl Sulfones

Removal of the sulfonyl group from vinylic carbon is usually carried out by reductive methods or addition-elimination processes where the sulfonyl group is replaced by a tributylstannyl substituent, which is then replaced by H.

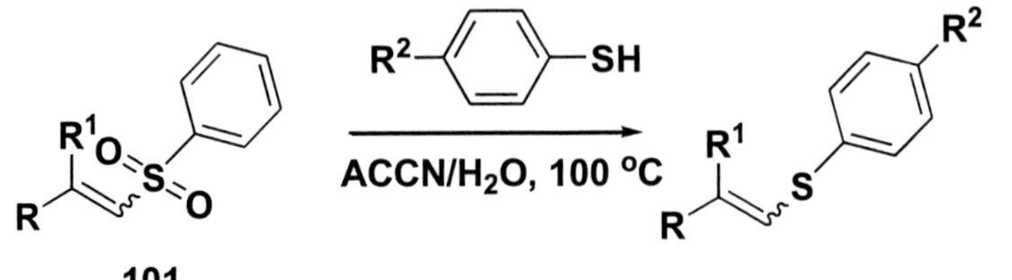

R^2–C6H4–SH

ACCN/H_2O, 100 °C

101

	R	R^1
a	C_6H_5	H
b	4-$(CH_3)C_6H_4$	H
c	4-$(CF_3)C_6H_4$	H
d	3,4-$(MeO)_2C_6H_3$	H
e	$C_6H_5CH_2CH_2$	H
f	o-C_6H_{11}	H
g	-$(CH_2)_5$-	

103 $R^2 = H$
104 $R^2 = CH_3$
105 $R^2 = NH_2$
110 $R^2 = CO_2Me$

Scheme 19.Thiodesulfonylation of vinyl Sulfones. Synthesis of vinyl sulfides.

Wnuk et al.[24] reported the stereoselective radical-mediated thiodesulfonylation of vinyl and (α-fluoro)vinyl sulfones in water. Such thiodesulfonylation provides a flexible alternative to the hydrothiolation of alkynes with thiols under radical or metal catalysis conditions. It also offers a convenient preparation of (α-fluoro)vinyl sulfides. This methodology can be envisaged as a reductive deoxygenation of sulfones to the corresponding sulfides.

Treatment of the sulfonyl-stabilized enolates generated from diethyl(phenylsulfonyl)methylphosphonate with aliphatic and aromatic aldehydes and ketones*100a-g* gave the corresponding *E*-vinyl sulfones *101a-f* and vinyl sulfone *101g* (72-95%, Scheme 18). Analogous treatment of *100a-h* with diethylfluoro(phenylsulfonyl)methylphosphonate produced (α-fluoro)vinyl sulfones *102a-h.*

Treatment of *E-101a-c* with benzenethiol afforded *103a-c* in 95%, 54% and 85% yields respectively, when the reaction was carried out in water, and initiated by ACCN (1,1′-azo*bis*(carbonitrilecyclohexane)), Scheme 18.

Treatment of *E-101a* with 4-methylbenzenethiol or 4-aminobenzenethiol in H_2O/ACCN,produced the corresponding vinyl sulfides *104a* (61%) and *105a* (55%). Analogously, *E-5c* was converted to *104c* and *105c* in 58 and 55% yields, respectively(Scheme 19).

Radical-mediated thiodesulfonylation of the vinylsulfones *101* occurred with retention of the *E* stereochemistry. In order to study the stereochemical outcome of the thiodesulfonylation reactions, *Z* –vinyl sulfones *101a* was prepared by anti-Markovnikov addition of PhSH/NaOH to phenylacetylene followed by the oxidation of the resulting (*Z*)-2-phenyl-1-phenylthioethene. Treatment of *Z-101a* with PhSH in aqueous medium produced sulfide *103a* in very good yields with inversion of stereochemistry (*E/Z*, 95:5). Thus the vinyl sulfides are formed predominantly with *E*-stereochemistry independently of the stereochemistry of the starting vinyl sulfones.

Thiodesulfonylation appears to be fairly general since sulfones *100b-d* with alkyl (methyl), electron withdrawing (CF_3) and electron donating substituents (CH_3O) on the phenyl ring attached to the double bond also produced(α-fluoro)vinyl sulfides.

R^2–C_6H_4–SH; ACCN/H_2O, 100 °C

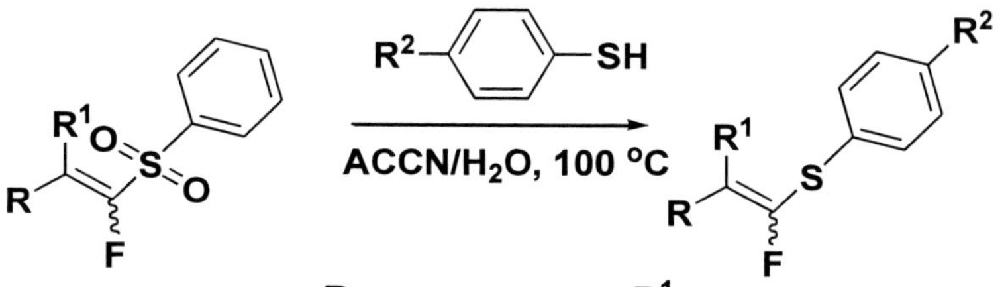

102

	R	R^1
a	C_6H_5	H
b	$4\text{-}(CH_3)C_6H_4$	H
c	$4\text{-}(CF_3)C_6H_4$	H
d	$3,4\text{-}(MeO)_2C_6H_3$	H
e	$C_6H_5CH_2CH_2$	H
f	$o\text{-}C_6H_{11}$	H
g	$-(CH_2)_5-$	
h	C_6H_5	CH_3

111 $R^2 = H$
112 $R^2 = CH_3$
113 $R^2 = NH_2$
114 $R^2 = CO_2Me$

Scheme 20. Thiodesulfonylation of (*E/Z*)- (α-fluoro)vinyl Sulfides.

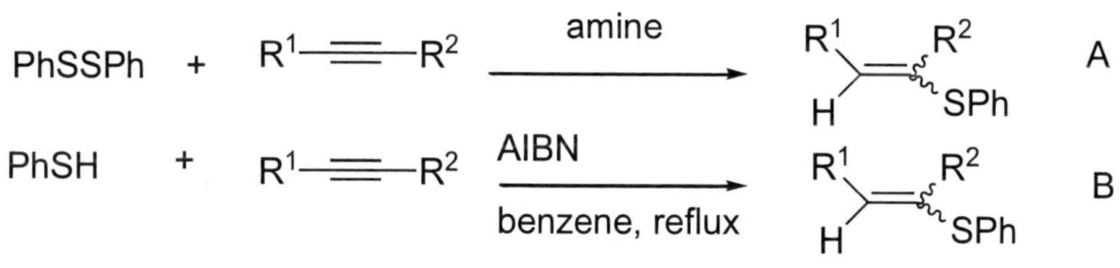

PhSSPh + R^1–≡–R^2 → (amine) R^1(H)C=C(R^2)SPh A

PhSH + R^1–≡–R^2 → (AIBN, benzene, reflux) R^1(H)C=C(R^2)SPh B

Scheme 21. Radical synthesis of 1-akenylsulfides.

Radical thiodesulfonylation permitted the synthesis of the sparsely developed (α-fluoro)vinyl sulfides *111-114* in high yields (Scheme 20).

Reaction of *102h* (*E/Z*, 57:43) with benzenethiol also afforded tetrasubstituted (α-fluoro)vinyl sulfide *111h* in 58% yield. Reactions of *102b-d* with benzenethiol produced *111b-d*in 82%, 60% and 73% yields, respectively, when the reactions are carried out in water, and initiated by ACCN. Thiodesulfonylation occurred with other aromatic thiols as well. In Scheme 20, a summary of the scope of this reaction is presented, with 4-methylbenzenethiol, and 4-aminobenzenethiol. It is noteworthy that hydrothiolation of alkynes is inapplicable for the synthesis of(α-fluoro)vinyl sulfides since the 1-fluoralkynes are unstable and virtually unknown. Desulfonylation occurred probably via β-elimination of the sulfonyl radical from the radical intermediates formed after addition of PhS• to vinyl sulfones (presumably via a radical addition-elimination mechanism).

8.4. Radical Hydrothiolation of Alkynes with Diphenyldisulfide and Tripropylamine

It is well-known that the benzenethiyl radical formed from thiophenol in the presence of AIBN undergoes an addition reaction with alkynes to give vinyl sulfides (Scheme 21, eq B).[25] In this reaction, the use of ill-smelling thiophenol and a hazardous radical initiator such as AIBN was required.

Although many reactions of alkynes with disulfides have been developed, disulfidation products are generally obtained.[26] However, reductive radical reaction of alkynes with disulfide to give 1-alkenyl sulfides has not been explored.

Ishibashi and collaborators[27] have developed a method for the formation of benzenethiyl radical from diphenyl disulfide with tripropylamine via a single electron transfer (SET) reaction in water. Inexpensive and environmentally friendly reagents were employed, and the experimental procedure is very simple and safe, according to Scheme 22.

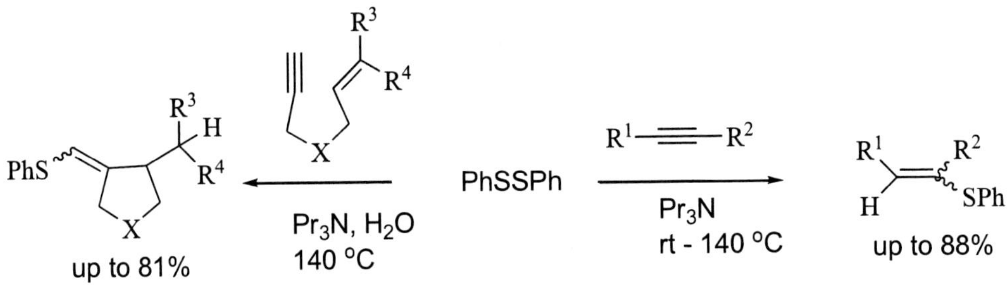

PhSSPh; R^1—≡—R^2; Pr_3N, rt - 140 °C; up to 88%; Pr_3N, H_2O, 140 °C; up to 81%

Scheme 22. Benzenethiyl radicals from diphenyldisulfide in water.

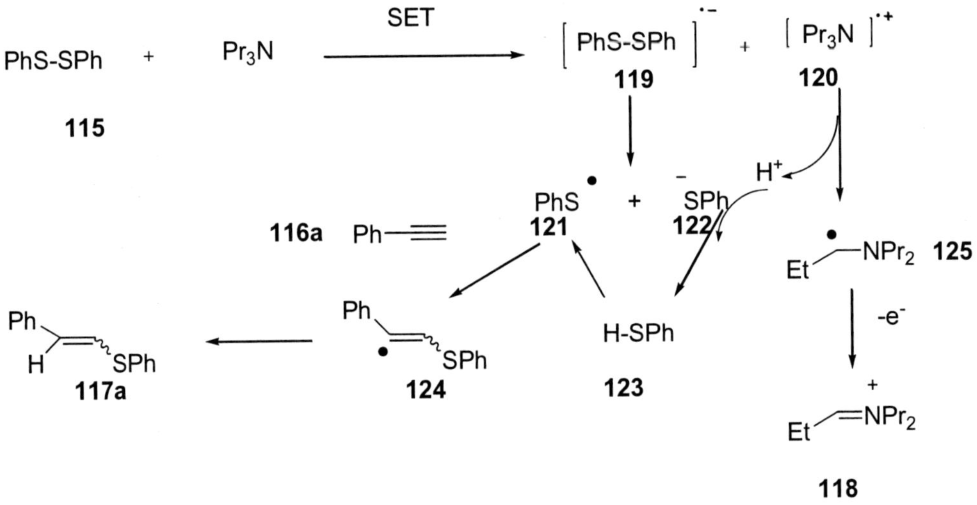

PhS-SPh 115 + Pr_3N —SET→ [PhS-SPh]$^{\bullet-}$ 119 + [Pr_3N]$^{\bullet+}$ 120; PhS$^\bullet$ 121 + $^-$SPh 122; H-SPh 123; 116a Ph—≡; 124; 117a; 125; $-e^-$; 118

Scheme 23. Mechanism for the radical synthesis of alkenylsulfides in water.

A plausible mechanism for the reaction is shown in Scheme 23. The reaction may be initiated by single electron transfer (SET) process of tripropylamine to disulfide *115* to generate an anion radical *119* and a cation radical *120*. The S-S bond cleavage of anion radical *119* generates benzenethiyl radical *121* and thiolate anion *122*. An attack of thiyl radical *121* on the alkyne *116a* gives the vinyl radical *124*, which then abstracts a hydrogen atom from benzenethiol *123* to give vinyl sulfide*117a* together with the benezenthiyl radical *121*. It should be noted that the formation of benzenethiol *123* might be a result of the removal of a proton of cation radical *120* (proton removal of *120* gives radical *125*). Therefore, a single electron transfer reaction between tripropylamine and diphenyl disulfide*117* to form cation radical *120* is important to give vinyl sulfides *117a*. Partial formation of compound *118a* might be a result of an attack of vinyl radical *124* on diphenyl disulfide *115* (Scheme 23).

9. Rates of Radical Reactions in Water

Solvent polarity can have tremendous effects on the kinetics of reactions involving charged species in solution. On the other hand, reactions of neutral radicals are less sensitive to solvent polarity effects, mainly because charged species are not involved, and there is not a significant change in dipole moment in the progression from reactants to transition state. However, other solvent properties such as viscosity or internal pressure can influence the rate of certain radical reactions; such solvent effects are much more difficult to detect in polar reactions because they are masked by the overwhelming effect of solvent polarity.

For example, solvent viscosity can affect the rate and product distribution when radical caged-pairs (geminate or diffusive) are involved. Internal pressure can influence rate if there is a difference in the volume of the reactants compared to the transition state ($\Delta V act \neq 0$) and can influence the relative rate of some radical reactions. However, these solvent effects are generally small, with changes in rate or product distribution not much greater than an order of magnitude. Consequently, instances where solvent *dramatically* affects the rate or selectivity of reactions involving neutral radicals are rare and noteworthy.

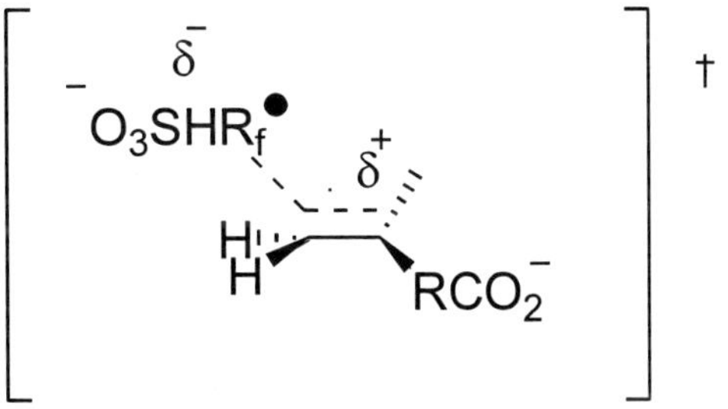

Figure 12. Coulombic effects postulated on the transition state for the radical addition of $^{\bullet}R_fSO_3^-$ onto alkenes.

The classic example of a significant solvent effect in a radical reaction involves free radical chlorinations of alkanes conducted in benzene solvent. The chain-carrying chlorine atom forms a complex with benzene, lowering its reactivity and increasing its selectivity (by nearly 2 orders of magnitude) in hydrogen atom abstractions. A more recent example of a significant solvent effect was reported by Ingold and co-workers, who found that rate constants for hydrogen atom abstractions from phenols were reduced in solvents where the phenol was stabilized by hydrogen bonding.[26c] In this case, it was the reactivity of the substrate, not the radical, that was diminished as a result of a solute/solvent interaction.

The fastest rate constants ever reported for this type of radical reaction were determined by Ingold and Lusztyk in water as sole solvent.[7] Aryl radicals generated from sodium 4-iodophenylsulfonate by laser flash photolysis reacted with non-activated olefins at rates of $k \approx$ 2–3 $x10^7$ M^{-1} s^{-1}, and the authors pointed out that the fast and selective aryl radical addition was largely due to the aqueous solvent.

For comparison, rate constants as low as $k \approx 5x10^5$ M^{-1} s^{-1} were derived in earlier experiments conducted in tetrachlorocarbon.[26b]

A key element to achieve selectivity in reactions of aryl radicals appears to be the use of water as a solvent or at least co-solvent.[7]

Ingold and Dolbier have initiated a laser flash photolysis (LFP) study[27] on the rates of perfluoroalkyl radical addition on to double bonds and found that the rates of addition in water are significantly larger than in perfluorocarbon solvents.

The authors[27] studied the reactions of the water-soluble radicals $^{\bullet}R_fSO_3^-$ on the addition of water-soluble alkenes of the type CH_2=$CRCO_2Na$ or other carboxylate-substituted terminal alkenes. They postulated that the observed rate enhancements in water almost certainly reflect the greater ability of this polar solvent to stabilize the polar transition state for addition of $^{\bullet}R_fSO_3^-$to the alkenes, such as that depicted in Figure 12.

The observed rate enhancements are all the more remarkable in view of the Coulombic repulsion that must be met in a transition state such as that depicted in Figure 12,involving two negatively charged ions. It was recognized as early as 1922 by Bronsted that for “ions of the same sign the repulsive forces will tend to keep them apart.

A general expectation is that the addition reactions between the negatively charged radical and negatively charged alkene substrates would be *inhibited* by electrostatic repulsion. However, the presence of a polar solvent or a medium of increased ionic strength should at least partially ameliorate the inherently detrimental electrostatic effects present in a reaction between two negatively charged ions and, hence, facilitate such reactions, mainly by hydrogen-bonding stabilization and charge dispersion.

The authors[27] also postulate the presence of ionic strength effects.

Considering the dilute solutions used in the LFP experiments the authors arrive at the conclusion that solvent effects provide the simplest and most reasonable explanation for the observed rate constant differences. If not for the negative impact of Coulombic repulsion in the transition state, these rate enhancements would most certainly be much larger.

Dolbier et al. [28] have found that perfluorinated radicals were much more reactive than their hydrocarbon counterparts in addition to normal, electron rich alkenes such as 1-hexene (40 000 times more reactive) in organic solvents, and that H transfer from $(Me_3Si)_3SiH$ to a perfluoro-*n*-alkyl radical such as *n*-$C_7F_{15}CH_2CH(\bullet)C_4H_9$ was 110 times more rapid than to the analogous hydrocarbon radicals (eq 21).

Barata-Vallejo and Postigo[14b] have reported the radical perfluoroalkylation of organic solvent-soluble alkenes in water, and found that the relative rates of radical addition / reduction in water were higher than those reported in organic solvents, for consecutive reactions.

According to what has been observed and measured by Dolbier et al.[28], in benzene-d_6, the ratio of products [*128*]/[*127*] (eq 21) should equal the ratio of rate constants for addition (of perfluorinated heptyl radical on 1-hexene) and rate constant for H abstraction from $(Me_3Si)_3SiH$ times the ratio of concentrations of alkene and silane (see also Chapter II).

$$n\text{-}C_7F_{15}I \xrightarrow{h\upsilon} n\text{-}C_7F_{15}^{\bullet} \begin{cases} \xrightarrow[k_{add}]{\text{CH}_2=\text{CHC}_4H_9} C_7F_{15}CH_2\dot{C}HC_4H_9 \ (\mathbf{126}) \xrightarrow{(Me_3Si)_3SiH} C_7F_{15}CH_2CH_2C_4H_9 \ (\mathbf{128}) \\ \xrightarrow[(Me_3Si)_3SiH]{k_H} n\text{-}C_7F_{15}H \ (\mathbf{127}) \end{cases} \quad (21)$$

According to the experimental conditions reported by Barata-Vallejo and Postigo[14b], employing equation 22, the authors would obtain a theoretical ratio of perfluoroalkylated alkane over reduced perfluoroalkane of *ca.* 1.3, which is not completely in agreement with the unobserved reduced perfluoroalkanes in their reaction systems (*i.e.*: $CHF_2(CF_2)_4CF_3$, when iodoperfluorohexane is employed).[14b]

$$\frac{[\mathbf{128}]}{[\mathbf{127}]} \quad \frac{k_{add}\,[\text{1-hexene}]}{k_H\,[(Me_3Si)_3SiH]} \qquad (22)$$

The electrophilicity of $R_f\bullet$ radicals are the dominant factor giving rise to their high reactivity. The stronger carbon-carbon bond which forms when $R_f\bullet$ versus $R\bullet$ radicals add to an alkene is a driving force for the radical addition (the greater exothermicity of the $R_f\bullet$ radical addition is expected to lower the activation energy).[29] It has been observed, in organic solvents, that the rates of addition of $R_f\bullet$ radicals onto alkenes correlate with the alkene IP (which reflects the *HOMO* energies).[30] Indeed, the major transition state orbital interaction for the addition of the highly electrophilic $R_f\bullet$ radical to an alkene is that between the *SOMO* of the radical and the *HOMO* of the alkene. Thus, the rates of $R_f\bullet$ radical addition to electron deficient alkenes are slower than those to electron rich alkenes (as observed in organic solvents). From the results in water,[14b] however, it becomes apparent, that in water the reactivity for both electron rich and electron deficient alkenes towards $R_f\bullet$ radical addition could be comparable. In order to clarify this subtle aspect of the reaction in water, the authors[14b] undertook a set of experiments designed to compare the ratios of $(k_H/k_{add})_{1\text{-hexene}}$ and $(k_H/k_{add})_{acrylonitrile}$for the addition reaction of n-$C_6F_{13}I$ to the electron rich 1-hexene and electron deficient acrylonitrile, respectively. These ratios of rate constants are obtained by plotting [n-$C_6F_{13}H$]/[C_6F_{13}-C_6H_{13}] *vs* [$(Me_3Si)_3SiH$]/[1-hexene] and [n-$C_6F_{13}H$]/[C6F13-CH_2CH_2CN] *vs* [$(Me_3Si)_3SiH$]/[acrylonitrile], respectively, when the reactions are initiated thermally, by using incremental amounts of $(Me_3Si)_3SiH$, and keeping the alkene and n-$C_6F_{13}I$ concentrations constant. They obtained slopes for both plots equal to 1.55±0.09 (r^2=0.998) and 1.88±0.19 (r^2=0.989) for $(k_H/k_{add})_{1\text{-hexene}}$ and $(k_H/k_{add})_{acrylonitrile}$ respectively. This seems to indicate that the reactivities of electron rich and electron deficient alkenes towards $R_f\bullet$ radicals in water are leveled off.

The ratio of rates constants $(k_H/k_{add})_{1\text{-hexene}}$ obtained in benzotrifluoride as solvent[28] for the reaction of n-$C_7F_{15}I$ and 1-hexene with $(Me_3Si)_3SiH$ as the hydrogen donor (eq 21, 22) is 6.32, while that same ratio of rate constants for the reaction of n-$C_6F_{13}I$ with 1-hexene in water is 1.55 (*vide supra*). Owed to the unavailability of the rate constant for R_f• radical addition onto double bonds*in water* makes comparisons difficult; however, the results from Barata-Vallejo and Postigo[14b] would seem to imply that the rate for hydrogen donation from $(Me_3Si)_3SiH$ to the R_f• radical relative to the addition reaction is four times slower in *water* than in benzotrifluoride as solvent (*i.e.*: $(k_H/k_{add})_{water}$ / $(k_H/k_{add})_{Benzotrifluoride}$ = 0.25). Indeed, it is likely that water does influence the H-transfer step from the silane to the C-radical. This is well established and important for phenols, however for silanes this is not known. In the phenol series H-bonding of the phenol with the solvent leads to a stabilization and consequently to less efficient H-transfer. H-bonding from the silane to water might have a similar effect, as the authors report on slower H-transfer in water.[14b] It can also be hypothesized that given the known hypervalency of silicon, coordination of a water molecule can diminish the hydrogen atom donation of the silicon hydride.

As mentioned above, by plotting [*n*-C_6F_{13}H]/[*1a*] *vs* [$(Me_3Si)_3SiH$]/[1-hexene], a straight line, whose slope represents $(k_H/k_{add})_{1\text{-hexene}}$ is obtained (with a value of 1.55±0.09, r^2=0.998). The intercept of this plot, shows, remarkably, no deviation from the ideal value of zero, purporting that the only source of *n*-C_6F_{13}• radical reduction (*i.e.*:*n*-C_6F_{13}H) is the silane, and not the solvent or the alkene.[30]

Newcomb, Horner, and collaborators[32] have measured the intramolecular rate constant for cyclization of radical *130b* in acetonitrile-water, and found a value of k = 4 x 10 ^{7} s^{-1} (eq 23).

129 **130** **131** (23)

a: X = CO_2Et; b: X = CO_2H; c: X = CO_2^-

Radical anion *130c*, from photolysis of *129c* in acetonitrile-water solution (eq 23), gave distinctly different results. The rate constant for cyclization of *130c* at 20 °C was k = 3.5 x 10 ^{6} s^{-1} at pH 7.4, an order of magnitude smaller than the rate constant for cyclization of *130b*. The pronounced effect of the carboxylate group in *130c* is noteworthy because the identity of the X group in other radicals*130* has little effect on the rates of cyclization; for radicals *130* with X =H, CH_3, OCH_3, CO_2Et, $CONEt_2$, and CO_2H, the rate constants for cyclization at 22 °C are in the range k = (2-5) x 10 ^{7} s^{-1} . A temperature-dependent kinetic study performed by the authors[32] showed that the reduction in the rate constant for cyclization of radical anion *130c* is due to an enthalpy effect in the transition state that slows the reaction, even though the entropy of activation for the reaction is more favorable than those for cyclizations of analogues with other X groups. Specifically, the Ea for cyclization of radical anion *130c* (eq 23) at pH 7.4 of 6.5 kcal/mol is ca. 3 kcal/mol greater than that found for any other radical *130* thus far studied. The authors[32] speculate that charge delocalization in radical anion *130c* and subsequent localization upon cyclization to *131c* is an important enthalpy feature.

The significant difference in rate constants for cyclizations of radical *130b* and radical anion *130c* permitted a kinetic titration study[32] that demonstrated the robust nature of the approach at various pH. Reactions were conducted in buffered acetonitrile-water solutions, and the observed rate constants were fit to eq 24, where k_A and k_B are rate constants for the cyclization of the acid*(130b)* and basic *(130c)* forms, respectively, and K_a is the acidity constant for *130b*.

$$k_{obs} = (k_A [H^+] + k_B K_a) (K_a + [H^+])^{-1} \qquad (24)$$

Regression analysis gave $k_A = (3.43 \pm 0.08) \times 10^{7}\ s^{-1}$, $k_B = (3.4 \pm 1.0) \times 10^{6}\ s^{-1}$, and Ka = $(2.3 \pm 0.4) \times 10^{-5}$. The apparent *p*Ka of α-carboxylic acid radical *130b* in water is 4.6 . The α-radicals from small alkanoic acids have *p*Ka values similar to those of the parent acids. Access to α- and β-carboxylate radicals could be important for studies of rearrangements catalyzed by coenzyme B12-dependent enzymes, as suggested by the authors[32] because it can produce model radical anions that closely resemble the reactive species in nature. The large effect of the carboxylate group in the kinetic parameters for cyclization of *130c*, a reaction that might be expected to have low sensitivity to charge effects, suggests that neutral ester and carboxylic acid radicals might be poor models for the radical anions. The authors[32] speculated that the negative charge in the radical anions is a critically important feature in rearrangements catalyzed by coenzyme B12-dependent enzymes, one that permits heterolytic fragmentation reaction pathways that cannot be accessed from neutral radicals.

The effects of water on radical reactivity can become of relevance in biological processes. Thus Barclay and collaborators[33] have discovered that bilirubin (BR) has strong antioxidant properties in water as compared to non-polar media, where the rates of BR with peroxyl radicals (5×10^{4}, $M^{-1}s^{-1}$)is comparable to that of vitamin E and trolox.

The antioxidant properties of BR depended markedly on the medium of the reaction. In contrast to its weak effect in a hydrocarbon-chlorobenzene solution, it displayed strong antioxidant activity in aqueous SDS micelles, phosphate buffer, pH = 7.4. In this medium, BR inhibited the oxidation of methyl linoleate initiated with azo-bis-(amidinopropane)dihydrochloride (ABAP) using the oxygen electrode or the pressure transducer system.[33] The authors propose an electron transfer mechanism to account for this large difference.[33]

The SET reaction with ionized bilirubin, *132* , and peroxyl radicals, *133*, known to be strongly (Scheme 24) polarized, is proposed as a plausible reaction mechanism (see Scheme 24). The initial SET would form an ion pair, *134*, and reactions of *134* such as direct proton transfer or via a separated ion pair would form a pyrrole radical, *135*, andthe hydroperoxide *136*. Since bilirubin possesses two pyrrole rings, it could deactivate two peroxyl radicals by SET. The proposed SET mechanism (Scheme 24) accounts for the remarkable effects of the reaction medium on the antioxidant activity of BR, and earlier reports on the antioxidant properties of BR in aqueous lipid dispersions probably also involved SET reactions.[33]

Although ordinary carbon-centered radicals have largely been referred to as rather insensitive to water effects, a distinct case is that given by hydroxyl radicals. An interesting example was given by Tanko and collaborators,[34] who found that the hydroxyl radical reacts through a polar transition state. As a consequence of this polarization, hydrogen bonding to water stabilizes the transition state, resulting in larger rate constants when water is the solvent. This explanation also predicts that the *magnitude* of the solvent effect will decrease when electronegative substituents are attached on the α-carbon of the substrate. The solvent effect is also significant when the substrate possesses electron-donating substituents, presumably because of competing inductive (electronwithdrawing) and resonance (electron-donating) effects.

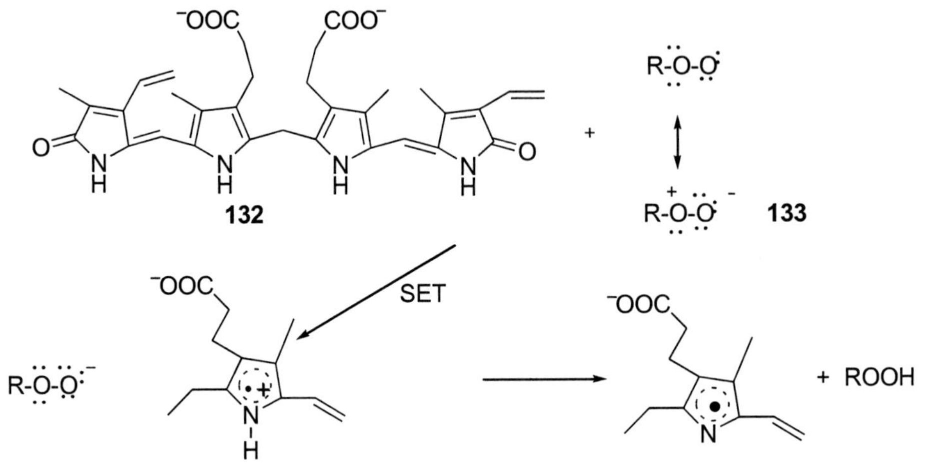

Scheme 24. Single electron transfer mechanism for the antioxidant activity of BR in water toward peroxyl radicals.

For the addition reactions of alkenes onto double bonds, calculations suggested that the hydroxyl moiety was nearly anionic in the transition state, providing a clear opportunity for water stabilization via hydrogen bonding.[34] The effect water has on modulating HO• reactivity has enormous implications. In biological systems, this means that HO• may be less reactive in the hydrophobic regions of a cell than previously believed, *i.e.*, hydroxyl radical does not necessarily react with the first molecule it encounters. Indeed, Nature may use this as a containment strategy when hydroxyl radical is produced naturally within cells.[34a] Abstraction of the 4′-hydrogen atom from the sugar-phosphate backbone of duplex DNA is a major reaction for hydroxyl radical (HO•)

A large number of base damage products arising from the reaction of hydroxyl radical with DNA have been characterized. These products arise via hydrogen atom abstraction or, more commonly, addition of hydroxyl radical to the π-bonds of the bases. Here, we consider the formation of several common DNA base damage products arising from reactions at thymidine and deoxyguanosine. The pathways described here produce some of the most prevalent oxidative base damage products and also illustrate the general types of reactions that commonly occur following the attack of radicals on the nucleobases.[34b,c]

There has been an increasing emphasis on the development of alternative hydrogen transfer agents due in large part to concerns about the toxicity of tin-containing compounds. Some of the more exciting advances in this direction involve the use of water and alcohols as safe "green" hydrogen atom transfer agents. The high O-H bond dissociation energies (BDEs) of alcohols (*ca.* 105 kcal/mol) and water (118 kcal/mol) suggest that hydrogen atom transfers from these sources to carbon-centered radicals will be too slow to be useful, but Lewis acid complexed alcohols and water have much reduced O-H BDEs and can react with alkyl radicals rapidly.[35]

Newcomb and collaborators[35b] have found that at room temperature the rate constants for reactions of triethylborane complexes of water and methanol as H-atom donors to alkyl radicals are only 2 orders of magnitude smaller than those for reactions of tin hydride reagents. These reactions are adequately fast at room temperature for some synthetic applications, and they are increasingly competitive with radical rearrangements at reduced temperatures. One anticipates increasing numbers of applications of borane complexes as radical reducing agents given the ease of removal of the boron-containing byproducts, especially if a mixed reagent system can be developed that permits the use of common alkyl halides as radical precursors.[35b]

Newcomb and collaborators [36] also determined the rate constants for reactions of $Cp_2Ti^{III}Cl$-complexed water and methanol with a secondary alkyl radicals. At ambient temperature in THF, the titanium(III) reagent complexed with deuterium oxide and water reacts with the secondary 1-dodecyl cyclobutyl radicalwith rate constants of 1.0×10 and 2.3×10^4 M^{-1} s^{-1}, respectively. In benzene containing 0.95 M methanol, the Cp_2TiIIICl-MeOH complex reacts with a rate constant of 7.5×10^4 M^{-1} s^{-1}.[36] The titanium(III) reagent apparently activates water and methanol more strongly than Et_3B with the result that the H-atom transfer reaction of the $Cp_2Ti^{III}Cl$-H_2O complex is 5 times as fast as the H-atom transfer reaction of Et_3B-H_2O at room temperature.[36]

The authors inform that the Arrhenius function for reaction of $Cp_2Ti^{III}Cl$-H_2O had a normal entropic term; however, the unusually low entropic term for the borane-water complex leads to more efficient hydrogen atom transfer trapping by this species at low temperatures. Radical reductions by H-atom transfer from water or alcohol complexes of $Cp_2Ti^{III}Cl$ will be useful when radicals are generated by reduction of epoxides or α,β-unsaturated ketones.[36]

It is noteworthy that the rate constants found in this work are about 1 order of magnitude smaller than those for radical reduction reactions of tin hydrides and similar to those for reactions of $((CH_3)_3Si)_3SiH$.[37] The reactions are fast enough to be used in many radical chain reaction sequences, and the kinetics illustrate a large degree of O-H bond activation possible by complexation of simple hydroxylic compounds with a strong Lewis acid.

More recently, Cuerva and Cárdenas[38] have demonstrated the initial assumption that titanocene(III) -water complexes are a unique class of HAT reagents. They are able to reduce efficiently carbon-centered radicals of diverse nature. The success of this transformation is based on two key features: (a) an excellent binding capabilities of water toward titanocene(III) complexes and (b) a low activation energy for the HAT step.[38] Therefore, the observed reactivity can be explained in the framework of an unprecedented HAT reaction involving water.[38]

REFERENCES

[1] Yorimitsu, H.; Shinokubo, H.; Matsubara, S.; Oshima, K. Triethylborane-Induced Bromine-Atom Transfer Radical Addition in Aqueous Media: Study of the Solvent Effect on Radical Addition Reactions. *J.Org.Chem.* 2001, *66*, 7776-7785.

[2] Kondo, J.; Shinokubo, H.; Oshima, K. From Alkenylsilanes to Ketones with Air as an Oxidant. *Angew.ChemieInt.Ed.* 2003, *42*, 825-827.

[3] Miyabe, H.; Ueda, M.; Naito, T. Free Radical Reactions of Imines in Water. *J.Org.Chem.* 2000, *65*, 5043-5047. Miyabe, H.; Fujii, K.; Goto, T.; Naito, T. A New Alternative to the Mannich Reaction: Tandem radical Addition-Cyclization Reaction. *Org.Lett.* 2000, *2*, 4071-4074.

[4] McNabb, S.B.; Ueda, M.; Naito, T. Addition of Electrophilic and Heterocyclic Carbon-Centered Radicals to GlyoxylicOxime Ethers. *Org.Lett.* 2004, *6*, 1911-1914.

[5] Miyabe, H.; Yamaoka, Y.; Takemoto, Y. Triethylborane-Induced Intermolecular Radical Additions to Ketimines. *J.Org.Chem.* 2005, *70*, 3324-3327.

[6] Tu, Z.; Lin, C.; Jang, Y.; Yang, Y-J.; Ko, S.;Fang, H.; Liu, J-T.; Yao, C-F. One-pot Synthesis of Malononitriles by Free Radical Reactions of Ylidenmalononitriles with Et_3B and Iodoalkane in water –ether Biphase medium.*TetrahedronLett.* 2006, *47*, 6133-6137.

[7] Garden, S. J.; Avila, D. V.; Beckwith, A. L. J.; Bowry, V. W.; Ingold, K. U.; Lusztyk, J. Absolute Rate Constant for the Reaction of Aryl Radicals with Tri-n-butyltin Hydride. *J.Org. Chem.* 1996, 61, 805-809.

[8] Heinrich, H.R.; Wetzel, A.; Kirschstein, M. Intermolecular Radical Carboaminohydroxylation of Olefins with Aryl Diazonium Salts and TEMPO. *Org.Lett.* 2007, *9*, 3833-3835

[9] Heinrich, M.; Blank, O.; Ullrich, D.; Kirschstein, M. Allylation and Vinylation of Aryl Radicals Generated from Diazonium Salts. *J.Org.Chem.* 2007, *72*, 9609-9616.

[10] Yorimitzu, H.; Nakamura, T.; Shinokubo,H.; Oshima, K. Triethylborane-mediated Atom Transfer Radical Cyclization in Water. *J.Org.Chem.* 1998, *63*, 8604-8605.

[11] a.-Wakabayashi, K.; Yorimitzu, H.; Shinokubo, H.; Oshima, K. Radical Cyclization of N-allyl-2-halo Amide in Water. *Bull.Chem.Soc.Jpn.* 2000, *73*, 2377-2378. b.-Yorimitsi, H.; Nakamura, T.; Shinokubo, H.; Oshima, K.; Omoto, K.; Fujimoto, H. Powerful Solvent Effect of Water in Radical Reaction: Triethylborane-Induced Atom-Transfer Radical Cyclization in Water. *J.Am.Chem.Soc.* 2000, *122*, 11041-11047.

[12] Kita, Y.; Nambu, H.; Ramesh, N.G.; Anilkumar, G.; Matsugi, M. A Novel and Efficient Methodology for the C-C Bond Forming Radical Cyclization of Hydrophobic Substrates in Water. *Org.Lett.* 2000, *3*, 1157-1160.

[13] a.-Yorimitsu, H; Wakabayashi, K.; Shinokubo, H.; Oshima, K. Radical Addition of 2-Iodoalkalamide or 2-Iodoalkanoic Acid to Alkenes with a Water-soluble Radical Initiator in Aqueous Media: Facile Synthesis of γ-Lactones. *Bull-Chem.Soc.Jpn.* 2001, *74*, 1963-1970. b.- Godineau, E.; Schenk, K.; Landais, Y. Synthesis of Fused Piperidinones through a Radical-Ionic Cascade. *J.Org.Chem.* 2008, *73*, 6983-6993. c.- Bella, A. F.; Slawin, A. M. Z.; Walton, J. C. Approach to 3-aminoindolin-2-ones via Oxime Ether Functionalized Carbamoylcyclohexadienes.*J. Org. Chem.* 2004, *69*, 5926–5933. d.- Boiteau, L.; Boivin, J.; Liard, A.; Quiclet-Sire, B.; Zard, S. Z. A Short Synthesis of (±) Matrine. *Angew.Chem., Int. Ed.* 1998, *37*, 1128–1131. e.- Roberts, B. P. Polarity-Reversal Catalysis of Hydrogen-Atom Abstraction Reactions: concepts and applications in organic chemistry*Chem. Soc. Rev.* 1999, *28*, 25–35.

[14] a.-Postigo, A.; Kopsov, S.; Ferreri, C.; Chatgilialoglu, C. Radical Reactions in Aqueous Medium Using $(Me_3Si)_3SiH$.*Org. Lett.* 2007, *9*, 5159-5162. b.- Barata-Vallejo, S.; Postigo, A. $(Me_3Si)_3SiH$-mediated Radical Perfluoroalkylation Reactions of Olefins in Water.*J.Org.Chem.* 2010, *75*, 6141-6148. c.-Khan, T.A.; Tripoli, R.; Crawford, J.L.; Martin, C.G.; Murphy, J.A. Diethylphosphine Oxide (DEPO): High-Yielding and

Facile Preparation of Indolones in Water.*Org.Lett.* 2003, *5*, 2971-2974. d.- Gonzalez Martin, C.; Murphy, J.A.; Smith, C.R. Replacing tin in radical chemistry: *N*-ethylpiperidine Hypophosphite in Cyclisation Reactions of Aryl Radicals. *TetrahedronLett.* 2000, *41*, 1833. e.-Leca, D.; Fensterbank, L.; Lacote, E.; Malacria, M. Recent Advances in the Use of Phosphorus-Centered Radicals in Organic Chemistry. *Chem.Soc.Rev.* 2005, *34*, 858-865. g.- Ceylan, M.; Fındık, E. Acetate—Mediated Addition of Dimedon to Chalcone Derivatives.*Synth. Commun.*2008, *38*, 1070. h. Wang, G.-W.; Dong, Y.-W.; Wu, P.; Yuan, T.-T.; Shen, Y.-B. Unexpected Solvent-Free Cycloadditions of 1,3-Cyclohexanediones to 1-(Pyridin-2-yl)-enones Mediated by Manganese(III) Acetate in a Ball Mill.*J. Org.Chem.* 2008, *73*, 7088-7095. i.- Ye, Y.; Wang, L.; Fan, R. Aqueous Iodine(III)-Mediated Stereoselective Oxidative Cyclization for the Synthesis of Functionalized Fused Dihydrofuran Derivatives. *J.Am.Chem.Soc.* 2010, *75*, 1760-1763.

[15] Lampard, C.; Murphy, J. A.; Lewis, N. Tetrathiafulvalene as a Catalyst for Radical–Polar Crossover Reactions. *J. Chem. Soc., Chem. Commun.* 1993, 295-297.

[16] Patro, B.; Merrett, M.C.; Markin, S.D.; Murphy, J.A.; Parkes, K.E.B. Radical-polar Crossover Reactions with a Water-Soluble Tetrathiafulvalene Derivative. *TetrahedronLett.* 2000, *41*, 421-424.

[17] a.-Lemoine, P.; Daoust, B. Water as an Efficient Medium for the Synthesis of Functionalized Enol Ethers. *TetrahedronLett.* 2008, *49*, 6175-6178.b.- Olivier, C.; Renaud, P. Organoboranes as a Source of Radicals.*Chem.Rev.* 2001, *101*, 3415-3434.

[18] Kirillova, M.V.Kirillov, A.M.; Pombeiro, A.J.L. Metal-Free and Copper-Promoted Single-Pot Hydrocarboxylation of Cycloalkanes to Carboxilic Acids in Aqueous Medium. *Adv.Synth.Catal.* 2009, *351*, 2936-2948.

[19] Panchaud, P.; Renaud, P. A convenient Tin-Free Procedure for Radical Carboazidation and Azidation. *J.Org.Chem.* 2004, *69*, 3205-3207.

[20] Opstal, T.; Verpoort, F. From Atom Transfer Radical Addition to Atom Transfer Radical Polymerisation of Vinyl Monomers Mediated by Ruthenium Indenylidene Complexes. *New J. Chem.* 2003, *27*, 257-262.

[21] a.-Nambu, H.; Hata, K.; Matsugi, M.; Kita, Y. Efficient Synthesis of Thioesters and Amides from Aldehydes by Using an Intermolecular Radical Reaction in Water. *Chem. Eur. J.* 2005, *11*, 719 – 727. b.-Nambu, H.; Hata, K.; Matsugi, M.; Kita, Y. The Direct Synthesis of Thioesters Using an Intermolecular radical Reaction of Aldehydes with Dipentafluorophenyldisulfide in Water. *Chem.Comm.* 2002, 1082-1083.

[22] Zhao, R.; Lind, J.; Mere′nyi, G.; Eriksen, T. E. Kinetics of One-Electron Oxidation of Thiols and Hydrogen Abstraction by Thiyl Radicals from .alpha.-Amino C-H Bonds.*J. Am. Chem. Soc.* 1994, *116*, 12010-12015

[23] Nauser, T.; Schöneik, C. Thiyl Radicals Abstract Hydrogen Atoms from the C-H Bonds in Model Peptides: Absolute Rate Constants and Effect of Aminoacid Structure. *J.Am.Chem.Soc.* 2003, *125*, 2042-2043.

[24] Sacasa, P.R.; Zayas, J.; Wnuk, S.F. Radical-mediated Thiodesulfonylation of the Vinyl Sulfones: Access to (α-fluoro)vinyl Sulfides. *TetrahedronLett.* 2009, *50*, 5424-5427.

[25] Benati, L.; Montevecchi, P. C.; Spagnolo, P. Free-radical Reactions of Benzenethiol and DiphenylDisulphide with Alkynes. Chemical Reactivity of Intermediate 2-(phenylthio)vinyl Radicals *J. Chem. Soc., PerkinTrans. 1* 1991, *9*, 2103-2109.

[26] a. Ananikov, V. P.; Kabeshov, M. A.; Beletskaya, I. P.; Khrustalev, V. N.; Antipin, M. Y. New Catalytic System for S–S and Se–Se Bond Addition to Alkynes Based on Phosphite Ligands.*Organometallics* 2005, *24*, 1275-1283. b.- Galli, C. Radical reactions of arenediazonium ions: An easy entry into the chemistry of the aryl radical.*Chem. Rev.* 1988, *88*, 765–792. c. Litwinienko, G.; Ingold, K. U. Solvent Effects on the Rates and Mechanisms of Reaction of Phenols with Free Radicals. *Acc. Chem. Res.* 2007, *40*, 222–230.

[27] Zhang, L.; Dolbier, W.R. Jr.; Sheeller, B.; Ingold, K.U.Absolute Rate Constants of Alkene Addition Reactions of a Fluorinated Radical in Water. *J.Am.Chem.Soc.* 2002, *124*, 6362-6366.

[28] Rong, X. X.; Pan. H.-Q.; Dolbier, W. R. *Jr*. Reactivity of Fluorinated Alkyl Radicals in Solution. Some Absolute Rates of Hydrogen-Atom Abstraction and Cyclization.*J.Am.Chem.Soc.* 1994, *116*, 4521-4522.

[29] (a) Scaiano, J. C.; Stewart, L. C. Phenyl radical kinetics.*J. Am. Chem.* Soc. 1983, *105*, 3609-3614. (b) Johnston, L. J.; Scaiano, J. C.; Ingold, K. U. Kinetics of Cyclopropyl Radical Reactions. 1. Absolute Rate Constants for some Addition and Abstraction Reactions.*J. Am. Chem.* Soc. 1984, *106*, 4877-4881.

[30] Delest, B.; Shtarev, A. B.; Dolbier, W. R.Jr. Reactivity of Perfluoro-n-alkyl Radicals with Reluctant Substrates. Rates of Addition to $R_FCH_2CH=CH_2$ and $R_FCH=CH_2$*Tetrahedron* 1998, *54,* 9273-9280.

[31] (a) Litwinienko, G.; DiLabio, G.A.; Mulder, P.; Korth, H.-G.; Ingold, K. U. Intramolecular and Intermolecular Hydrogen Bond Formation by Some Ortho-Substituted Phenols: Some Surprising Results from an Experimental and Theoretical Investigation.*J.Phys.Chem .A* 2009, *113*, 6275-6288 (b) Nielsen, M. F.; Ingold, K. U. Kinetic Solvent Effects on Proton and Hydrogen Atom Transfers from Phenols. Similarities and Differences. *J.Am.Chem.Soc.* 2006, *128*, 1172-1182. (c) Snelgrove, D. W.; Lusztyk, J.; Banks, J. T.; Mulder, P.; Ingold, K. U. Kinetic Solvent Effects on Hydrogen-Atom Abstractions: Reliable, Quantitative Predictions via a Single Empirical Equation.*J.Am.Chem.Soc.* 2001, *123*, 469-477. (d) Avila, D. V.; Ingold, K. J. Dramatic Solvent Effects on the Absolute Rate Constants for Abstraction of the Hydroxylic Hydrogen Atom from tert-Butyl Hydroperoxide and Phenol by the Cumyloxyl Radical. The Role of Hydrogen Bonding.*J.Am.Chem.Soc.* 1995, *117*, 2929-2930

[32] Miranda, N.; Daublain, P.; Horner, J.H.; Newcomb, M. Carboxylate-Substituted Radicals from Phenylselenide Derivatives. Designs on Models for Coenzyme B12-Dependent Enzyme-Catalyzed Rearrangements. *J.Am.Chem.Soc.* 2003, *125,* 5260-5261.

[33] Hatfield, G.L.; Barclay, R.L.C.Bilirubin as an Antioxidant: Kinetic Studies of the Reaction of Bilirubin with Peroxyl Radicals in Solution, Micelles, and Lipid Bilayers. *Org.Lett.* 2004, *6*, 1539-1542.

[34] a.-Mitroka, S.; Zimmeck, S.; Troya, D.; Tanko, J.M. How Solvent Modulates Hydroxyl Radical Reactivity in Hydrogen Atom Abstractions. *J.Am.Chem.Soc.* 2010, *132*, 2908-2913. b.-Gates, K.S. An Overview of Chemical Processes That Damage Cellular DNA:Spontaneous Hydrolysis, Alkylation, and Reactions with Radicals. *Chem.Res.Tox.* 2009, *22*, 1747-1760. c.-Wagner, J.R.; Cadet, J. Oxidation Reactions of Cytosine DNA Components by Hydroxyl Radical and One-Electron Oxidants in Aerated Aqueous Solutions. *Acc.Chem.Res.* 2010, *43*, 564-571.

[35] a.- Tantawy, W.; Zipse, H. *Eur. J. Org. Chem.* 2007, 5817–5820. b.- Jin, J.; Newcomb, M. Rate Constants for Reactions of Alkyl Radicals with Water and Methanol Complexes of Triethylborane. *J.Org.Chem.* 2007, *72*, 5098-5103.

[36] Jin, J.; Newcomb, M. Rate Constants for Hydrogen Atom Transfer Reactions from Bis(cyclopentadienyl)titanium(III) Chloride-Complexed Water and Methanol to an Alkyl Radical. *J.Org.Chem.* 2008, *73*, 7901-7905.

[37] Chatgilialoglu, C.; Newcomb, M. Hydrogen donor abilities of the Group XIV hydrides. *Adv. Organomet. Chem.* 1999, *44*, 67–112.

[38] Paradas, M.; Campaña, A.G.; Jimenez, T.; Robles, R.; Oltra, J.E.; Buñuel, E.; Justicia, J.; Cárdenas, D.J.; Cuerva, J.M. Understanding the Exceptional Hydrogen-Atom Donor Characteristics of Water in Ti(III)-Mediated Free-Radical Chemistry. *J.Am.Chem.* 2010, *132*, 12748-12756.

In: Organic Radical Reactions in Water and Alternative Media ISBN 978-1-61209-648-3
Editor: J. Alberto Postigo, pp. 45-78

Chapter 2

SILYL RADICALS IN WATER

J. Alberto Postigo[*]

Department of Chemistry-Faculty of Science, University of Belgrano,
Villanueva 1324 CP 1426-Buenos Aires-Argentina

ABSTRACT

This Chapter is focused on highlighting the recent advances on synthetically-useful organic reactions employing silicon-centered radicals in water, and presenting new reactions in water, mediated by silyl radicals. In doing so, several types of organic radical transformations will be discussed, such as reduction of organic halides utilizing nontoxic organosilane reducing agents in water, transformation of azides into amines, synthesis of protecting silyl ethers in water, hydrosilylation reactions of cabon-carbon double and triple bonds, and radical cyclization reactions in water induced by silicon-centered radicals. More recently, intermolecular radical carbon-carbon bond formation reactions mediated by silyl radicals have allowed the synthesis of perfluoroalkyl-substituted compounds in water, widening the scope for the syntheses of fluorophors. These silicon radical-mediated chain reactions in water are initiated through different methods, among which, thermal, photochemical, and dioxygen initiations are reported to be the most successful methods in water. A versatile aspect of the radical methodology employed in water will be presented in terms of dealing with water-soluble and organic solvent-soluble substrates in these silicon radical-mediated reactions in water. In this regard, for an efficient chain process to take place in water, a chain carrier must be used when water-soluble substrates are employed, whereas organic solvent-soluble materials do not require a chain transporter when silyl radicals are used in water.

Keywords: Silyl radicals • $(Me_3Si)_3SiH$ • silyl radicals in water • radical carbon-carbon bond formation in water • chain carrier.

---

[*] Corresponding Author: Dr. Al Postigo. Fax: (+)54 11 4511-4700. E-mail: alberto.postigo@ub.edu.ar

1. INTRODUCTION

The majority of organic radical reactions employing Si-centered radicals take place in organic solvents, neat media, and on surface chemistry, and only recently has water been used as a convenient reaction medium for these radicals.[1,2]

The development of novel water-soluble organosilane compounds and their applications to radical reactions in water is constantly being paid attention to, as a means to producing effective radical organic transformations in water, such as reductions of hydrophilic organic halides.

In the radical chain processes, the initial silyl radicals are generated by some initiation. The most popular initiator is 2′,2′-azobis*iso*butyronitrile (AIBN), with a half-life of 1 h at 81 °C, generating the incipient radicals that commence the radical chain reaction. Other azo-compounds are used from time to time[3] as well as the thermal decomposition of di-*tert*butyl peroxide[4] depending on the reaction conditions. Triethyl borane (Et_3B) in the presence of very small amounts of oxygen is an excellent initiator for low temperature reactions (down to – 78 °C). Also air-initiated reactions have recently been reported (eq 1).[4b]

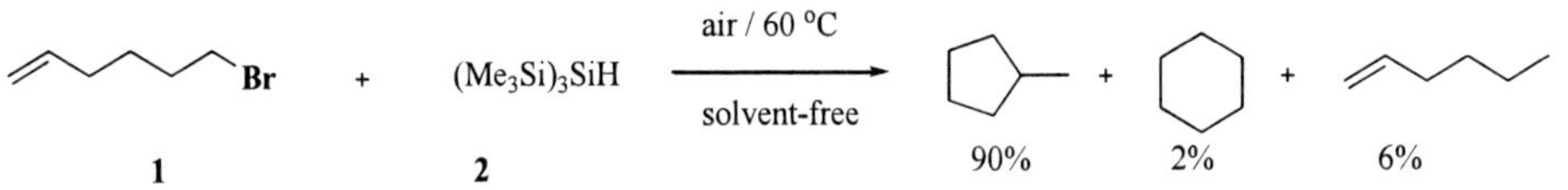

Br (1) + $(Me_3Si)_3SiH$ (2) —air / 60 °C, solvent-free→ 90% + 2% + 6% (1)

Togo and Yokoyama[5] have dealt with the necessity of performing reduction reactions in water of water-soluble halides. They have shown that the reactivity of water-soluble arylsilanes, as reducing agents, is much higher than that of alkylsilanes, and that the reactivity of diarylsilanes was higher than that of monoarylsilanes and triarylsilanes. Thus, the solubility of bis[4-(2-methoxyethoxy)phenyl]silane*3* and bis{4-[2-(2-methoxyethoxy)ethoxy]phenyl}silane *4* were about 1.0 x10^{-2} M and 1.5 x 10^{-2} M, respectively.

The reduction of potassium*o*-bromobenzoate and potassium *o*-iodobenzoate, which form the sp^2 carbon radicals, showed that the bromine atom abstraction is somewhat difficult, whereas the iodine atom of potassium *o*-iodobenzoate is easily removed by the silyl radical. These results and others obtained by the same authors showed that organosilanes *3* and *4* (Fig. (1)) can promote radical reductions of alkyl bromides, alkyl iodides, and aryl iodides, initiated by Et_3B, in aqueous media under aerobic conditions.

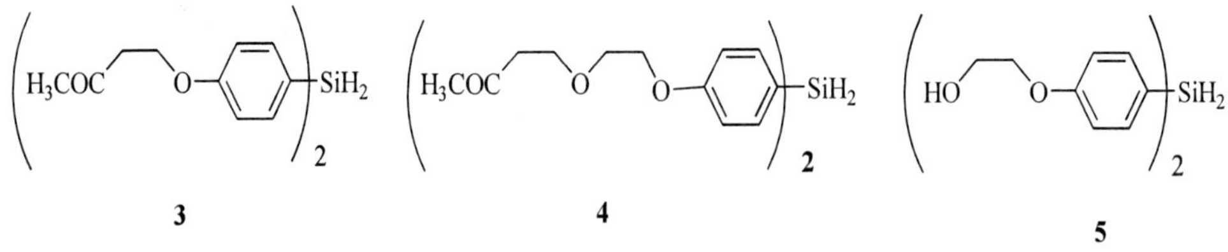

H_3COC O SiH_2 2 (3) H_3COC O O SiH_2 2 (4) HO O SiH_2 2 (5)

Figure 1. Water-soluble arylsilanes utilized in radical reductions of hydrophilic organic halides.

Radical cyclizations in water using organosilanes *4* and *5* have also been studied. The radical cyclization of potassium 7-bromo-2-heptenoate with *4* and *5* was carried out to afford 48 and 82% yields of cyclopentyl acetic acid, respectively, while no direct reduction product, *i.e.* 2-heptenoic acid, was formed. This clearly demonstrated the radical nature of the process.[5]

Water is also a choice of solvent for free-radical polymerization. The high heat capacity of water allows effective transfer of the heat from polymerization. Compared to organic solvents, the high polarity of water distinguishes remarkably the miscibility of many monomers from polymers. Today, aqueous free-radical polymerization is applied in industries.[6]

Several interesting photoinitiators based on the silyl radical chemistry have been proposed as a means to effecting polymerization in aqueous suspensions.[7] Among these compounds, (4-tris(trimethylsilyl)silyloxy)benzophenone generates silyl radicals under light irradiation that produce high rates of polymerization. A water-soluble poly(methylphenylsilylene) derivative has been used as a photoinitiator of radical polymerization of hydrophilic vinyl monomers with great success.[8] Following, we shall describe different radical triggering events that have recently been used for generating silicon-centered radicals in water that will be used throughout this chapter.

1.A. Initiation by Thermal Decomposition of an Azo Compound

The water-insoluble radical initiator 1,1′-azobis(cyclohexanecarbonitrile) (ACCN; half-life of 2.33 h at 100 °C) has been found to give the best performance for both hydrophobic and hydrophilic substrates in initial studies and this trend has been confirmed by successive experiments. The procedure is the following: In a 5 mL Wheaton-vial®, provided with a stir bar, a heterogeneous aqueous mixture of the substrate (10 mM), $(Me_3Si)_3SiH$ (1.2 – 2.0 equiv., or other hydrosilane) and ACCN (0.3 equiv.) is flushed with Ar for ten minutes before heating at 100° C for 2-4 hours or otherwise indicated. After the reaction time elapsed, addition of pentane and extraction, the organic-phase is analyzed.

1.B. Photochemical Radical Initiation

The initiation of the hydrosilylation reactions in water can be accomplished directly with light (low pressure Hg lamp, 254 nm) in the absence of a radical chemical precursor (*e.g.* peroxide), where most of the light is absorbed by $(Me_3Si)_3SiH$. To this effect, for all substrates studied, the absorption of $(Me_3Si)_3SiH$ (12 mM) at the irradiation wavelength (254 nm) should represent *ca.* 95%-97% of the total absorption of the mixture (substrate and reagents). Nevertheless, as $(Me_3Si)_3SiH$ is not soluble in water, it is assumed that its local droplet-concentration could be much higher, and therefore results in higher local UV-absorbances.

A volume of Ar-degassed water (3 mL) is placed in a quartz cell provided with a stir bar, with subsequent addition of $(Me_3Si)_3SiH$ (3 x 10-5 moles) and the substrate (3 x 10-5 moles) by syringe. The cell is mounted on a stir plate very near the lamp (1 cm) and stirred vigorously throughout the irradiation (1.5 – 2 h). The temperature is controlled thermostatically at 20 °C. At the working concentrations, most of the light (254 nm) is absorbed by $(Me_3Si)_3SiH$.

1.C. Radical Initiation by Dioxygen

A balloon filled with pure oxygen connected to the vessel where no apparent bubbling resulted, allowed dioxygen to be introduced up to its solubility limits in water. The dioxygen-initiated radical-induced reactions in Ar-degassed water is carried out by adding subsequently $(Me_3Si)_3SiH$(6×10–5 moles) and the substrate (5×10–5 moles). The vessel is tight-sealed, connected with a balloon filled with 99.99 % dioxygen, and vigorously stirred at 20 °C (24 h). As a slight positive oxygen-pressure is exerted on the reaction vessel, air does not leak in the system. For hydrophilic alkynes, $HOCH_2CH_2SH$ (0.3 equiv.) is employed as the chain propagating agent (*vide infra*).

2. Reduction of Organic Compounds andHydrosilylation Reactions of Double Bonds

Interestingly, Et_3B/O_2 initiation can be performed in aqueous solution. For instance, a wide range of aryl and alkyl halides are reduced in water by water-soluble organosilanes using Et_3B/O_2 initiation (Figure 2).

It is known that in organic solvents, *tris*(trimethylsilyl)silane, $(Me_3Si)_3SiH$, is an efficient reducing agent for organic halides. Also, the reported methodology of polarity–reversal catalysis is well documented in organic solvents. The thiol/silane couple shows not only an efficient synergy of radical production and regeneration, but could also provide for the use of an amphiphilic thiol, in order to enhance the radical reactivity at the interface. For the reduction of an organic halide (RX) by the couple $(Me_3Si)_3SiH$ / $HOCH_2CH_2SH$ under radical conditions, the propagation steps depicted in Scheme 1 are expected. That is, the alkyl radicals abstract hydrogen from the thiol and the resulting thiyl radicals abstract hydrogen from the silane, so that the thiol is regenerated along with the chain carrying silyl radical for a given RX.[9,10].

The proposal of $(Me_3Si)_3SiH$ in water is attractive from the point of view of its commercial availability. Recently, Postigo and Chatgilialoglu[11] tested the reducing agent $(Me_3Si)_3SiH$ in water and observed its high stability in deareated aqueous media and high temperatures. They subjected a series of organic halides to reduction with $(Me_3Si)_3SiH$ in water with different initiators, azo compounds and Et_3B. The initiators studied that afforded the best reduction yields with $(Me_3Si)_3SiH$ were the water soluble 2,2′-azobis(2-amidinopropane) dihydrochloride (AAPH) and 1,1′-azo*bis*(cyclohexanecarbonitrile) (ACCN, organic-solvent soluble, see section 1.A.-). The half-life of ACCN at 100 °C is 2.33 h, while that of AAPH is *ca.* 1.1 h at 73 °C.

The reduction of hydrophilic 4-iodobutyric acid and hydrophobic 5-iodouracil, afforded the corresponding reduced products in yields >90 %, with both initiators.

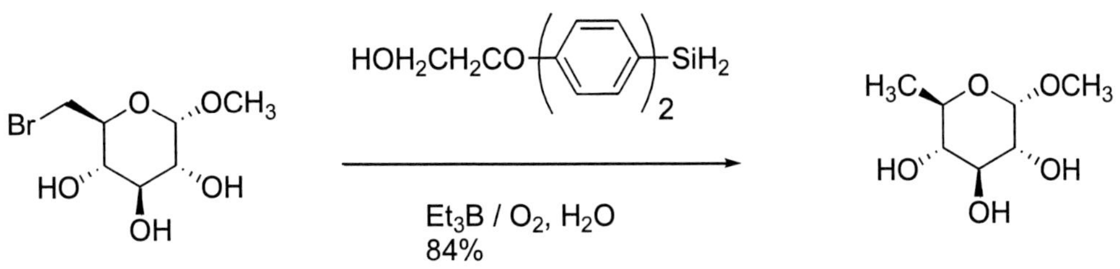

Figure 2. Reduction of bromosugars by hydrosoluble organosilanes.

$$R^{\bullet} + HOCH_2CH_2SH \longrightarrow RH + HOCH_2CH_2S^{\bullet}$$

$$HOCH_2CH_2S^{\bullet} + (TMS)_3SiH \rightleftharpoons HOCH_2CH_2SH + (TMS)_3Si^{\bullet}$$

$$(TMS)_3Si^{\bullet} + RX \longrightarrow (TMS)_3SiX + R^{\bullet}$$

$$RX + (TMS)_3SiH \longrightarrow RH + (TMS)_3SiX$$

Scheme 1. Polarity Reversal Catalysis of Silanes with Thiols.

The reduction of hydrophilic (1*S*)-bromocamphor-10-sulfonic acid and 5-bromouridine were also considered under similar reaction conditions. Using the water-soluble AAPH initiator no reaction occurred for the camphor derivative, whereas 5-bromouridine afforded uridine in 82% yield (based on 17% converted substrate). However, when 3 mM ACCN is used as initiator, both substrates afforded 90% yields of the corresponding reduction products, although the conversion of the starting material was as low as 10%. By increasing the amount of ACCN, however, the disappearance of starting material increased in favor of reduction product.[11] In this work, the relevance of 2-mercaptoethanol in the reduction process was revealed. For the reduction of 5-bromouridine, the optimal ratio of substrate / 2-mercaptoethanol was found to be 3-3.5.[11]

Other substrates of biological relevance bearing halogen atoms such as 8-bromoadenosine and 8-bromoguanosine were also subjected to reduction with $(Me_3Si)_3SiH$ / $HOCH_2CH_2SH$ in water initiated by ACCN. Very high yields of reduced products were obtained under these reaction conditions (>80 %).

Later on, Postigo and Chatgilialoglu[12] reported on two methods for the use of $(Me_3Si)_3SiH$ in water, depending on the hydrophilic or hydrophobic character of the substrates.

The reduction of water-insoluble organic substrates proceeded in a heterogeneous mixture of substrate, $(Me_3Si)_3SiH$, and ACCN , in water which is previously de-oxygenated with Ar and heated at 100 °C for 4 h. For water-soluble substrates, $HOCH_2CH_2SH$ is used. Thus, reduction of 5-bromo-nicotinic acid, 5′-iodo-5′-deoxyadenosine, and other hydrophilic halides do proceed by the couple $(Me_3Si)_3SiH$ / $HOCH_2CH_2SH$, where the alkyl or aryl radicals (R) abstract hydrogen from the thiol in the water phase, and the resulting thiyl radicals migrate into the lipophilic dispersion of the silane and abstract a hydrogen atom, thus regenerating the thiol along with the chain-carrying silyl radical for a given RX (Scheme 1). The reaction of the silyl radical is expected to occur at the interface of the organic dispersion with the aqueous phase. It is worth mentioning that the reaction of thiyl radicals with silane is estimated to be exothermic by *ca.* –3.5 Kcal mol^{-1}.[13]

The same reaction conditions were also applied to the radical cyclization of 1-allyloxy-2-iodobenzene derivative *6*, as shown in eq 2, but in this case, 2-mercaptoethanol was not needed.

Me I O —$(Me_3Si)_3SiH$, ACN / water, 4h, 100 °C→ O (2)

6 **7, 85%**

The reaction afforded 85 % yield of cyclized product *7*.

There is little knowledge on the reduction of *gem*-dichlorides by $(Me_3Si)_3SiH$ in organic solvents, and this knowledge was limited to some stereoselective examples.[14a] Reduction of *gem*-dichlorides *8* and *9* in water (eq 3) with $(Me_3Si)_3SiH$ and ACCN under the usual experimental conditions (see section 1.A.-, this time at 70 °C, 2 h) proceeded smoothly affording thc corresponding monochloride derivatives, in quantitative yields, as a diastereoisomeric mixture of compounds*10/11* in a 1.7:1 ratio for both cases (*8* and *9*).

R_1, Cl, R_2 Cl —$(Me_3Si)_3SiH$, H_2O→ R_1, Cl, R_2 H + R_1, H, R_2 Cl (3)

8 R_1= C_5H_{11}, R_2=H
9 R_1=Me, R_2= Ph

10 **11**

The diastereoselectivity outcome of this reaction is likely due to the influence of the substituents on the rate of the cyclopropyl radical and on the shielding of the two faces of the cyclopropyl ring.[14a,15,16]

Another successful class of radical reductions in water has been obtained by the transformation of azides into primary amines under the same experimental conditions. The results are reported in Table 1.

Again, no reaction was observed in the absence of amphiphilic 2-mercaptoethanol. The mechanistic steps of this reaction are shown in Scheme 2, in analogy with the pathways reported for the radical reduction of aromatic azides with triethylsilane in toluene.[12,17,18]

Table 1.Reduction of water-soluble azides

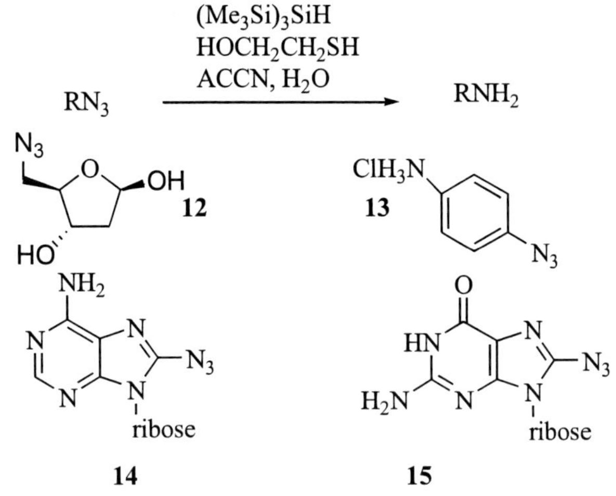

$$RN_3 \xrightarrow[\text{ACCN, } H_2O]{(Me_3Si)_3SiH,\ HOCH_2CH_2SH} RNH_2$$

12, 13, 14, 15

RN_3	conversion (%)	RNH_2, (%)
12	70	99
13	>99	90
14	>99	95
15	>99	99

De-oxygenation reactions in water (Barton-McCombie reaction) have also been attempted using $(Me_3Si)_3SiH$ as reducing agent.[12] As observed in organic solvents, the reaction is independent of the type of the thiocarbonyl derivative (e.g.: *O*-arylthiocarbonate, *O*-thiocarbamate, thiocarbonyl imidazole or xanthate). On the other hand, the water-soluble material does require the presence of the chain-carrier 2-mercaptoethanol.

The radical-base hydrosilylation reactions are generally performed in organic solvents, or under solvent free conditions.[14] More recently, these reactions are also performed in continuous-flow microreactors.[15] Lately, these reactions are performed in water.[12]

$$RN_3 + (Me_3Si)_3Si^\bullet \longrightarrow R\text{-}\dot{N}\text{-}Si(SiMe_3)_3 + N_2$$

$$R\text{-}\dot{N}\text{-}Si(SiMe_3)_3 + HOCH_2CH_2SH \longrightarrow R\text{-}NH\text{-}Si(SiMe_3)_3 + HOCH_2CH_2S^\bullet$$

$$HOCH_2CH_2S^\bullet + (Me_3Si)_3SiH \longrightarrow HOCH_2CH_2SH + (Me_3Si)_3Si^\bullet$$

$$R\text{-}NH\text{-}Si(SiMe_3)_3 + H_2O \longrightarrow RNH_2$$

Scheme 2. Proposed reaction steps for the reduction of azides in water.

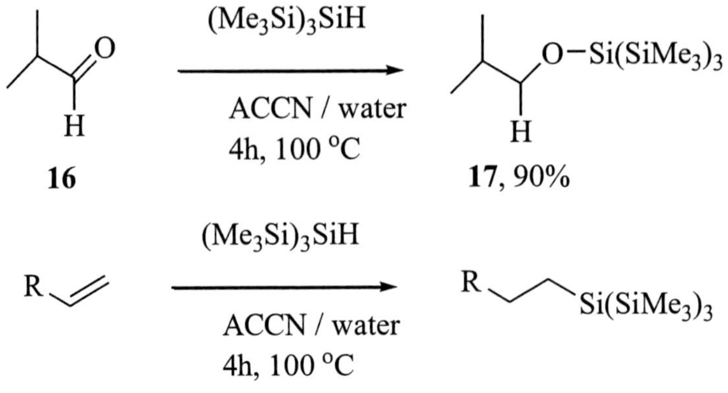

$$\mathbf{16} \xrightarrow[\text{ACCN / water, 4h, 100 °C}]{(Me_3Si)_3SiH} \mathbf{17},\ 90\%\ (O-Si(SiMe_3)_3)$$

$$R\text{-CH=CH}_2 \xrightarrow[\text{ACCN / water, 4h, 100 °C}]{(Me_3Si)_3SiH} R\text{-CH}_2\text{CH}_2\text{-Si(SiMe}_3)_3$$

18 R = $(CH_2)_4OH$ **19**, 99%
20 R = $(CH(OH)(CH_2)_4CH_3$ **21**, 99%

Scheme 3. Hydrosilylation reactions of multiple bonds Using $(Me_3Si)_3SiH$ in water.

Postigo *et al.*[12] effected the hydrosilylation reaction of unsaturated bonds in water, using different hydrophobic compounds, such as aldehydes (*16*), alkenes (*18,20*), under the same reaction conditions reported previously (section 1.A.-, and Scheme 3).

The efficiency of the reaction was very good, and in all cases, good to quantitative formation of the hydrosilylation products was achieved. These results showed that the nature of the reaction medium does not play an important role either in influencing the efficiency of the radical transformation or in the ability to dissolve the reagents. The authors attribute the success of the radical transformations of all water-insoluble material suspended in the aqueous medium to the vigorous stirring that creates an efficient vortex and dispersion. Probably, the radical initiation benefits from the enhanced contact surface of tiny drops containing $(Me_3Si)_3SiH$ and ACCN.[12]

For water –soluble material, hydrosilylation of multiple bonds in water is reported by the same authors to vary slightly. In this case, as referred to before, the reducing system $(Me_3Si)_3SiH$ / $HOCH_2CH_2SH$ in water is used. The amphiphilic thiol is successfully employed for radical reactions in the heterogeneous system of vesicle suspensions.[16] Excellent results of hydrosilylation of multiple bonds were achieved by adding this amphiphilic thiol to the system. The treatment of hydrophilic substrates in water has the additional advantage of an easy separation of the silane by-products by partition between water and organic phases.

More recently, Calandra et al.[21] undertook the hydrosilylation reaction of a series of alkenes (Scheme 4) in water with $(Me_3Si)_3SiH$ (*e.g.*: 3-chloroprop-1-en, prop-2-en-1-ol, prop-2-en-1-amine, *tert*-butyl vinyl ether, and *n*-butyl vinyl ether) initiated thermally and by light (*vide infra*, sections 1.A.- and 1.B.-) and found very good yields of hydrosilylated alkanes. In Table 2, columns 3 and 4, yields of the hydrosilylation products derived from alkenes obtained by ACCN initiation and photochemical initiation are given for the series of alkenes (Scheme 4) studied. Interestingly, under thermal initiation, the hydrosilylated product derived from hydrophobic 3-chloroprop-1-ene (*22*) is obtained in 65% relative yield (Table 2, column 3, entry 1, product *22a*, 2-(3-chloropropyl)-1,1,1,3,3,3-hexamethyl-2-(trimethylsilyl)trisilane, without chlorine atom reduction). There is also a product derived from the chlorine atom substitution by the $(Me_3Si)_3Si$ group (product *22b*, 2-allyl-1,1,1,3,3,3-hexamethyl-2-(trimethylsilyl)trisilane), which is obtained in 35% relative yield. These relative yields were calculated from 1H NMR integration. The hydrosilylated product derived from *tert*-butyl vinyl ether is also obtained in high yield (substrate*23*, 74%). High yields of hydrosilylation are obtained from *n*-butyl vinyl ether (*24*)[12] (isolated in 99% yield) in water under ACCN initiation.[21]

Hydrosilylation products derived from methylenecyclobutane (*25*), acrylonitrile[9,12] (*26*), and vinyl butyrate (*27*), are obtained in water in yields ranging from 68% to quantitative, when the reactions are initiated by ACCN (entries 4,5, and 6, column 3, Table 2).[21]

When the water soluble prop-2-en-1-ol (*28*), prop-2-en-1-amine (*29*), and organic solvent-soluble crotonaldehyde (*30*) are treated with $(Me_3Si)_3SiH$ in the presence of a thiol under ACCN initiation in water, the corresponding hydrosilylated products are obtained in higher than 75% yields (Table 2, column 3, entries 7, 8, and 9). In the absence of the thiol 2-mercaptoethanol, poor yields of hydrosilylated products are obtained. A notable case is that from substrate*29* (isolated global product yield 75%), which renders both a product derived from simple radical hydrosilylation of the C-C double bond (product *29a*, 15 % relative yield) and a ring-closed product (*29b*, *i.e.*: 2-(1,1,1,3,3,3-hexamethyl-2-(trimethylsilyl)trisilan-2-yl)ethanamine 85 % relative yield). The formation of product *29b* is accounted for in Scheme 5, where the incipient carbon centered radical derived from the addition of the $(Me_3Si)_3Si$ radical to the C-C double bond of the allylamine abstracts a hydrogen atom from the NH_2 group, rendering the aminyl radical that attacks the Si atom of the ancillary $(Me_3Si)_3Si$ group and undergoes intramolecular radical cyclization with loss of Me_3Si radical.[21]

From Table 2, column 4, the yields of hydrosilylation reactions of alkenes initiated by light in water are reported. Interestingly, the hydrosilylated product derived from hydrophobic 3-chloroprop-1-ene (*22*) is not obtained under photoinitiation. The hydrosilylated product derived from *tert*-butyl vinyl ether (*23*, Table 2, column 4, entry 2) is obtained in moderate yields (60%). The hydrosilylation product from *n*-butyl vinyl ether[12] is obtained in rather low yield (substrate*24*, hydrosilylated product isolated in 31 % yield) in water.[21]

When the water soluble prop-2-en-1-ol (*28*), prop-2-en-1-amine (*29*), and organic solvent-soluble crotonaldehyde (*30*) are treated with $(Me_3Si)_3SiH$ in the presence of a thiol under light initiation in water, the corresponding hydrosilylated products are obtained in high yields (Table 2, column 4, entries 7 , 8, and 9). As observed in the thermal initiation (*vide supra*), substrate*29* (isolated global product yield 65 %), renders both a product derived from simple radical hydrosilylation of the C-C double bond (product *29a*, 50% relative yield) and a ring-closed product (*29b*, *i.e.*: 2-(1,1,1,3,3,3-hexamethyl-2-(trimethylsilyl)trisilan-2-yl)ethanamine 50% relative yield). The formation of product *29b* is accounted for in Scheme 5. This behavior has been observed before in allyl and homoallyl alcohols.[22] The authors subjected the open chain hydrosilylated product derived from substrate *29*, *i.e.*: 3-(1,1,1,3,3,3-hexamethyl-2-(trimethylsilyl)trisilan-2-yl)propan-1-amine (*29a*), to photolysis (254-nm) and thermal treatment (ACCN, 100 °C) in water, as indicated in sections 1.B.- and 1.A.-, respectively. After two-hour reaction, no cyclic product *29b* was formed. This experiment was performed in order to rule out the formation of *29b*, as from a secondary reaction pathway (*i.e.:29a*).[21]

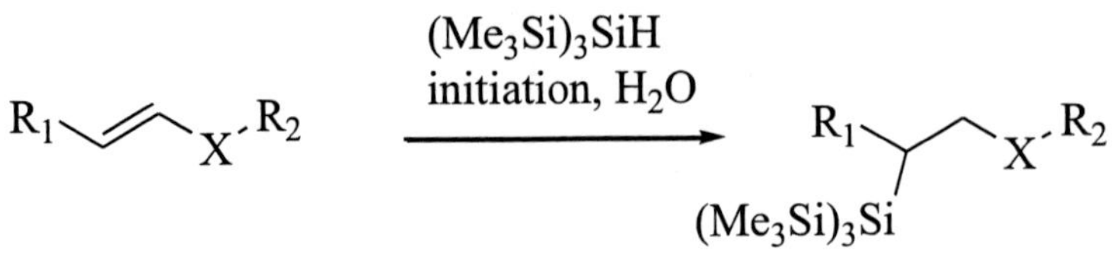

$$R_1\text{-}CH{=}CH\text{-}X\text{-}R_2 \xrightarrow[\text{}]{(Me_3Si)_3SiH,\ \text{initiation, } H_2O} R_1\text{-}CH(Si(SiMe_3)_3)\text{-}CH_2\text{-}X\text{-}R_2$$

X=CH_2, **22** R_1=H, R_2=Cl
28 R_1=H, R_2=OH
29 R_1=H, R_2=NH_2
X=O, **23** R_1=H, R_2=^{t}Bu
24 R_1=H, R_2=^{n}Bu
27 R_1=H, R_2=COC_3H_7
X=CH **30** R_1=CH_3, R_2=O
X=C 26 R_1=H, R_2=N

Scheme 4. Hydrosilylation reactions of alkenes in water By various initiating techniques.

In order to give further support to the mechanism suggested in Scheme 4, the isolated open chain product *29a* (4 mM) was treated in water (2 mL) with *t*-butylhydroperoxide (*t*-BuOOH, 10mM). The reaction was initiated by light (266 nm). After 40 min-reaction, product *29b* was obtained in *ca.*45% yield. This simple experiment shows that the aminyl radical II (Scheme 5) produced in this latter reaction from H-abstraction from *29a* by *t*-BuO• radical, can trigger an intramolecular cyclization with displacement of the TMS group to render product *29b*. Therefore, 1,3-H atom migration to render II (Scheme 5) could be suggested to take place along the reaction co-ordinate. In other works, the intramolecular substitution reaction at silicon, S_Hi mechanism[23], has been suggested to yield ring-closure products.[24]

Notably, the hydrosilylation product derived from substrate*28*, allylic alcohol, only affords an open chain product (99% isolated yield), as opposed to allylamine*29* where products *29a* and *29b* are observed. This difference could be related to the difference in the nucleophilicity of oxygen- and nitrogen-centered radicals in water, as opposed to organic solvents.[22]

When the reaction is carried out in the absence of the thiol 2-mercaptoethanol, poor yields of hydrosilylated products are obtained from water-soluble substrates*28* and *29*. Water-soluble hydrosilylated products are isolated by mixing the aqueous layer with pentane, the pentane layers discarded (to exclude excess of silane) and the aqueous phase lyophilized.

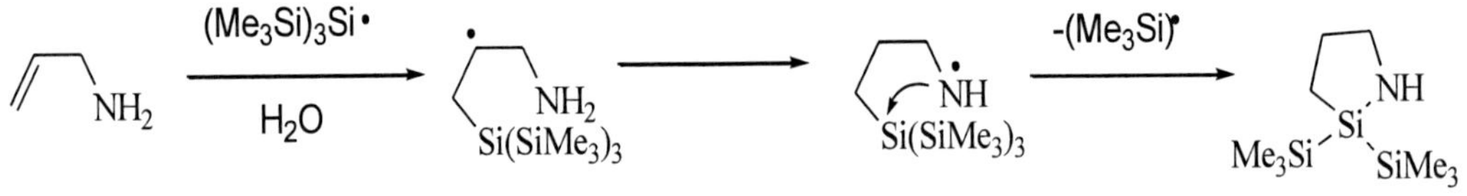

$(Me_3Si)_3Si•$ / H_2O; NH_2; NH_2, $Si(SiMe_3)_3$; NH, $Si(SiMe_3)_3$; $-(Me_3Si)•$; NH, Si, Me_3Si, $SiMe_3$

Scheme 5. Proposed mechanism for formation of cyclic product *29b* from the hydrosilylation reaction of allyl amine*(29)* in water.

Table 2. Hydrosilylation product yields from C-C multiple-bonded organic solvent-soluble substrates and hydrophilic substrates(10 mM) in de-oxygenated water with $(Me_3Si)_3SiH$ (12 mM) under different initiation conditions (ACCN and $h\nu$) [21]

Entry	Substrate	ACCN (100 °C, 2h) Yield, %	$h\nu$ (254 nm) Yield, %
1	3-chloroprop-1-ene (22)	(22a)65 [b] (22b)35 [b]	-
2	*Tert*-butyl vinyl ether (23)	74 [a]	60 [a]
3	*n*-Butyl vinyl ether(24)	99 [a]	31 [a]
4	Methylenecyclobutane (25)	68 [a]	54 [a]
5[ref.9,11]	Acrylonitrile (26)	80 [a]	82 [a]
6	Vinyl butyrate (27)	99 [a]	88 [a]
7	Prop-2-en-1-ol (28)	99 [a,c]	99 [a,c]
8	Prop-2-en-1-amine (29)	(29a)15 [b,c,d] (29b)85 [b,c,d]	(29a)50[b,c,d] (29b)50 [b,c,d]
9	Crotonaldehyde (30)	75 [a,c]	90 [a,c]

[a] Isolated Yield.

[b] Relative product yield.

[c] $HOCH_2CH_2SH$ (0.3 equiv.) was added as chain carrier.

[d] global Isolated yield; 65%.

Hydrosilylation products derived from methylenecyclobutane (*25*), acrylonitrile[9,12] (*26*), and vinyl butyrate (*27*), are obtained in water in yields ranging from 54% to 88%, when the reactions are initiated photochemically (entries 4,5, and 6, column 4, Table 2). The authors suggest that UV-irradiation of $(Me_3Si)_3SiH$ generates silyl radicals from homolysis of the Si-H bond. The silyl radicals in turn add to the unsaturated C-C bonds generating carbon-centered radicals that abstract hydrogen atoms from $(Me_3Si)_3SiH$, thus regenerating the silyl radicals. The hydrosilylation of prop-2-en-1-ol *28* and 3-chloroprop-1-ene *22* in the absence of solvent[10,11] and in the presence of air has also been attempted. A neat mixture (50×10^{-5} moles of substrate and 50×10^{-5} moles of $(Me_3Si)_3SiH$) was allowed to stir for 48 hrs. in an open vessel. After the reaction time elapsed, no hydrosilylation product is observed, and the substrate is recovered unaltered. From these results, water can be regarded as a good support medium for carrying out the radical chain transformation, as the solubility of oxygen in the neat substrate is compromised. In a recent report[19], the radical chain hydrosilylation reaction of C-C multiple-bonded compounds in water has been initiated with dioxygen, and found that this radical initiation is very sensitive to the dioxygen concentration. However, yields of hydrosilylated products derived from alkenes under dioxygen radical initiation are rather low.[19]

In the present study,[21] it was found that substrate*22*, forms products *22a* and *22b* under thermal initiation. Formation of product*22a* and product *22b* can be accounted for from Scheme 6.[21] Product *22a* arises from an ordinary hydrosilylation of the C-C double bond of *22*, while product *22b* is the result of a β-elimination of the chlorine atom from the intermediate radical species I (Scheme 6).

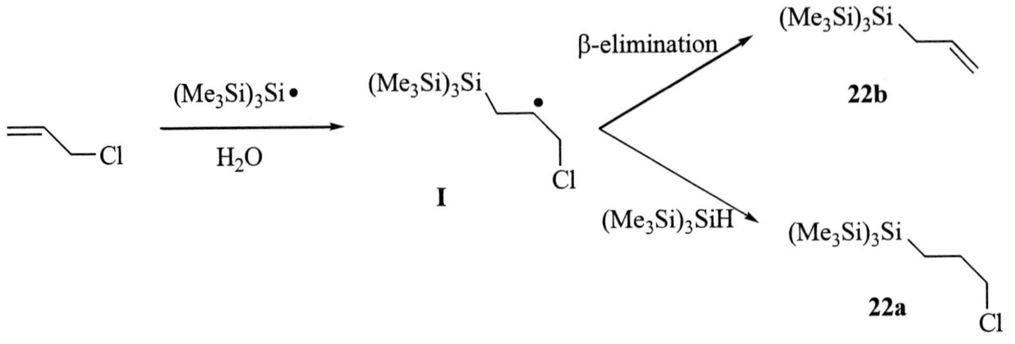

$(Me_3Si)_3Si\bullet$, H_2O; I; β-elimination → 22b; $(Me_3Si)_3SiH$ → 22a

Scheme 6.Proposed mechanism for formation of products *22a* and *22b* in water.

In alkenyl-substituted 1,1-dichlorocyclopropane ring systems[19], the hydrosilylation of C-C double bonds under radical thermal initiation precedes the chlorine atom reduction in water; that is, the hydrosilylation of C-C double bonds is faster than the chlorine atom removal by the $(Me_3Si)_3Si$ radical and that the one-chlorine atom reduction occurs under forced reaction conditions. Thus the authors[19] investigated the competition reaction between the *gem*-dichlorocyclopropane moiety reduction and hydrosilylation of double bonds in alkenyl-substituted *gem*-dichlorocyclopropane derivatives in water. The authors found that the hydrosilylation of the double bonds in alkenyl-substituted *gem*-dichlorocyclopropane derivatives precedes the chloro atom reduction in water. The substrate studied was*31* (eq 4). Reaction of *31* in water according to the usual protocol (section 1.A.- at 70 °C) afforded the hydrosilylated product *32* with retention of both chlorine atoms. Prolonged reaction times showed the formation of monochloride*33*.

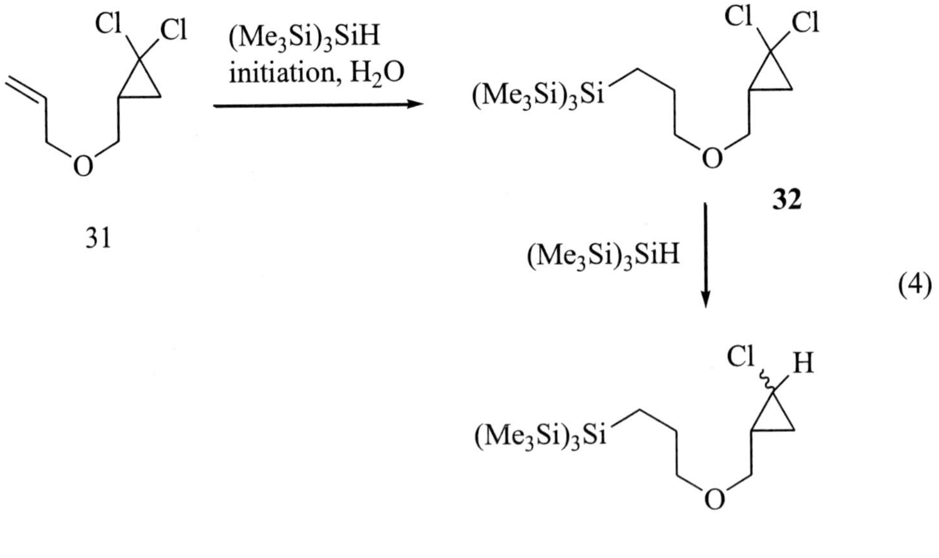

31 → $(Me_3Si)_3SiH$, initiation, H_2O → 32 → $(Me_3Si)_3SiH$ → 33, cis/trans=1.5/1 (4)

3. HYDROSILYLATION REACTIONS OF ALKYNES IN WATER

Recently, Zhou *et al*[4b] reported that phenylacetylene is hydrosilylated in water in 1 hour through air initiation to render the hydrosilylated product in 85 % yield. It is well-known that the hydrosilylation of alkynes[10] in organic solvents takes place efficaciously under thermal initiation (AIBN), with the hydrosilylation products being obtained in high isolated yield, albeit in *E:Z* varying isomeric ratios. However, hydrosilylation of alkynes in water is very efficient and stereoselective under thermal initiation, as has been reported before.[12] For example, the *Z:E* ratio for the hydrosilylated product derived from 1-cyclohexylacetylene changes from 45:55 in toluene (80 °C) to 74:26 in water (100 °C).[16] Thus, in organic solvents the initiation with AIBN takes place efficaciously albeit with poor stereoselectivity, while in water, ACCN is preferably used as initiator, and the stereoselectivity is much higher.[12,19]

$(Me_3Si)_3SiH$; ACCN/water; 4h, 100 °C; **34** → $Si(SiMe_3)_3$ product, 95 % (5)

The hydrosilylation of water-soluble propiolic acid*34* (eq 5) [12,19] was tested under thermal-initiation conditions (section 1.A.-). The reaction proceeded efficiently giving the *Z*-alkene with optimal yield (99%, eq 5). It is worth noting that the hydrosilylation has recently been reported to be efficient also under neat conditions, where the initiation was linked to the presence of adventitious oxygen.[18] The authors[19] postulate that the higher stereoselectivity in favor of the *Z* isomer, under the experimental conditions, suggests that additional factors are playing a role in water. It could be hypothesized that the hindrance of approach of the bulky silane to the radicals may also be influenced by the organization of the organic material dispersed in water.

$(Me_3Si)_3Si$, H, R ⇌ $(Me_3Si)_3Si$, R, H (6)

The hydrosilylation reaction of propiolic acid (*34*) and 3-chloroprop-3-yne (*35*) have also been tested under photochemical initiation conditions (Scheme 7), although in the former case, 2-mercaptoethanol is added as the chain carrier. The reactions proceed efficaciously affording the respective hydrosilylated alkenes in optimal reaction yields (95 %, Scheme 7, Table 3, column 5, entries 1 and 2). These hydrosilylated alkenes are obtained with varying *Z:E* isomeric ratios higher than 93:7. The retention of the chlorine atom is observed in this hydrosilylated alkene product derived from substrate*35*.[12]

The hydrosilylation reaction of organic solvent-soluble 3-chloroprop-1-yne *35* and water-soluble prop-2-yn-1-ol *36* initiated by ACCN have been tested under the conditions described above (using in this latter case, 2-mercaptoethanol as the chain carrier). The reactions proceed efficaciously affording the respective hydrosilylated alkenes in optimal reaction yields (89 %

and 98% , respectively, Scheme 7, Table 3, column 4, entries 2, and 3). These alkenes are obtained with *Z:E* isomeric ratios varying from 89:11 (that from alkyne *35*) and 88:12 (that from alkyne *36*), determined by ^{1}H NMR spectroscopy.[12]

O_2 (R.T.) or ACCN, 100 °C, 2h or *h*u, 2 h

$(Me_3Si)_3SiH$/water

X–R Si(SiMe$_3$)$_3$ + R–X Si(SiMe$_3$)$_3$

X=CO	**34** R= OH
X=CH_2	**35** R= Cl
X=CH_2,	**36** R= OH
X=$(CH_2)_6$	**37** R= H
X=C_6H_{11}	**38**
X=C_6H_4	**39** R= H
X=CH_2	**40** R= NH_2
X=4-C_6H_4	**41** R= Cl
X=4-C_6H_4	**42** R= Me
X=3.4-C_6H_3	**43** R= Me
X=4-C_6H_4	**44** R= F
X=4-C_6H_4	**45** R= MeO

Scheme 7. Hydrosilylation ($(Me_3Si)_3SiH$) reactions of triple-bonded substrates in water by different initiation methods.

$(Me_3Si)_3SiH$ as a pure material or in solution reacts spontaneously and slowly at ambient temperature with molecular oxygen from air, to form the siloxane. The mechanism of this unusual process has been studied in some detail.[14] Absolute rate constants for the spontaneous reaction of $(Me_3Si)_3SiH$ with molecular oxygen (reaction 7) have been determined to be 3.5 ×10 -5 M -1 s –1 at 70 °C, and theoretical studies elucidated the reaction co-ordinates.[25]

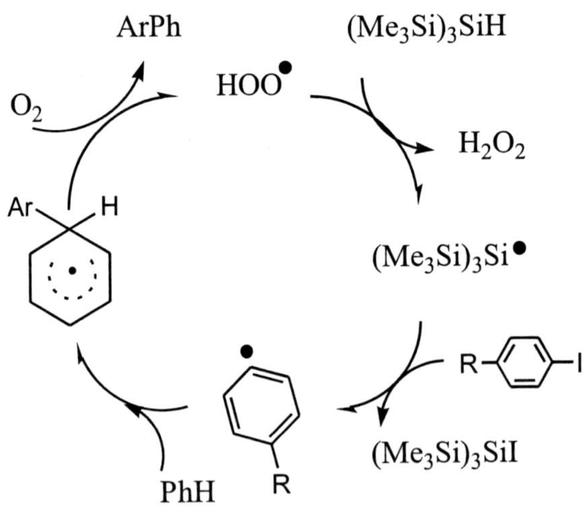

Scheme 8. Dioxygen initiation in $(Me_3Si)_3SiH$-mediated reactions in benzene.

$$(Me_3Si)_3SiH + O_2 \longrightarrow (Me_3Si)_3Si^{\bullet} + HOO^{\bullet} \quad (7)$$

In organic solvents (benzene), Curran et al.[26] reported the $(Me_3Si)_3SiH$ / dioxygen-mediated radical initiation reaction in the absence of an azo compound or light. Thus it was shown that $(Me_3Si)_3SiH$ mediates in the radical addition reaction of aryl iodides to benzene, and the rearomatization is achieved through oxygen. The mechanism of this reaction in benzene is described in Scheme 8.

Silyl radicals, generated from reaction of $(Me_3Si)_3SiH$ with oxygen, abstract the iodine atom from iodobenzene generating aryl radicals that suffer intramolecular addition to benzene (solvent) to form the cyclohexadienyl radical. Oxygen-induced rearomatization affords products along with hydroperoxyl radicals (HOO•). This radical abstracts hydrogen from the silane yielding $(Me_3Si)_3Si$• radicals completing the chain reaction.

Though it has been established that the thermal decomposition of ACCN is an excellent method for initiating the hydrosilylation reaction of unsaturated organic compounds in water (*vide supra*), the high temperature needed for the decomposition of the initiator (ACCN, 70 – 100 °C) precludes the treatment of thermally labile substrates and compromises the stability of products. Also, the photochemical initiation is very sensitive to the nature of the alkyne studied, since only alkynes with low absorptivities at 254 nm can be employed. The photostability of the hydrosilylated alkenes at the irradiation wavelength is also a compromising and limiting factor in choosing this latter initiation technique.

Alkynes, are more effectively hydrosilylated in water by $(Me_3Si)_3SiH$ through dioxygen initiation than alkenes are, where this condition affords only moderate to low yields of hydrosilylated alkanes (*vide supra*). Remarkably, the hydrosilylation of alkynes in water by $(Me_3Si)_3SiH$ proceeds with the highest *Z*-stereoselectivity when the reactions are initiated by dioxygen (Table 3, column 3). In a recent report[19] alkynes such as 1-octyne (*37*), 1-cyclohexylacetylene (*38*), 1-phenylacetylene (*39*) and propiolic acid (*34*) are treated with $(Me_3Si)_3SiH$ in water under dioxygen (see Section 1.C.-) initiation and yield the respective alkenes stereoselectively in high yields. Normally *Z* alkenes (*Z:E* ratios >99:1, isolated alkene yields >95 %) are formed. The hydrosilylation reactions of alkynes in water under dioxygen initiation have shown the highest degree of stereoselectivity achieved under this milder initiation technique.[19,27] For example, the *Z:E* ratio from hydrosilylation of 1-cyclohexylacetylene varies from 96:4 in toluene (25 °C, with Et_3B/air as initiator) to >99:1 in water (22 °C), under dioxygen initiation. More recently, these reactions are also performed in continuous-flow microreactors.[14]. Comparison of these data with the analogous reactions carried out in toluene at 80-90 °C and AIBN as initiator[20], not only shows better product yields but also a higher stereoselectivity in favor of the *Z* isomer. Unconjugated vinyl radicals are known to be sp^2 hybridized and to invert with a very low barrier (eq 6). Therefore, *Z*-hydrosilylated alkene products are preferentially obtained.

A new series of light and thermal sensitive alkyne substrates and hydrosilylated products were studied through the dioxygen initiation, as this latter was deemed the most convenient radical triggering event[27], providing further mechanistic evidence for the dioxygen-radical initiation in water.

The water-soluble prop-2-yn-1-ol[21](*36*) , and prop-2-yn-1-amine (*40*) have been tested under dioxygen initiation conditions employing the amphiphilic 2-mercaptoethanol as the chain carrier. The reactions proceed efficaciously (24 hrs) affording the respective hydrosilylated alkenes in optimal reaction yields (98%, Scheme 9, Table 3, column 3, entries 3 and 5, respectively). These hydrosilylated alkenes are obtained with *Z:E* isomeric ratios equal to 99:1 and 96:4 respectively. Water-soluble hydrosilylated alkene products are rinsed with pentane (to discard silane excess), lyophilized, and column-chromatographed by reverse-phase silica gel.

The hydrosilylation of water-insoluble 3-chloroprop-1-yne (*35*) with $(Me_3Si)_3SiH$ initiated by dioxygen (24 hrs) affords a high yield of the corresponding hydrosilylated alkene product (97 %), in a *Z :E* ratio equal to 99:1 (Table 3, column 3, entry 2). Retention of the chlorine atom is observed in this hydrosilylated alkene product. It should be pointed out that a small minor by-product (*ca.* 5%) detected from the oxygen-initiated reactions in water corresponds to a compound observed by mass spectrometry of mass 280 and formula minima $C_9H_{28}Si_3O_2$ whose structure has been assigned to $(Me_3SiO)_2Si(H)SiMe_3$, arising from the autoxidation of silane.[20] This by-product was confirmed by GC-co-injection with an authentic sample which has been synthesized according to a reported procedure.[20]

When hydrosilylation reactions of substituted phenylacetylenes (*41-45*) are attempted under dioxygen initiation (Table 3, column 3, entries 6-10) in water, the respective hydrosilylated styrene derivatives are obtained in high yields, ranging from 88% yield (that derived from 1-chloro-4-ethynylbenzene (*41*)) to 96 % yield (that from 4-ethynyltoluene (*42*)), and 75% yield from 3,4-dimethyl-ethynylbenzene (*43*). Notably, the stereoselectivity observed in all *tris*(trimethylsilyl)silyl-substituted styrenes are $\geq$ 95% in favor of the *Z* isomers. For comparison, the hydrosilylation in water of the parent phenylacetylene (*39*) initiated by dioxygen is shown in Table 3, where the hydrosilylated styrene is obtained quantitatively in the *Z* geometric isomer exclusively (Table 3, column 3, entry 4).[27].

When substituted phenylacetylenes (*41-45*) are subjected to hydrosilylation reactions in water initiated by ACCN, the respective hydrosilylated styryl derivatives are obtained in good yields albeit with lower stereoselectivities than those observed under dioxygen initiation (Table 3, column 4, entries 6-10). The stereoselectivities are nevertheless in favor of the *Z* alkenes, ranging from 71:29 to 87:13. Notably, the stereoselectivities are lower than those obtained under dioxygen initiation. (*cf.* Table 3, columns 3 and 4). For comparison, the hydrosilylation in water of the parent phenylacetylene (*39*) initiated by ACCN is reported in Table 3 (entry 4, column 4) from previous studies, where the hydrosilylated styrene is obtained almost quantitatively (93%) with a *Z:E* stereoisomeric ratio equal to 70:20.[10]

The hydrosilylation reaction of water-soluble (*36*), and (*40*) have also been tested under photochemical initiation conditions employing 2-mercaptoethanol as the chain carrier. The reactions proceed efficaciously affording the respective hydrosilylated alkenes in optimal reaction yields (99 %, Scheme 9, Table 3, column 5, entries 3 and 5, respectively). These hydrosilylated alkenes are obtained with *Z:E* isomeric ratios equal to 95:5 and 91:9, respectively, determined by 1H NMR spectroscopy.[19,21]

When substituted phenylacetylenes (*41-45*) are subjected to hydrosilylation reactions in water initiated by light, the respective hydrosilylated styryl derivatives are obtained in good yields albeit with poorer stereoselectivities as compared to those obtained under dioxygen (Table 3, column 5, entries 6-10). The stereoselectivities in favor of the *Z* alkenes range from 66:34 to 85:15. For comparison, the hydrosilylation yield in water of the parent phenylacetylene (*39*) induced by light is shown in Table 3 (entry 4, column 5), where the hydrosilylated styrene is obtained almost quantitatively (98%) with a Z:E stereoisomeric ratio equal to 75:25.[27]

Table 3. Hydrosilylation reactions of C-C triple bonds. Organic solvent-soluble substrates and hydrophilic substrates (10 mM) in water with $(Me_3Si)_3SiH$ (12 mM) under different initiation conditions [27a]

Entry	Substrate	O_2 (R.T.) Isolated Yield, % (Z:E ratio)	ACCN Isolated Yield, % (Z:E ratio)	$h\nu$ Isolated Yield, % (Z:E ratio)
1.ref. [10,11]	Propiolic acid (34)	99 (99:1)	95 (99:1)	95 (97:3)
2	3-chloroprop-1-yne (35)	97 (99:1)	89 (89:11)	95 (93:7)
3[a], ref, [10]	Prop-2-yn-1-ol (36)	98 (99:1)	98 (88:12)	99 (95:5)
4 ref [10,11]	Phenylacetylene (39)	99 (99:1)	93 (70:30)	98 (75:25)
5[a]	Prop-2-yn-1-amine (40)	98 (96:4)	89 (88:12)	99 (91:9)
6	1-chloro-4-ethynylbenzene (41)	88 (95:5)	90 (71:29)	91 (79:21)
7r[ef.16]	4-ethynyl toluene (42)	96 (99:1)	90 (85:15)	83 (79:21)
8b	3,4-dimethyl-ethynyl benzene (43)	75 (99:1)	71 (80:20)	67 (66:34)
9[ref.17]	1-ethnyl –4-fluorobenzene (44)	89 (95:5)	85 (87:13)	80 (85:15)
10	4-ethynyl anisole (45)	95 (99:1)	89 (85:15)	90 (79:21)

[a]$HOCH_2CH_2SH$(0.3 equiv.) was added as chain carrier.
[b] 24 mM $(Me_3Si)_3SiH$ used, 4h-reaction .Isolated yield.

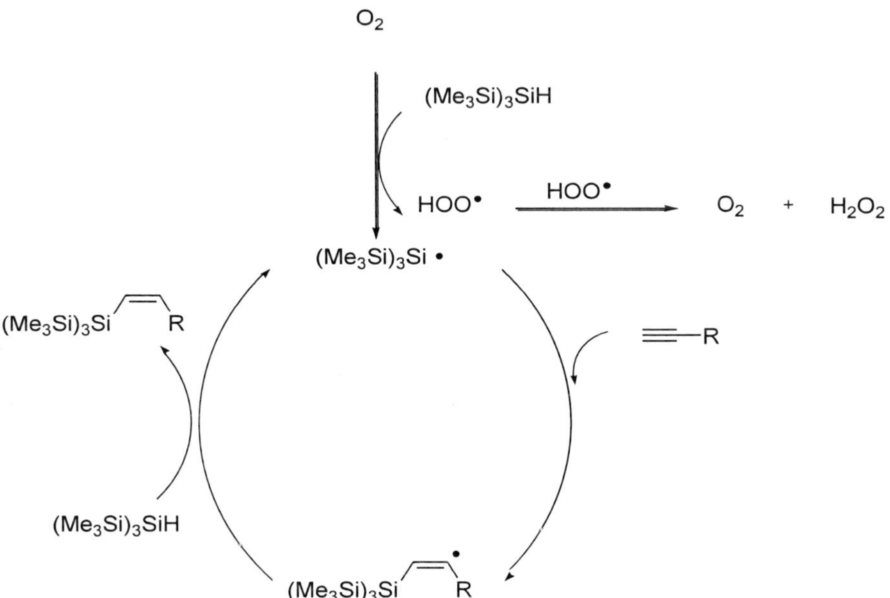

Scheme 9. Mechanism for the dioxygen-initiated hydrosilylation reactions of C-C triple bonds in water.

For the dioxygen initiation, the authors postulate a mechanism as shown in Scheme 9 as operative in water for the hydrosilylation of C-C triple bonds. Upon reaction of dioxygen with $(Me_3Si)_3SiH$ (in water), silyl radicals along with hydroperoxyl radicals are generated. Hydroperoxyl radicals are known to be poorly reactive towards closed-shell compounds, and probably expected to recombine to form oxygen and hydrogen peroxide (or else abstract hydrogen from the silane to render $(Me_3Si)_3Si\bullet$ radicals that add to C-C triple bonds). No product derived from an addition reaction of hydroperoxyl radicals on C-C triple bonds has been detected under the reaction conditions. Nevertheless, hydrogen peroxide is expected to be a by-product of the reaction.[27] This latter was detected by adding to the aqueous reaction mixture (24 hour-reaction of 2-chloro-1-propyne or phenylacetylene, $(Me_3Si)_3SiH$, dioxygen and water, where the hydrosilylated alkene product was previously detected by GC) a fresh colorless solution of Fe(SCN)+ (10 mM) which turned immediately light orange due to the formation of Fe(SCN)+ +. A blank experiment considering a freshly-prepared oxygen-saturated aqueous mixture of the respective alkyne and $(Me_3Si)_3SiH$, did not lead to a change in color when Fe(SCN)+ was added to the aqueous mixture, purporting that the chain reaction initiates very slowly. As a matter of fact, the radical chain reaction under oxygen initiation is known to have a poor initiation yield, and the effectiveness of the overall radical hydrosilylation transformation relies on a very efficient propagation step. Another experiment carried out by adding H_2O_2 (5%) to a 10 mM solution of Fe(SCN)+ , isolated hydrosilylated alkene product (10 mM) and $(Me_3Si)_3SiH$ (10 mM) produced an orange solution.[27]

The silyl radicals perform the well-known addition reaction to C-C triple bonds. Unconjugated σ vinyl radicals are known to be sp2 hybridized and to invert with a very low barrier (eq 6). Although $(Me_3Si)_3Si\bullet$ radicals have been shown to isomerize alkenes, the post-isomerization of the hydrosilylation adduct is not observed due to steric hindrance. The higher Z-stereoselectivity observed upon addition of $(Me_3Si)_3Si\bullet$ to triple bonds is rationalized in terms of the $E\sigma$ vinyl silylated radical being more hindered to abstract hydrogen from $(Me_3Si)_3SiH$ than the $Z\sigma$ vinyl silylated radical (eq 6). It is evident that the higher *Z* stereoselectivity observed upon dioxygen initiation in water (as compared to light-induced or thermal) is related to the stability of substrates and products under the experimental conditions. The milder dioxygen-initiated hydrosilylation reaction in water, as compared to the thermal and light-induced methods, circumvents issues associated with stability of substrates and products under the reaction conditions. In the photochemical initiation, light is partially absorbed by the substrate (*e.g.*: phenylacetylenes).

It is worth mentioning that under the experimental procedure, initiation with oxygen does not take place under O_2 purging of the solution, nor with slow, continuous bubbling of O_2 or air. Oxygen or air introduced slowly by syringe-pump techniques resulted in a less efficient initiation of the chain radical reaction. Probably, the slow oxygen addition into the solution with the balloon technique described above allows for sufficient oxygen to be dissolved in the water environment, thus initiating the radical chain (the solubility of oxygen in water is 1.34 ×10 -3 M at 22 °C; taking into account that the substrate and $(Me_3Si)_3SiH$ concentrations are *ca.* 10 times higher, limiting working initiator concentrations could be reached by the methodology).

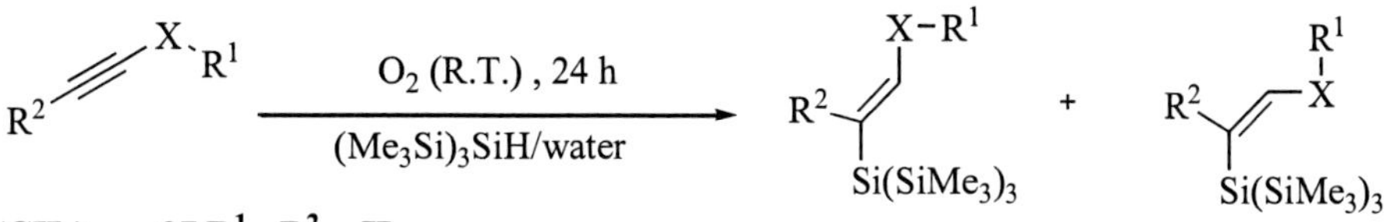

X=$(CH_2)_6$, **37** $R^1 = R^2 = H$
X=C_6H_4 **39** $R^1 = R^2 = H$
X=$(CH_2)_3$, **46** $R^1 = OH$, $R^2 = H$
X=CH(OH) **47** $R^1 = C_6H_5$, $R^2 = H$
X-R^1= ^tBu **48** $R^2 = H$
X=CO_2 **49** $R^1 = CH_2CF_3$, $R^2 = H$
X=CO_2 **50** $R^1 = CH_2CH_2OH$, $R^2 = H$
X=3,4-C_6H_3 **51** $R^1 = CH_3$,F $R^2 = H$
X=4-C_6H_4 **52** $R^1 = CF_3$, $R^2 = H$
X=C_6H_4 **53** $R^1 = H$, $R^2 = CN$

O_2 (R.T.) , 24 h; $(Me_3Si)_3SiH$/water

Scheme 10. Hydrosilylation ($(Me_3Si)_3SiH$) reactions of triple-bonded substrates in water by the dioxygen initiation method.

The water-soluble pent-4-yn-1-ol(*46*), and 2-hydroxyethyl propiolate (*50*) have been tested under dioxygen initiation conditions employing the amphiphilic 2-mercaptoethanol as the chain carrier (Table 4). The reactions proceed efficaciously (24 hrs) affording the respective hydrosilylated alkenes in optimal reaction yields (98% and 96%, respectively, Scheme 10, Table 4, column 3, entries 3 and 7, respectively). These hydrosilylated alkenes are obtained with *Z:E* isomeric ratios equal to 99:1.

For comparison, the hydrosilylation reaction of propiolic acid (*34*)with $(Me_3Si)_3SiH$ in water initiated by dioxygen is reported, where a high *Z:E* stereoselectivity ratio is obtained (Table 3, column 3, entry 1).[19,12]

Table 4. Hydrosilylation reactions of C-C triple bonds. Organic solvent-soluble substrates and hydrophilic substrates(10 mM) in water with $(Me_3Si)_3SiH$ (12 mM) under di-oxygen initiation conditions

Entry	Substrate	O_2 (R.T.) Isolated Product Yield, % (*Z:E* ratio)
1 [ref.3]	Oct-1-yne (37)	97 (99:1)
2	Phenylacetylene (39)	99 (99:1)
3 [ref.2]	Pent-4-yn-1-ol [a] (46)	98 (99:1)
4 [ref.4b]	1-phenylprop-2-yn-1-ol (47)	75 (99:1)
5 [ref.20]	3,3-dimethylbut-1-yne (48)	98 (96:4)
6. [ref.27b]	2,2,2-trifluoroethylpropilate (49)	99 (99:1)
7 [ref.27b]	2-hydroxyethyl propiolate [a] (50)	96 (99:1)
8 [ref.4b]	4-ethynyl-1-fluoro-3-methylbenzene (51)	89 (95:5)
9 [ref. 27c]	1-ethynyl-4-trifluoromethylbenzene (52)	95 (99:1)
10 [ref.20]	3-phenylpropiolonitrile [b] (53)	88 (95:5)

a HOCH2CH2SH (0.3 equiv.) was added as chain carrier.
b 24 mM (Me3Si)3SiH used. Isolated yield.

The hydrosilylation of water-insoluble oct-1-yne (*37*) with $(Me_3Si)_3SiH$ initiated by dioxygen (24 hrs) affords a high yield of the corresponding hydrosilylated alkene product (97 %), in a *Z:E* ratio equal to 99:1, (Table 4, column 3, entry 1).

When hydrosilylation reactions of substituted phenylacetylenes (*51-53*) are attempted under dioxygen initiation (Table 4, column 3, entries 8, 9, and 10, respectively) in water, the respective hydrosilylated styrene derivatives are obtained in high yields, ranging from 89% yield (that derived from 4-ethynyl-1-fluoro-3-methylbenzene (*51*)) to 95 % yield (that from 1-ethynyl-4-trifluoromethylbenzene (*52*)), and 88% yield from 3-phenylpropiolonitrile (*53*). Notably, the stereoselectivity observed in all *tris*(trimethylsilyl)silyl-substituted styrenes are ≥ 95% in favor of the *Z* isomers.

4. $(Me_3Si)_3SiH$ Carbon-Carbon Bond Formation in Water

Perfluoroalkyl compounds have been the subject of intense studies during the past twenty years for their wide applications in different fields of chemistry. The development of fluorous combinatorial techniques have driven the search for convenient ways for introducing fluorous tags[28] containing perfluoroalkyl groups into various organic compounds.[29] The synthesis of these compounds cannot be achieved through classical nucleophilic substitutions on perfluoroalkyl halides, R_fX, as these substrates are thwarted from reacting by the S_N1 mechanism, on account of the low stability of carbocations, and impeded to undergo S_N2 substitutions due to repulsion of the lone electron pairs of the fluorine atoms to the backside attack by the nucleophile.[30]

Compounds bearing the perfluoroalkyl moiety R_f-C bond, however, have been synthesized by different routes. One such route involves addition of R_f• radicals to double bonds.[31] Perfluoroalkyl iodides and bromides are convenient sources of perfluoroalkyl radicals in the presence of radical initiators.[32]

The photoinitiation based on the homolytic dissociation of perfluoroalkyl iodides, R_f-I, (the CF_2-I bond) is also applicable for the iodoperfluoroalkylation of unsaturated compounds with R_fI.[33] Ogawa et al. undertook an iodoperfluoroalkylation of unsaturated carbon-carbon double and triple bonds in benzotrifluoride as solvent.[33] These authors also utilized non-conjugated dienes, conjugated dienes, allenes, vinylcyclopropanes, and isocyanides as radical-acceptor substrates for the radical iodoperfluoroalkylation reactions in benzotrifluoride, affording good yields of the corresponding iodoperfluoroalkylated derivatives. Another route to the synthesis of compounds with perfluoroalkyl moieties is through the $S_{RN}1$ mechanism, which involves radicals and radical ions as intermediates.[30,34]

On the other hand, intermolecular radical carbon-carbon bond formation reactions, *i.e.:* consecutive reactions, demand a careful synthetic planning to achieve carbon-carbon coupling products in fairly good yields. The key step in these consecutive reactions generally involves the intermolecular addition of R• radicals to a multiple-bonded carbon acceptor. When a hydride chain carrier is involved, care has to be exercised in order to ensure that the effective rate of the radical addition is higher than the rate of H atom transfer.

When silicon-centered radicals are used,[35] for an efficient chain process it is important that (i) the R'_3Si• radical reacts faster with RZ (the precursor of radical R•) than with the alkene, and (ii) that the alkyl radical reacts faster with the alkene (to form the radical adduct) than with the silicon hydride. This process has in due course been termed "*disciplined intermediates*".[36] The hydrogen donation step controls the radical sequence and the concentration of silicon hydride often serves as the variable by which the product distribution can be influenced. The majority of sequential radical reactions using silanes as mediators for the intermolecular carbon-carbon bond formation deals with tris(trimethylsilyl) silane,$(Me_3Si)_3SiH$, in organic solvents.[37] The need to resort to more environmentally friendly solvents opened the scope of radical carbon-carbon bond formation reactions in water, and other aqueous mixtures. Atom transfer intermolecular carbon-carbon bond formation reactions in water have been investigated in detail by many authors.[38,39] It has been reported[40] the intermolecular carbon-carbon bond formation reaction in water from ethyl bromoacetate and 1-octene affording ethyl 4-bromodecanoate in 80% yield when the reaction is initiated by triethylborane (Et_3B) / air (eq 8).

C_2H_5O O Br + nC_6H_{13} — Et_3B/air, water → C_2H_5O O n-C_6H_{13} Br (8)

80%

Other radical precursors such as bromomalonate and bromoacetonitrile give excellent results of bromine atom transfer products in water. A tandem radical addition-oxidation sequence in water which converts alkenyl silanes into ketones has been elegantly described by Oshima et al.[40] Other types of intermolecular radical carbon-carbon bond formation reactions in water have been lately reported, describing radical additions to radical acceptors such as imines and their derivatives.[41,42,43]These latter consecutive radical reactions are also initiated by BEt_3 / air.

As has been shown in section *3.-*, the dioxygen-radical initiation has recently been successfully applied in water for the hydrosilylation of alkynes employing $(Me_3Si)_3SiH$ / oxygen, rendering high yields of the respective hydrosilylated alkene products with excellent *Z* stereoselectivity.[19a,27] In sections *2.-* and *3.-* the attempted radical reactions in water mediated by silyl radicals have been illustrated, among them, reduction of organic halides[11,12], reduction of azides[12], hydrosilylation reactions of multiple-bonded substrates[12,19a,21], intramolecular carbon-carbon bond formation reactions,[12] etc., either by azo compounds decomposition, photochemically or through dioxygen initiation.[44]

Dolbier et al.[45] have found that perfluorinated radicals were much more reactive than their hydrocarbon counterparts in addition to normal, electron rich alkenes such as 1-hexene (40 000 times more reactive) in organic solvents, and that H transfer from $(Me_3Si)_3SiH$ to a perfluoro-n-alkyl radical such as n-$C_7F_{15}CH_2CH(\bullet)C_4H_9$ was 110 times more rapid than to the analogous hydrocarbon radicals (eq 9).

$$n\text{-}C_7F_{15}I \xrightarrow{h\nu} n\text{-}C_7F_{15}^{\bullet} \begin{cases} \xrightarrow[K_{add}]{\diagup\!\!\diagdown C_4H_9} C_7F_{15}\diagdown\!\!\diagup\!\!\overset{\bullet}{\diagdown} C_4H_9 \ (\mathbf{54}) \xrightarrow{(Me_3Si)_3SiH} C_7F_{15}\diagdown\!\!\diagup\!\!\diagdown C_4H_9 \ (\mathbf{56}) \\ \xrightarrow[K_H]{(Me_3Si)_3SiH} n\text{-}C_7F_{15}H \ (\mathbf{55}) \end{cases} \quad (9)$$

Thus the authors determined that k_{add} (eq 9) has a value of 7.9 x 106 $M^{-1}s^{-1}$ in benzene-d_6 at 298 K, and the value of k_H is *ca.* 50 x 106 $M^{-1}s^{-1}$ in benzene-d_6 at 303 K.

Given the known rate acceleration effects of radical reactions in water[42,44], it became worthwhile studying intermolecular carbon-carbon bond formation radical reactions in this medium, mediated by silyl radicals. Barata-Vallejo and Postigo[46] embarked on a study of the perfluoroalkylation reactions of electron rich alkenes and alkenes with electron withdrawing groups, to explore the scope and limitations of the intermolecular addition reactions of perfluoroalkylated radicals on alkenes in water mediated by $(Me_3Si)_3SiH$. The initiators employed are azo compounds and dioxygen which upto then had given the best results in previous studies.[11,12,19a, 21,27]

A preliminary set of experiments was conducted[46] in order to adjust the right stoichiometry of the radical addition reactions of $R_f\bullet$ radicals to alkenes so as to favor the radical addition product over the reduction product in water (as in products *55* and *56*, equation 9). For this preliminary experiment, 1-hexene was used as the radical acceptor, and n-$C_6F_{13}I$, as the source of $R_f\bullet$ radicals. Reduction product n-$C_6F_{13}H$ and addition product C_6F_{13}-C_6H_{13} were both obtained under different reaction conditions (by incremental amounts of $(Me_3Si)_3SiH$, and keeping alkene and n-$C_6F_{13}I$ concentrations constant), under thermal initiation. The most favorable reaction conditions were obtained by using a molar ratio of *alkene* : $(Me_3Si)_3SiH$: R_fI equal to 25:2.5:5 and this ratio was chosen as optimal.[47]

When 1-hexene is allowed to react with n-1-iodopefluorohexane, n-$C_6F_{13}I$, in water, initiated by $(Me_3Si)_3SiH$ / dioxygen, 1,1,1,2,2,3,3,4,4,5,5,6,6-tridecafluorododecane *57a*[47,48] is obtained in 65% yield (Scheme 11, Table 5, entry 1). Analogously, 1-octene and 1-decene afford, upon reaction with n-$C_6F_{13}I$ under the same radical conditions, products *58a*[48a,b] and *59a* [48b] in 71 and 74% yields (isolated yields), respectively (Scheme 11,Table 5, entries 2 and 3).

Styrene and *p*-methylstyrene when reacted with n-$C_6F_{13}I$ in water under $(Me_3Si)_3SiH$ / dioxygen initiation, give products *60a*[49] and *61a*[50] in 54 and 63% yields respectively (Scheme 11, Table 5, entries 4 and 5). In these latter cases, reduction of the iodo compound (*i.e.*: 1,1,1,2,2,3,3,4,4,5,5,6,6-tridecafluorohexane, $CHF_2(CF_2)_4CF_3$) was concomitantly obtained, as determined from the 1H and 19F NMR spectra of the crude reaction mixtures (centered triplet at 6.29 ppm, J = 22 Hz, from terminal HCF_2, in the 1H NMR spectrum and from the 19F NMR spectrum, the CF_2H peak at δ = -114.05 ppm). Alkenes with electron withdrawing groups such acrylonitrile (*26*), crotonaldehyde (*30*), methylacrylate, and vinyl methyl ketone also react with n-$C_6F_{13}I$ under the same radical conditions, to afford the respective perfluoroalkylated products *62a* [48c,51], *63a*, *64a*, and *65a* [51-54] in yields ranging from 60 to 77% (Scheme 11, Table 5, entries 6-9).

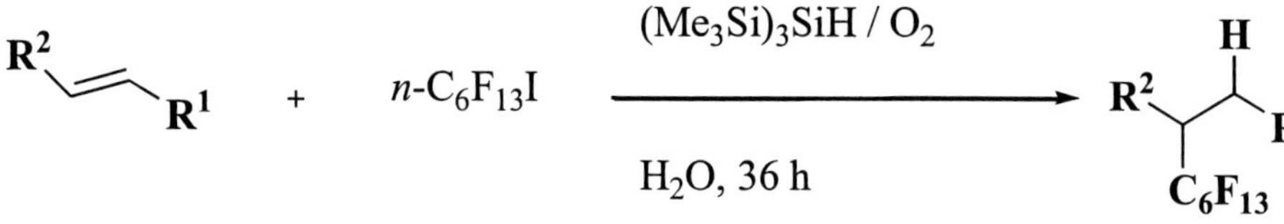

$$R^2\text{CH=CH}R^1 + n\text{-}C_6F_{13}I \xrightarrow[H_2O,\ 36\ h]{(Me_3Si)_3SiH\ /\ O_2} R^2CH(C_6F_{13})CH(H)R^1$$

$R^1 = C_4H_9$, $R^2 = H$	**57a, 65%**
$R^1 = C_6H_{13}$, $R^2 = H$	**58a, 71%**
$R^1 = C_8H_{17}$, $R^2 = H$	**59a, 74%**
$R^1 = Ph$, $R^2 = H$	**60a, 54%**
$R^1 = p\text{-}CH_3C_6H_4$, $R^2 = H$	**61a, 63%**
$R^1 = CN$, $R^2 = H$	**62a, 77%**
$R^1 = CHO$, $R^2 = CH_3$	**63a, 59%**
$R^1 = COOMe$, $R^2 = H$	**64a, 75%**
$R^1 = COMe$, $R^2 = H$	**65a, 76%**

Scheme 11. Intermolecular radical carbon-carbon bond formation in water.reactions of different alkenes with *n*-1-iodoperfluorohexane initiated by $(Me_3Si)_3SiH$ / dioxygen.

Encouraged by the above results obtained from reactions in water with n-$C_6F_{13}I$ and different alkenes initiated by $(Me_3Si)_3SiH$ / dioxygen, the same authors proceeded with the perfluoroalkylation reactions of electron rich 1-hexene with an array of perfluoroalkyl halides (iodides and bromides) under the same radical conditions in water.

When 1-hexene reacts with *n*-1-iodo-perfluorobutane, n-C_4F_9I, under the reaction conditions described above, 1,1,1,2,2,3,3,4,4-nonafluorodecane *57b*[53] is obtained in 88% yield (Scheme 12, Table 6, entry 1). The reaction of 1-hexene with 1,4-diiodo-1,1,2,2,3,3,4,4-octafluorobutane, n-IC_4F_8I, affords 1,1,2,2,3,3,4,4-octafluoro-1-iodo-decane *57c,* obtained in 75% yield (Scheme 12, Table 6, entry 2), and 15% of reduced 1,1,2,2,3,3,4,4-octafluoro-decane, as observed from the 19F NMR spectrum, the CF_2H peak at $\delta = -114.15$ ppm.

Table 5. Intermolecular radical carbon-carbon bond reactions in water. Reactions of different alkenes (40-50 mM) with *n*-1-iodoperfluorohexane (10 mM), initiated by $(Me_3Si)_3SiH$ (5 mM) / dioxygen

Entry	Alkene	Product, (%)[a]
1 ref.48c	1-Hexene	57a, (65)
2 ref.50c	1-Octene	58a, (71)
3 ref.48b	1-Decene	59a, (74)
4 ref.51	Styrene	60a[b], (54)
5 ref.48b	4-Methylstyrene	61a[b], (63)
6 ref.48c,51	Acrylonitrine	62a, (77)
7	Crotonaldehyde	63a, (59)
8 ref.51	Methylacrylate	64a, (75)
9 ref.48b	Vinyl methyl ketone	65a, (76)

[a] Isolated yield after purification, based on $C_6F_{13}I$.

[b] 1,1,1,2,2,3,3,4,4,5,5,6,6-tridecafluorohexane, $CHF_2(CF_2)_4CF_3$was also obtained in 30-40 %.

When *n*-1-bromo perfluorohexane, *n*-$C_6F_{13}Br$, is allowed to react with 1-hexene, product *57a* is obtained in poor yield (21%, Scheme 12, Table 6, entry 3 *cf.* with the same reaction carried out with *n*-$C_6F_{13}I$, Scheme 11, Table 5, entry 1). Analogously, the reaction of 1-hexene with *n*-1-bromo perfluorooctane, *n*-$C_8F_{17}Br$, progresses poorly to give product *57d* in 14% yield, while the same reaction employing *n*-$C_8F_{17}I$ affords *57d* in 72% yield (Scheme 12, Table 6, entries 3 and 4, respectively). In the same fashion, *n*-1-iodoperfluorodecane, *n*-$C_{10}F_{21}I$, reacts with 1-hexene to afford *57e* in 70% yield (Scheme 12, Table 6, entry 7), but with *n*-$C_{10}F_{21}Br$, product *57e* is obtained in 9% yield (Table 6, entry 6). This could be in agreement with a stronger BDE of the bond R_f-Br than that of R_f-I.

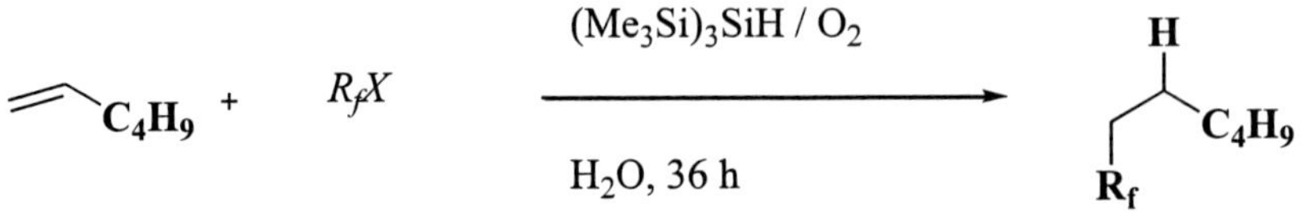

$$\text{CH}_2\text{=CH–C}_4\text{H}_9 + R_fX \xrightarrow[\text{H}_2\text{O, 36 h}]{(\text{Me}_3\text{Si})_3\text{SiH / O}_2} R_f\text{CH}_2\text{–CH(H)–C}_4\text{H}_9$$

$R_f = C_4F_9$, X = I — **57b, 88%**
$R_f = (CF_2)_4I$, X = I — **57c, 75%**
$R_f = C_6F_{13}$, X = Br — **57a, 21%**
$R_f = C_8F_{17}$, X = I — **57d, 72%**
$R_f = C_8F_{17}$, X = Br — **57d, 14%**
$R_f = C_{10}F_{21}$, X = Br — **57e, 9%**
$R_f = C_{10}F_{21}$, X = I — **57e, 70%**

Scheme 12. Intermolecular radical carbon-carbon bond formation in water. Reactions of 1-hexene with haloperfluoroalkanes in water, initiated by $(Me_3Si)_3SiH$ / dioxygen.

In order to cast some light into the efficiency of the chain reaction with dioxygen, and see whether the low yields obtained from 1-hexene and R_fBr and $(Me_3Si)_3SiH$ / dioxygen are due either to a slow initiation step or the retardation in the propagation step on account of the different BDE of bromo and iodo perfluoroalkanes, the authors undertook the radical chain initiation with the azo compound ACCN, at 70 °C in water.

Table 6. Intermolecular radical carbon-carbon bond formation in water. Reactions of different iodoperfluoroalkanes (10 mM) with 1-hexene (50 mM), initiated by $(Me_3Si)_3SiH$ (5 mM)/ dioxygen

Entry	Perfluoroalkane	Product, (%) [a]
1 [ref.55]	C_4F_9I	57b, (88)
2	$I(CF_2)_4I$	57c[b], (75)
3 [ref.48c]	$C_6F_{13}Br$	57a, (21)
4 [ref.48c]	$C_8F_{17}I$	57d, (72)
5 [ref.48c]	$C_8F_{17}Br$	57d, (14)
6 [ref.48b]	$C_{10}F_{21}Br$	57e, (9)
7 [ref.48b]	$C_{10}F_{21}I$	57e , (70)

[a] Isolated yield based on R_fX.

[b] 15% of reduced 1,1,2,2,3,3,4,4-octafluoro-decane was also obtained, as observed from the 19F NMR spectrum, the CF_2H peak at δ = -114.15 ppm.

When a series of alkenes are allowed to react with bromo perfluoroalkanes (*n*-1-bromo perfluorohexane, *n*-1-bromo perfluorooctane, and *n*-1-bromo perfluorodecane) in water as described above. for the thermal radical initiation reaction, the respective perfluoroalkylated products are obtained in yields ranging from 50 to 77% (Scheme 13, Table 7).

Thus, 1-hexene reacts with *n*-$C_6F_{13}Br$, *n*-$C_8F_{17}Br$, and *n*-$C_{10}F_{21}Br$ to afford the corresponding perfluoroalkylated hexanes *57a*[51,47], *57d* [50,46], and *57e* [50,46] in 75, 70, and 62% yields, respectively (Scheme 13, Table 7, entry 1). 1-Octene (Table 7, entry 2) and 1-decene (Table 7, entry 3) also afford perfluoroalkylated alkanes in fairly good yields (50-77%). Styrene and 4-methylstyrene react with *n*-$C_6F_{13}Br$, *n*-$C_8F_{17}Br$, and *n*-$C_{10}F_{21}Br$ to afford the alkylated products in good yields, ranging from 46 to 81% (Scheme 13, Table 7, entries 4 and 5).

Acrylonitrile, upon reaction with the series $C_nF_{2n+1}Br$ (n = 6, 8, and 10) affords the perfluoroalkylated-substituted propiononitriles *62a* [54], *62d*[54], and *62e [*54,50] in 51, 50, and 44% yields, respectively (Table 7, entry 6). Crotonaldehyde, upon reaction with *n*-$C_6F_{13}Br$, *n*-$C_8F_{17}Br$, and *n*-$C_{10}F_{21}Br$ under the reaction conditions described above, gives the perfluoroalkylated-substituted butyraldehydes *63a*, *63d*, and *63e* in 62, 60, and 55% yields, respectively (Scheme 13, Table 7, entry 7).

Methyl acrylate upon reaction with the series $C_nF_{2n+1}Br$ (n = 6, 8, and 10) affords the perfluoroalkylated-substituted methyl propionates *64a*, *64d*[55], and *64e* in 55, 53, and 43% yields, respectively (Scheme 13, Table 7, entry 8), while vinyl methyl ketone when reacted with *n*-$C_6F_{13}Br$, *n*-$C_8F_{17}Br$, and *n*-$C_{10}F_{21}Br$ in water under ACCN / $(Me_3Si)_3SiH$ thermal initiation (Section 1.C.-), affords the 5-perfluoroalkylated-substituted 2-butanones *65a* [54,55], *65d* [56,54] , and *65e*[54] in 67, 70, and 66% yields, respectively (Scheme 13, Table 7, entry 9).

The same authors also undertook the ACCN-radical initiated perfluoroalkylation of 1-hexene with $C_nH_{2n+1}I$ (n = 8, and 10), and obtained compounds*57d* and *57e* [48d], respectively, in yields ranging from 90-95%, while the same yields under dioxygen-radical initiation were in the 70% range (*cf.* O_2-initiated radical yields of *57d* and *57e* from Table 6, entries 4 and 7, respectively).

These sets of experiments would reveal that the lower yields obtained with the bromo perfluoroalkanes and alkenes in water under dioxygen initiation than those obtained under ACCN initiation could be attributed to a slower initiation in the former rather than a retardation in the propagation step due to differences in BDE of R_f-I and R_f-Br bonds; however, some involvement of the BDE of R_f-I *versus* R_f-Br should also be considered.

It is observed that the yields of products *57a-65a* are much better under ACCN initiation (Table 7) than under dioxygen initiation (Table 5).

It can be deduced that all water-insoluble material (substrates and reagents) suspended in the aqueous medium can interact due to the vigorous stirring that creates an efficient vortex and dispersion. In the dioxygen initiation, the chain mechanism probably benefits from the enhanced contact surface of tiny drops containing $(Me_3Si)_3SiH$ and dioxygen. The mechanism of the $(Me_3Si)_3SiH$-mediated intermolecular perfluoroalkylation of alkenes in water is depicted in Scheme 14.

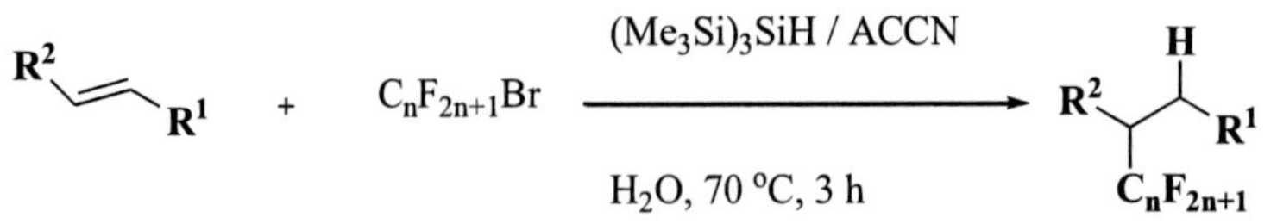

$$R^2CH{=}CHR^1 + C_nF_{2n+1}Br \xrightarrow[H_2O,\ 70\ ^\circ C,\ 3\ h]{(Me_3Si)_3SiH\ /\ ACCN} R^2CH(C_nF_{2n+1})CH(H)R^1$$

$R^1 = C_4H_9$, $R^2 = H$	**n=6, 57a, 75%; n=8, 57d, 70%; n=10, 57e, 62%**
$R^1 = C_6H_{13}$. $R^2 = H$	**n=6, 58a, 77%; n=8, 58d, 63%; n=10, 58e, 59%**
$R^1 = C_8H_{17}$, $R^2 = H$	**n=6, 59a, 65%; n=8, 59d, 59%; n=10, 59e, 48%**
R^1 = Ph, R^2 = H	**n=6, 60a, 80%; n=8, 60d, 81%; n=10, 60e, 77%**
R^1 = *p*-$CH_3C_6H_4$, R^2 =H	**n=6, 61a, 66%; n=8, 61d, 55%; n=10, 61e, 46%**
R^1 = CN, R^2 = H	**n=6, 62a, 51%; n=8, 62d, 50%; n=10, 62e, 44%**
R^1 = CHO, $R^2 = CH_3$	**n=6, 63a, 62%; n=8, 63d, 60%; n=10, 63e, 55%**
R^1 = COOMe, R^2 = H	**n=6, 64a, 55%; n=8, 64d, 53%; n=10, 64e, 43%**
R^1 = COMe, R^2 = H	**n=6, 65a, 67%; n=8, 65d, 70%; n=10, 65e, 66%**

Scheme 13. Intermolecular radical carbon-carbon bond formation in water. Reactions of alkenes with *n*-1-bromoperfluoroalkanes ($C_nF_{2n+1}Br$, n = 6, 8, and 10) in water, initiated by $(Me_3Si)_3SiH$ / ACCN.

Table 7.Intermolecular radical carbon-carbon bond formation in Water. Reactions of different *n*-1-bromoperfluoroalkanes, $C_nF_{2n+1}Br$ (n = 6, 8, and 10) (10 mM), with alkenes (40-50 mM), initiated by $(Me_3Si)_3SiH$ (5 mM)/ ACCN (3 mM) at 70 °C.

Entry	Bromoperfluoroalkane			
	1-Alkene	*n*-C_6F_{13}Br	*n*-C_8F_{17}Br	*n*-$C_{10}F_{21}$Br
		Perfluoroalkylated Product, (%product) a		
1	1-Hexene	57a[ref.48c], (75)	57d, (70)	57e, (62)
2	1-Octene	58a[ref.50], (77)	58d, (63)	58e , (59)
3	1-Decene	59a[ref.48c], (65)	59d, (59)	59e, (48)
4	Styrene	60a[ref.51], (80)	60d, (81)	60e, [ref.51] (77)
5	4-Methylstyrene	61a[ref.48b], (66)	61d, (55)	61e, (46)
6	Acrylonitrile	62a[ref.48c,51], (51)	62d, (50)	62e[ref 53], (44)
7	Crotonaldehyde	63a, (62)	63d, (60)	63e, (55)
8	Methylacrylate	64a,[ref.51] (55)	64d, (53)	64e, (43)
9	Vinyl methyl ketone	65a[ref.48b], (67)	65dref.53, (70)	65e[ref 53], (66)

[a]Isolated yield after purification.

The $(Me_3Si)_3Si•$ radical, produced by some kind of radical initiation (either by dioxygen or thermal decomposition of ACCN initiator) in water, abstracts the halogen atom (iodine or bromine) from R_fX. This $R_f•$ radical reacts faster with the alkene than with the silicon hydride, affording the perfluoroalkylated radical adduct *67*. Radical adduct *67* abstracts hydrogen from the silane, affording the perfluoroalkyl-substituted alkane*68*, and regenerating the silyl radical, thus propagating the chain.

According to what has been observed and measured by Dolbier et al.[53], in benzene-d_6, the ratio of products [*56*]/[*55*] (eq 9) should equal [47] the ratio of rate constants for addition (of perfluorinated heptyl radical on 1-hexene) and rate constant for H abstraction from $(Me_3Si)_3SiH$ times the ratio of concentrations of alkene and silane [47].

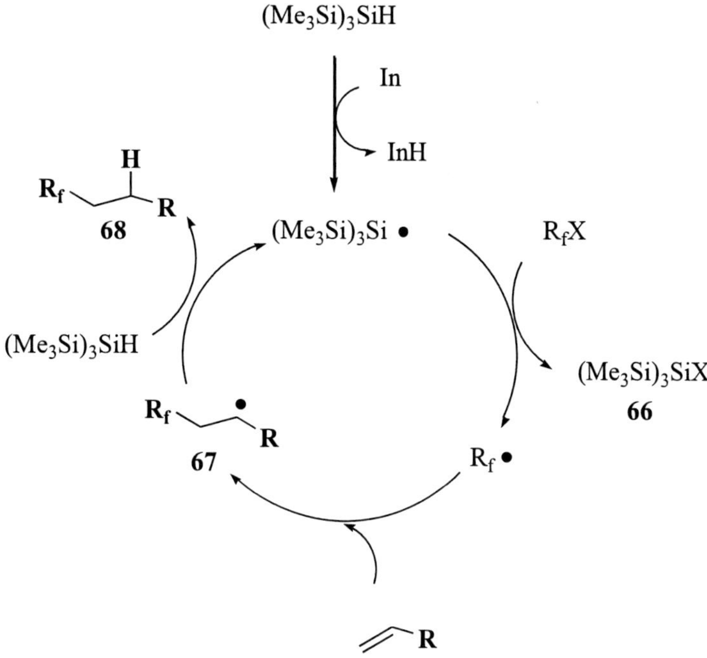

Scheme 14. Mechanism for radical carbon-carbon bond formation reactions from alkenes and perfluoroalkyl iodides in Water, mediated by $(Me_3Si)_3SiH$.

According to the experimental conditions, employing equation in reference 47, a theoretical ratio of perfluoroalkylated alkane over reduced perfluoroalkane of *ca.* 1.3 would be obtained, which is not completely in agreement with the unobserved reduced perfluoroalkanes in these reaction systems in water (*i.e.*: $CHF_2(CF_2)_4CF_3$, when iodoperfluorohexane is employed). Transition state polar effects (in water) must be playing a decisive role in such radical addition reactions onto alkenes in water exerting an acceleration effect at the expense of the hydrogen atom transfer reaction.[57]

The electrophilicity of R_f• radicals are the dominant factor giving rise to their high reactivity. The stronger carbon-carbon bond which forms when R_f• versus R• radicals add to an alkene is a driving force for the radical addition (the greater exothermicity of the R_f• radical addition is expected to lower the activation energy). It has been observed, in organic solvents, that the rates of addition of R_f• radicals onto alkenes correlate with the alkene IP (which reflects the *HOMO* energies).[53] Indeed, the major transition state orbital interaction for the addition of the highly electrophilic R_f• radical to an alkene is that between the *SOMO* of the radical and the *HOMO* of the alkene. Thus, the rates of R_f• radical addition to electron deficient alkenes are slower than those to electron rich alkenes (as observed in organic solvents). From these results, however, it becomes apparent, that in water the rates for R_f• radical addition to both electron rich and electron deficient alkenes could be comparable, given the similar reaction yields (*cf.* in Table 7, entries 1-3 and 6-9, respectively). In order to clarify this subtle aspect of the reaction in water, a set of experiments designed to compare the ratios of $(K_H/K_{add})_{1\text{-hexene}}$ and $(K_H/K_{add})_{acrylonitrile}$for the addition reaction of n-$C_6F_{13}I$ to the electron rich 1-hexene and electron deficient acrylonitrile, respectively was undertaken. These ratios of rate constants are obtained by plotting [n-$C_6F_{13}H$]/[*57a*] *vs* [$(Me_3Si)_3SiH$]/[1-hexene] and [n-$C_6F_{13}H$]/[*62a*] *vs* [$(Me_3Si)_3SiH$]/[acrylonitrile], respectively, when the reactions are initiated thermally, by using incremental amounts of $(Me_3Si)_3SiH$, and keeping the alkene and n-$C_6F_{13}I$ concentrations constant. The authors obtained slopes for both plots equal to 1.55±0.09 (r^2=0.998) and 1.88±0.19 (r^2=0.989) for $(K_H/K_{add})_{1\text{-hexene}}$ and $(K_H/K_{add})_{acrylonitrile}$ respectively. This seems to indicate that the reactivities of electron rich and electron deficient alkenes towards R_f• radicals in water are leveled off.

The ratio of rates constants $(K_H/K_{add})_{1\text{-hexene}}$ obtained in benzotrifluoride as solvent [43] for the reaction of n-$C_7F_{15}I$ and 1-hexene with $(Me_3Si)_3SiH$ as the hydrogen donor (eq 9) is 6.32, while that same ratio of rate constants for the reaction of n-$C_6F_{13}I$ with 1-hexene in water is 1.55 (*vide supra*). Owed to the unavailability of the rate constant for R_f• radical addition onto double bonds*in water* makes comparisons difficult; however, the results would seem to imply that the rate for hydrogen donation from $(Me_3Si)_3SiH$ to the R_f• radical relative to the addition reaction is four times slower in *water* than in benzotrifluoride as solvent (*i.e.:* $(K_H/K_{add})_{water}$ / $(K_H/K_{add})_{Benzotrifluoride}$ = 0.25).

The intermolecular radical carbon-carbon bond formation reactions presented herein in water take advantage of the halophilicity of silyl radicals towards iodine and bromine atoms. In this case, the halogen atom transfer reaction from *66* towards the perfluoroalkylated radical adduct *67* (Scheme 14) is slower than the hydrogen atom abstraction from the $(Me_3Si)_3SiH$, thus providing the reduced perfluoroalkylated product *68*. As a support to the mechanism proposed (Scheme 14), 2-iodo-1,1,1,3,3,3-hexamethyl-2-(trimethyl)trisilane (*66*, X= I, Scheme 14) was isolated and characterized from the reaction mixtures described in Scheme 11. In the same fashion, from reactions described in Scheme 13, 2-bromo-1,1,1,3,3,3-hexamethyl-2-(trimethyl)trisilane (*66*, X = Br, Scheme 14) was detected by gas chromatography, isolated, and compared with an authentic sample. As mentioned above, by plotting [n-$C_6F_{13}H$]/[*57a*] *vs* [$(Me_3Si)_3SiH$]/[1-hexene], a straight line, whose slope represents $(K_H/K_{add})_{1\text{-hexene}}$ is obtained (with a value of 1.55±0.09, r^2=0.998). The intercept of this plot, shows, remarkably, no deviation from the ideal value of zero, purporting that the only source of n-C_6F_{13}• radical reduction (*i.e.:n*-$C_6F_{13}H$) is the silane, and not the solvent or the alkene.[53]

It is interesting to point out that the scope of the radical dioxygen initiation reactions with silanes in water has been extended from the use in hydrosilylation reactions of C-C multiple bonds[12,45] (section 1.2..-), to the intermolecular radical carbon-carbon bond formation. From results obtained in Tables 5, 6, and 7, the thermal ACCN-initiation is deemed a better radical initiating technique than the dioxygen-initiation, as opposed to what has been observed in the radical hydrosilylation of alkynes in water.[19a,27] However, should thermally labile alkenes need be employed, the radical dioxygen initiation methodology is a good option of choice, as fairly good yields of carbon-carbon coupling products can be obtained in water under the mild dioxygen initiation technique, in the absence of reduced perfluorinated halides. Interesting and notorious solvent effects can be invoked in water on the rates of reactions of perfluoroalkyl radicals towards alkenes and silicon hydrides, respectively, as opposed to those found in benzene, where the latter rates are *ca.* six times faster at equal concentrations of silanes and alkenes. This account[46] provides also a convenient method to achieve perfluoroalkylation reactions of alkenes in water to render perfluoroalkylated alkanes as key intermediates in the synthesis of fluorophors and other fluorinated materials.

In a previous study,[12] the radical cyclization in water of 1-((*E*)-but-2-enyloxy)-2-iodobenzene (*6*) to afford 3-ethyl-2,3-dihydrobenzofuran (*7*) in 85% yield (eq 2) has been reported, when $(Me_3Si)_3SiH$ and an azo initiator is employed. In this account, 6-bromo-3,3,4,4,5,5,6,6-octafluoro-1-hexene[45] (12 mM) *69* was subjectedto reaction (24 h) with $(Me_3Si)_3SiH$ (8 mM) and dioxygen in water (5 mL), as described in Section 1.C., and obtained the *exo-trig* cyclization product 1,1,2,2,3,3,4,4-octafluoro-5-methylcyclopentane *70* (eq 10) in 76% yield (isolated).[58]

O_2 (R.T.) , 24 h
$(Me_3Si)_3SiH$/water
69 **70** (10)

Though the measurement of the rate constant for cyclization in the heterogeneous water system is difficult to be obtained, the cyclohexane cyclized product has not been observed in water under the reaction conditions reported. No uncyclized-reduced product is either observed.[58]

Analogously, cyclization of 5-bromo-1,1,2,3,3,4,4,5,5-nonafluoro-pent-1-ene (12 mM) *71* in water triggered by $(Me_3Si)_3SiH$ (8 mM) / dioxygen leads to nonafluorocyclopentane, the *exo-trig* cyclization product in 68% yield (isolated). No reduced product could be isolated from the reaction mixture. The reaction carried out in benzene-d_6 does not lead to cyclization product (eq 11).[58]

O_2 (R.T.) , 24 h
$(Me_3Si)_3SiH$/water
71 68% (11)

When 3,3,4,4-1,5-hexadiene (40 mM) *72* is allowed to react (24 h) in water with $(Me_3Si)_3SiH$ (5 mM) / dioxygen and C_2F_5I (10 mM), product *73* is obtained in 61% yield, based on C_2F_5I (eq 12).

72 + C_2F_5I O_2 (R.T.) , 24 h
$(Me_3Si)_3SiH$/water
73 (12

Summary and Outlook

The synthetically useful silyl radical-mediated transformations that can be accomplished in water encompass simple organic halide reductions, azide reductions, deoxygenation reactions, hydrosilylation reactions of carbon-carbon multiple bonds and carbon-oxygen double bonds, intramolecular carbon-carbon bond formation, and intermolecular consecutive radical reactions, and more recently, perfluoroalkylation reactions of multiple bonds. Future studies should be directed at the syntheses of more complex structures by intermolecular carbon carbon and carbon-heteroatom transformations in water mediated by silyl radicals.

References

[1] Yorimitsu, T.; Nakamura, H.; Shinokubo, K.; Oshima, K.; Fujimoto, H. Powerful Solvent Effect of Water in Radical Reaction: Triethylborane-induced Atom-Transfer Radical Cyclization in Water. *J.Am.Chem.Soc.* 2000, *122,* 11041-11047.

[2] Yorimitsu, H.; Shinokubo, H.; Oshima, K. Radical Reaction by a Combination of Phosphinic Acid and a Base in Aqueous Media. *Bull.Chem.Soc.Jpn.* 2001, *74*, 225-235.

[3] Yasuda, H.; Uenoyama, Y.; Nobutta, O.; Kobayashi, S.; Ryu, I. Radical Chain Reactions Using THP as a Solvent. *TetrahedronLett.* 2008, *49*, 367-370.

[4] (a) Naka, A.; Ohnishi, H.; Ohsita, J.; Ikadai, J.; Kunai, A.; Ishikawa M. Silicon-Carbon Unsaturated Compounds. 70. Thermolysis and Photolysis of Acylpolysilanes with Mesitylacetylene. *Organometallics* 2005, *24*, 5356-5363. (b) Wang, J.; Zhu, Z.; Huang, W.; Deng, M.; Zhou, X.J. Air-initiated Hydrosilylation of Unactivated Alkynes and Alkenes and Dehalogenation of Halohydrocarbons by Tris(trimethylsilyl)silane under Solvent-free Conditions. *J.Organomet.Chem.* 2008, *693*, 2188-2192.

[5] Yamazaki, O.; Togo, H.; Nogami, G. ; Yokoyama, M. Radical Addition of 2-iodoalkanamide or 2-iodoalkanoic Acid to Alkenols using a Water-soluble Radical Initiator in Water. A Facile Synthesis of γ-lactones. *Bull.Chem.Soc.Jpn.* 1997, *70,* 2519-2523.

[6] (a) Mecking, S.; Held, A.; Bauers, F.M. Aqueous Catalytic Polymerization of Olefins. *Angew.Chem.Int.Ed.* 2002, *41*, 544-561. (b) Li, C-J.; Chen, L. Organic Chemistry in Water. *Chem.Soc.Rev.* 2006, *35*, 68-82.

[7] Lalevée, J.; Blanchard, N.; El-Roz, M.; Graff, B.; Allonas, X.; Fouassier, J.P. New Photoinitiators Based on the Silyl Radical Chemistry: Polymerization Ability, ESR Spin Trapping and Laser Flash Photolysis Investigation. *Macromolecules* 2008, *41*, 4180-4186.

[8] Kminek, I.; Yagci, Y.; Schnabel, W. A Water-soluble Poly(methylphenylsilylene) Derivative as a Photoinitiator of Radical Polymerization of Hydrophilic Vinyl Monomers. *PolymerBull.* 1992, *29*, 277-282.

[9] Roberts, B.P. Polarity-reversal Catalysis of Hydrogen-atom Abstraction Reactions: Concepts and Applications in Organic Chemistry.*Chem.Soc.Rev.* 1999, *28,* 25-35.

[10] Chatgilialoglu, C. *Organosilanes in Radical Chemistry*, Wiley-VCH, Chichester, 2004, pp 1-142.

[11] Postigo, A.; Ferreri, C.; Navacchia, M.L.; Chatgilialoglu, C. The Radical-based Reduction with $(TMS)_3SiH$ on Water. *Synlett* 2005, 2854-2856.

[12] Postigo, A.; Kopsov, S.; Ferreri, C.; Chatgilialoglu, C. Radical Reactions in Aqueous Medium Using $(Me_3Si)_3SiH$. *Org.Lett.* 2007, *9,* 5159-5162.

[13] Chatigilialoglu, C. The Tris(trimethylsilyl)silane/Thiol Reducing System: A Tool for Measuring Rate Constants for Reactions of Carbon-Centered Radicals with Thiols.*Helv.Chim.Acta* 2006, *89*, 2387-2398.

[14] (a) Apeloig, Y.; Nakash, M. Reversal of Stereoselectivity in the Reduction of Gem-dichlorides by Tributyltin Hydride and Tris(trimethylsilyl)silane. Synthetic and Mechanistic Implications.*J.Am. Chem.Soc.* 1994, *116*, 10781–10782. (b) Liu, Y.; Yamazaki, S.; Izuhara, S. Modification and Chemical Transformation of Si(111) Surface. *.J.Organomet.Chem.* 2006, *691*, 5809-5824.

[15] Odedra, A.; Geyer, K.; Gusstafsson, T.; Gimour, R.; Seeberger, P.H. Safe, FacileRadical-based Reduction and Hydrosilylation Reactions in a Microreactor Using Tris(trimethylsilyl)silane.*Chem.Commun.* 2008, 3025-3027.

[16] Chatgilialoglu, C.; Ferreri, C.; Mulazzani, Q.C.; Ballestri, M.; Landi; L. *Cis−trans* Isomerization of Monounsaturated Fatty Acid Residues in Phospholipids by Thiyl Radicals. *J.Am.Chem.Soc.* 2000, *122*, 4593-4601.

[17] Benati, L.; Bencivenni, G.; Leardini, R.; Minozzi, M.; Nanni, D.; Scialpi, R.; Spagnolo, P.; Zanardi, G. Radical Reduction of Aromatic Azides to Amines with Triethylsilane. *J.Org.Chem.* 2006, *71*, 5822-5825.

[18] Liu, Y.; Yamazaki, S; Yamabe, S. Regioselective Hydrosilylations of Propiolate Esters with Tris(trimethylsilyl)silane. *J.Org.Chem.* 2005, *70,* 556-561.

[19] Postigo, A.; Kopsov, S.; Zlotsky, S.S.; Ferreri, C.; Chatgilialoglu, C. Hydrosilylation of C-C Multiple Bonds Using $(Me_3Si)_3SiH$ inWater. Comparative Study of the Radical Initiation Step.*Organometallics* 2009, *28,* 3282-3287.

[20] Kopping, B.; Chatgilialoglu, C.; Zehnyder, M.; Giese, B. Tris(trimethylsilyl)silane: an Efficient Hydrosilylating Agent of Alkenes and Alkynes. *J.Org.Chem.* 1992, *57*, 3994-4000.

[21] Calandra, J.; Postigo, A.; Russo, D.; Sbarbati-Nudelman, N.; Tereñas, J.J. Photochemical Initiation of Radical Hydrosilylation Reactions of C-C Multiple Bonds in Water. *J.Phys.Org.Chem.* 2010, *23,* 944-949.

[22] Trost, B.M.; Ball, Z.T.; Laemmhold, K.M. An Alkyne Hydrosilylation-oxidation Strategy for the Selective Installation of Oxygen Functionality. *J.Am.Chem.Soc.* 2005, *127,* 10028-10038.

[23] Studer, A.; Steen, H. The S_Hi Reaction at Silicon - A New Entry into Cyclic Alkoxysilanes. *Chem.Eur.J.* 1999, *5,* 759-773.

[24] Indeed, upon silyl radical attack on the C-C double bond of 29, the 1-amino-3-(1,1,1,3,3,3-hexamethyl-2-(trimethylsilyl)trisilan-2-yl)propan-2-yl radical would be formed. This latter radical, if S_Hi mechanism took place, would suffer bond cleavage and homolytic substitution at silicon to render, upon ulterior hydrogen abstraction, 3-(1,1,1,3,3,3-hexamethyl-2-(trimethylsilyl)trisilan-2-yl)butan-1-amine product, which was not found in the reaction mixture.

[25] Zaborovskiy, A.B.; Lutsyk, D.S.; Prystansky, R.E.; Kopylets, V.I.; Timokhin, V.I.; Chatgilialoglu, C. A Mechanistic Investigation of $(Me_3Si)_3SiH$ Oxidation. *J.Organomet.Chem.* 2004, *689,* 2912-2919.

[26] Curran, P.; Keller, A.I. Radical Additions of Aryl Iodides to Arenes are Facilitated by Oxidative Rearomatization with Dioxygen. *J.Am.Chem.Soc.* 2006, *128,* 13706-13707.

[27] (a) Postigo, A.; Sbarbati Nudelman, N. Different Radical Initiation Techniques of Hydrosilylation Reactions of Multiple Bonds in Water: Dioxygen Initiation. *J.Phys.Org.Chem.* 2010, *23*, 910-914. (b) Liu; Y.; Yamazaki; S.; Yamabe, S.; Nakato, Y. A Mild and Efficient Si (111) Surface Modification via Hydrosilylation of Activated Alkynes.*J.Mat.Chem.* 2005; *15*, 4906-4913.(c) Wang, Z.; Pittelaud, J-P.; Montes, L.; Rapp, M.; Derane D.; Wnuk, S.F. Vinyl Tris(trimethylsilyl)silanes: Substrates for Hishama coupling. *Tetrahedron* 2008; *64,* 5322-5327.

[28] (a) Zhang, W. Synthetic Applications of Fluorous Solid-phase Extraction (F-SPE).*Tetrahedron* 2003, *59*, 4475-4489. (b) Curran, D.P. *Pharm.News,* 2002, *9*, 179-188. (c) Villard, A.L., Warrington, B.H.; Ladlow, M. A Fluorous Tag, Acid-labile Protecting Group for the Synthesis of Carboxamides and Sulfonamides. *J.Comb.Chem.* 2004, *6,* 611-622. (d) Zhang, W.; Luo, Z.; Chen, C.H.-T.; Curran, D.P. Solution-phase Preparation of a 560-compound Library of Individual Pure Mappicine Analogues by Fluorous Mixture Synthesis.*J.Am.Chem. Soc.* 2002, *124*, 10443-10450. (e) Lindsley, C.W.; Zhao, Z.; Newton, R.C.; Leister, W.H.; Strauss, K.A. A general Staudinger Protocol for Solution Phase Parallel Synthesis. *Tetrahedron Lett.* 2002, *43*, 4467-4470. (f) Lindsley, C.W., Zhao, Z.; Leister, W.H.; Strauss, K.A. Florous-tethered Amine Bases for Organic and Parallel Synthesis, Scope and Limitations. *Tetrahedron Lett.* 2002, *43*, 6319-6323.

[29] Park, G.; Ko, K.-S.; Zakharova, A.; Pohl, N.L. Mono- versus di Fluorous Tagged Glucosamines for Iterative Oligosaccharide Synthesis. *J.Fluor.Chem.* 2008, *129,* 978-982.

[30] Vaillard, S.; Postigo, A.; Rossi, R.A. Novel Perfluoroalkyl Diphenylphosphine Compounds. Synthesis and Reaction Mechanism. *Organometallics* 2004, *23*, 3003-3007.

[31] Bravo, A.; Bjorvski, H.-R.; Fontana, F.; Liguori, L.; Mele, A.; Minischi, F. New Methods of Free-Radical Perfluoroalkylation of Aromatics and Alkenes. Absolute Rate Constants and Partial Rate Factors for the Homolytic Aromatic Substitution by *n*-perfluorobutyl Radical. *J.Org.Chem.* 1997, *62*, 7128-7136.

[32] Baasner, B.; Hagemann, H.; Tatlow, J.C. *Houben-WeylOrgano-Fluorine Compounds*, Thieme: Stuttgart, New York, 2000.

[33] Tsuchii, K.; Imura, M.; Kamada, N.; Hirao, T.; Ogawa, A. An Efficient PhotoinducedIodoperfluoroalkylation of Carbon−Carbon Unsaturated Compounds with Perfluoroalkyl Iodides. *J.Org.Chem.* 2004, *69*, 6658-6665.

[34] (a) Rossi, R.A.; Postigo, A. Recent Advances on Radical Nucleophilic Substitution Reactions. *Curr.Org.Chem.* 2003, *7*, 747-763. (b) Rossi, R.A.; Pierini, A.B.; Penñeñory, A.B. Nucleophilic Substitution Reactions by Electron Transfer. C*hem. Rev.* 2003, *103*, 71-168.

[35] Postigo, A. in "*Photochemical Generation of Silicon-centered Radicals and their Reactions*". Chapter VI in "*Progress in Photochemistry*", Nova Science Publications. Hauppauge, New York.2008, pp. 193-226.ISBN 978-1-60456-568-3.

[36] Barton, D.H.R. in *Half a Century of Free Radical Chemistry*, Cambridge University Press, Cambridge, 1993 pp 1-195.

[37] Chatgilialoglu, C. *Organosilanes in Radical Chemistry*, John Wiley & Sons, Ltd. 2004, pp 143-155.

[38] Yorimitsu, H.; Shinokubo, H.; Matsubara, S.; Oshima, K. Triethylborane-induced Bromine Atom-Transfer Radical Addition in Aqueous Media. Study of the Solvent Effect on Radical Addition Reactions.*J.Org.Chem.* 2001, *66*, 7776-7785.

[39] Miyabe, H.;Fujii, K.; Goto, T.; Naito T. A New Alternative to the Mannich Reaction: Tandem Radical Addition−cyclization Reaction for Asymmetric Synthesis of γ-butyrolactones and β-amino Acids. *Org.Lett.* 2000, *2*, 4071-4074.

[40] Kondo, J.; Shinokubo, H.; Oshima, K. From Alkenylsilanes to Ketones with Air as the Oxidant.*Angew.ChemieInt.Ed.* 2003, *42,* 825-827.

[41] Miyabe, H.; Ueda, M.; Naito, T. Free-radical Reaction of Imine Derivatives in Water.*J.Org.Chem.* 2000, *65*, 5043-5047.

[42] McNabb, S.B.; Ueda, M.; Naito, T.Addition of Electrophilic and Heterocyclic Carbon-centered Radicals to GlyoxylicOxime Ethers.*Org.Lett.* 2004, *6,* 1911-1914.

[43] Miyabe, H.; Yamaoka, Y.; Takemoto, Y. Triethylborane-induced Intermolecular Radical Addition to Ketimines.*J.Org.Chem.* 2005, *70*, 3324-3327.

[44] Postigo, A. Synthetic Organometallic Radical Chemistry in Water. *Curr.Org.Chem.* 2009, *13,* 1683-1704.

[45] Rong, X.X.; Pan, H.Q.; Dolbier, W.R.Jnr. Reactivity of Fluorinated Alkyl Radicals in Solution. some Absolute Rates of Hydrogen-atom Abstraction and Cyclization. *J.Am.Chem.Soc.* 1994, *116*, 4521-4522.

[46] Barata-Vallejo, S.; Postigo, A. $(Me_3Si)_3SiH$-mediated Intermolecular Perfluoroalkylation Reactions of Olefins in Water. *J.Org.Chem.* 2010, *75*, 6141-6148.

[47] where k_{add} is the rate constant for of addition of perfluorinated radical to 1-hexene, k_H is the rate constant for H atom abstraction from the silane, and 56 and 55 the reduced perfluorinated alkane, and the addition product, respectively, shown in eq 2.

[48] (a) Davis, C.R.; Burton, D.J.; Yang, Z-Yu. Titanium-catalyzed Addition of Perfluoralkyl Iodides to Alkenes. *J.Fluor.Chem.* 1995, *70*, 135-140. (b) Le, T.D.; Arlauskas, R.A.; Weers, J.G. Characterization of the Lipophilicity of Fluorocarbon Derivatives Containing Halogens or Hydrocarbon Blocks. *J.Fluor.Chem.* 1996, *78*, 155-163. (c) Huang, X-T.; Chen, Qing-Yun. Nickel(0)-catalyzed Fluoroalkylation of Alkenes, Alkynes, and Aromatics with Perfluoroalkyl Chlorides. *J.Org.Chem.* 2001, *66*, 4651-4656.

[49] Yang, G.S.; Xie, X-J.; Zhao, G.; Ding, Y. Palladium Catalyzed Cross Coupling Reaction of Organoboronic Acids with β-(perfluoroalkyl)ethyl iodides. *J.Fluor.Chem.* 1999, *98*, 159-161.

[50] Chen, W.; Xu, L.; Xiao, J. A general Method to Fluorous Ponytail-substituted Aromatics. *TetrahedronLett.* 2001, *42*, 4275-4278.

[51] Hu, Chan-Ming; Qiu, Yao-Ling. Cobaloxime-catalyzed Hydroperfluoroalkylation of Electron-Deficient Alkenes with Perfluoroalkyl Halides: Reaction and Mechanism. *J.Org.Chem.* 1992, *57*, 3339-3342.

[52] Kyoko, N.; Oshima, K.; Utimoto, K. Trialkylborane as an Initiator and Terminator of Free Radical Reactions. Facile Routes to Boron Enolates via α-carbonyl Radicals and Aldol Reaction of Boron Enolates. *Bull.Chem.Soc.Jpn.* 1991, *64,* 403-409.

[53] Delest, B.; Shtarev, A.B.; Dolbier, W.R. Reactivity of Perfluoro-*n*-alkyl Radicals with Reluctant Substrates. Rates of Addition of $R_fCH_2CH=CH_2$ and $R_fCH=CH_2$. *Tetrahedron* 1998, *54,* 9273-9280.

[54] Briza, T.; Kvicala, J.; Paleta, O.; Cermak, J. Preparation of Bis(polyfluoroalkyl)cyclopentadienes, New Highly Fluorophilic Ligands for FluorousBiphase Catalysis. *Tetrahedron* 2002, *58*, 3841-3846.

[55] Sturz, G.; Clement, J.C.; Daniel, A.; Molinier, J.; Lenzi, M. *Bull.Soc.Chemie France,* 1981, *1,* 161-165.

[56] Umemoto, T.; Kuriu, T.; Nakayama, S-ichi. Reactions of $R_fPh(OSO_2)CF_3$ with Alkenes and Alkadienes. *TetrahedronLett.* 1982, *23,* 1169-1172.

[57] Tedder, J.M. The Importance of Polarity, Bond Strength and Steric Effects in Determining the Site of Attack and the Rate of Free Radical Substitution in Aliphatic Compounds. *Tetrahedron* 1982, *38*, 313-329.

[58] Calandra, J.; Postigo, A. Synthetically-useful Dioxygen-initiated Radical Chain Reactions in Water. Unpublished results.

In: Organic Radical Reactions in Water and Alternative Media ISBN 978-1-61209-648-3
Editor: J. Alberto Postigo, pp. 79-169

Chapter 3

METAL-CENTERED RADICALS IN WATER

J. Alberto Postigo*

Department of Chemistry-Faculty of Science. University of Belgrano.
Villanueva 1324 CP 1426-Buenos Aires-Argentina

ABSTRACT

In this chapter metal-centered radicals are used in water to accomplish several synthetically-useful organic transformations such as Reformatzky reactions, alkylation and allylation reactions of carbonyl compounds and imine derivatives, radical conjugate additions, metal-mediated radical cyclizations, pinacol and other coupling reactions, oxidation and reduction reactions in water, etc. In doing so, the array of metal radicals used encompass elements from the main group (IIIB, IVB, V and VIB groups), as well as transition metals.

INTRODUCTION

The array of radical reactions of synthetic utility encompasses several areas of chemical sciences, among which, organometallic radical chemistry has become a valid option for the synthetic chemist during the past 50 years. More recently, chemists have successfully started to explore the use of water in radical reactions, and the involvement of metal-centered radicals to carry out organic synthetic transformations in this medium.

Most metal-mediated radical reactions in water are presented in the literature under a certain metal-centered radical chemistry or category. However, as a large group of metal-centered radicals have been found to show similar reactivity in water enabling common groups of organic transformations such as carbon-carbon bond formation and carbon-heteroatom bond formation reactions, these will be presented under "types of reactions in water " rather than under a single metal-centered radical heading or subtitle. This is the case

* Corresponding Author: Dr. Al Postigo. Fax: (+) 54 11 4511-4700. E-mail: alberto.postigo@ub.edu.ar

for a group of metal-centered radicals, particularly transition metal-centered radicals, which due to their reduction potential are likely to react readily as one-electron oxidants, and radicals and radical ions are found as intermediates in these reactions.

In this Chapter metal-centered radicals participate in water to accomplish several synthetically-useful organic transformations such as Reformatzky reactions, alkylation and allylation reactions of carbonyl compounds and imine derivatives, radical conjugate additions, metal-mediated radical cyclizations, pinacol and other coupling reactions, oxidation and reduction reactions in water, etc. In doing so, the group of metal radicals used involve elements from the main group (IIIB, IVB, V and VIB groups), as well as main transition metals and lanthanides.

In section 2.1.- transition metals such as Ce are shown to act as mediators in the addition of electrophilic carbon radicals to alkenes. In section 3.1.- Reformatzky reactions (reaction between a 2-halo ester and a carbonyl compound) are shown to proceed in water efficiently through the mediation of zinc.

In sections 3.2.- and 3.3.- radical alkylation reactions of carbonyl compounds and imine derivatives are illustrated to proceed in water through an array of different metals, such as indium, titanium, and silver, and mixtures of In / CuI / $InCl_3$, as well as Zn, Al, Sn

In section 3.4.- radical alkylation of electron-deficient alkenes in water are shown to be accomplished by the use of In, phosphorous-centered radicals, Zn / CuI, Zn / $AlCl_3$.

In sections 3.5.- and 3.6.- radical allylation reactions of carbonyl compounds and imine derivativesare performed through the intervention of such diverse metals as Zn, Bi, Sn, Mg, Mn, Sb. Pb, Hg, Cu, Ga, and In to yield interesting all-carbon addition compounds.

In section 3.7.- radical conjugate addition reactions of halo compoundsto α,β-unsaturated carbonyl compounds in water are shown to proceed in high yields with the use of Zn(Cu), Zn / CuI, Hg, In / CuI / $InCl_3$ and B. In section 3.8.- the radical synthesis of β,γ-unsaturated ketones in water is illustrated to take place by the reaction of α,β-unsaturated ketones and a regular ketone compound, mediated by In / $InCl_3$.

In section 3.9.- Mannich-type reactions in water are shown to take place through the use of different metals, such as Cu, and Ti.

For the synthetic chemist, metal-mediated radical cyclization reactions in water are depicted in section 3.10.- The metals that allow these transformations encompass In, Zn, Zn / CuI, Ag, Ti. Other types of carbon-carbon bond forming reactions in water mediated by metal radicals include pinacol coupling reactions, described in section 3.11.-. Dehalogenation reactions (section 3.12.-), and oxidation and reduction reactions (section 3.13) are also described in water with participation of metals.

Undoubtedly, metal-mediated radical reactions has become an active area of research, which has seen a rebirth with the use of water and other aqueous systems, lowering the environmental impact of transition metal chemistry.

Of the principal methods developed for the generation of radicals, redox processes based on electron-transfer deserve special mention. Chemical methods for electron-transfer oxidation involve the use of salts of high-valence metals such as Mn(III), Ce(IV), Cu(II), Ag(I), Co(III), V(V), and Fe(III). Among these, Mn(III) has received the most attention. In spite of the well-established and frequent use of this reagent for the generation of electrophilic carbon-centered radicals from enolic substrates, particularly in intramolecular processes, procedural problems associated with it have prompted the development of other oxidants of choice. The availability of Ce(IV) reagents as suitable one-electron oxidants assumes importance in this context.

2. Lanthanide Radicals in Water

2.1. Cerium-Mediated Radical Reactions in Water

The earliest report on the use of Ce(IV) salts for the generation of carbon-centered radicals dates back to the pioneering work of Heiba and Dessau in 1971, as the addition of acetone to 1-octene mediated by cerium(IV) acetate leading to keto-derivatives.[1]

Baciocchi focused on the oxidative addition of simple ketones and 1,3-dicarbonyl compounds to activated alkenes such as silyl enol ethers, enol ethers, and enol acetates. Representative examples include the synthesis of 4-ketaldehyde dimethyl acetals and 3-acyl furans by cerium ammonium nitrate, the CAN-mediated addition of carbonyl and dicarbonyl compounds, respectively, to enol acetates (Scheme 1).[2]

Nair et.al. have dealt with relevant synthetic transformations employing cerium ammonium nitrate, CAN [3] The array of useful reactions that utilize CAN involve carbon-carbon bond formation reactions leading to one-pot synthesis of dihydrofurans,[4] tetrahydrofurans, and aminotetralins. Also, carbon-heteroatom bond formation reactions mediated by CAN comprise oxidative addition of soft anions to alkenes.

As might be expected of very powerful one-electron oxidants (E_{red}*ca.* 1.6 eV), the chemistry of cerium(IV) oxidation of organic molecules is dominated by radical and radical cation chemistry.

All the reactions studied are based on the CAN-mediated generation of electrophilic carbon-centered radicals and their trapping with electron-rich substrates, either in methanol or acetonitrile as solvents. In addition, these radicals are also susceptible to dimerizations, and such processes have been investigated. For instance, Nicolaou has reported the synthesis of racemic hybocarpone 1 by CAN-mediated dimerization of naphthazirin 2 (Scheme 2).[5]

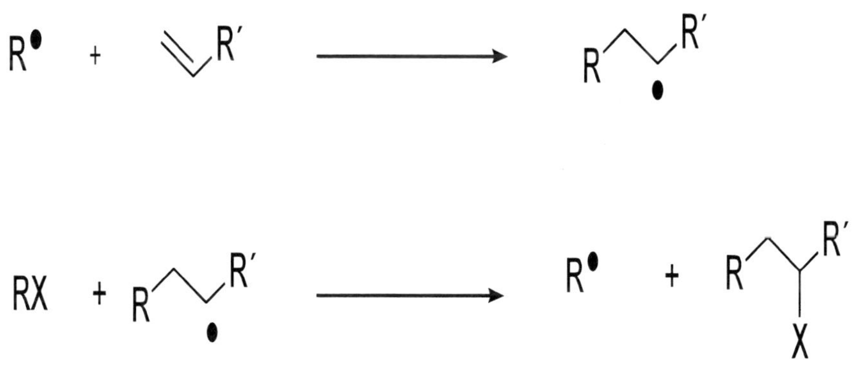

Scheme 1. CAN-mediated addition of carbonyl compounds.

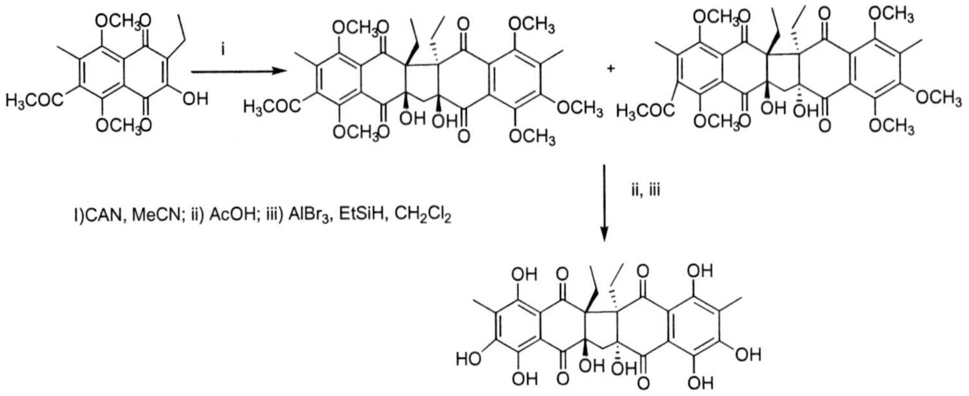

I)CAN, MeCN; ii) AcOH; iii) $AlBr_3$, EtSiH, CH_2Cl_2

Scheme 2.Synthesis of hybocarpone mediated by CAN.

The reactions proceed by regioselective addition of electrophilic carbon (or heteroatom)-centered radicals to alkenes, to generate stable radicals, such as benzylic, which are oxidized by a second equivalent of CAN to the corresponding cation, or add to another double bond, or ultimately, react with molecular oxygen to form ketone end-products. In case cations are formed, these react with nucleophilic solvents such as methanol. An example of such mechanistic versatile transformations is given by the synthesis of tetralone 3, in Scheme 3A from 3,4-dimethoxystyrene.

The utility of CAN in carbon-heteroatom bond formation, particularly C-S, C-N, C-Se, C-Br, and C-I bonds, is noteworthy in this connection. Mostly, these reactions involve the oxidative addition of heteroatom-centered radicals, formed by the oxidation of anions by CAN, to alkenes or alkynes. This topic has recently been reviewed.[6].

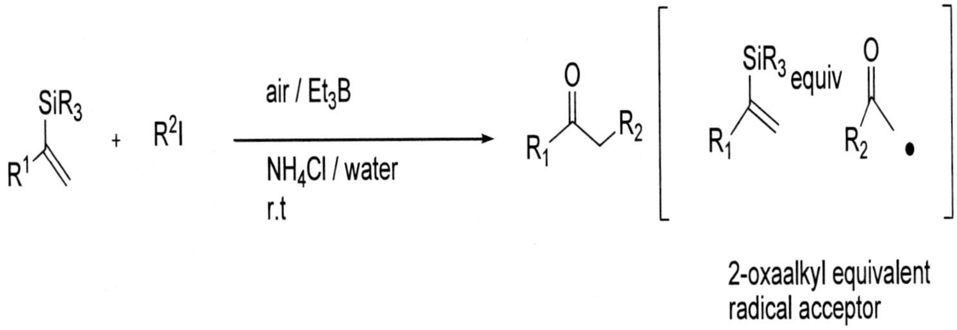

air / Et_3B

NH_4Cl / water

r.t

2-oxaalkyl equivalent
radical acceptor

4 R^1 = Me, R^2 = CH_2COOBn	**5, 80%**
6 R^1 = Ph, R^2 = CH_2COOBn	**7, 77%**
6 R^1 = Ph, R^2 = $CH_2COOnBu$	**8, 74%**
6 R^1 = Ph, R^2 = CH_2COPh	**9, 84%**
6 R^1 = Ph, R^2 = C_6F_{13}	**10, 81%**
11 R^1 = CO_2Me, R^2 = CH_2COOBn	**12, 61%**

Scheme 3A. Synthesis of tetralone 3 from 3,4-dimethoxystyrene.

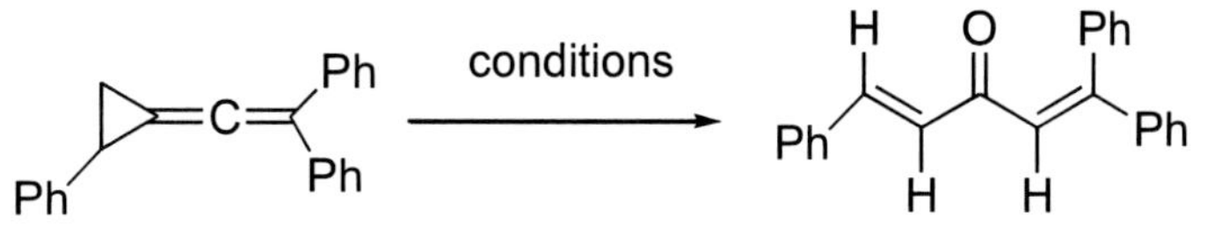

Scheme 3B. Radical synthesis of divinylketones in water.

In his seminal work, which marks the beginning of the use of CAN for the construction of carbon-heteroatom bonds, Trahanovsky demonstrated that CAN promotes a facile addition of azide radicals to olefins such as stilbene and acenaphthalene, to afford *trans*-β-azido-α-nitratoalkanes.[7] Lemieux exploited this protocol in the synthesis of azido sugars, which serve as convenient precursors of the corresponding amino sugars.[8]

However, in water or water mixtures, this reagent has been used less frequently.

A rapid and efficient conversion of styrenes, cycloalkanes, and α,β-unsaturated carbonyl compounds to di-bromides in good yields, using KBr and CAN in a biphasic system of water-dichloromethane, has been reported before. The major advantage of this transformation is that it offers a convenient alternative to the direct use of bromine.[9] Interestingly, acetylenes and arylcyclopropanes afforded the corresponding vicinal dibromoalkenes and 1,3-dibromides, respectively.[9,10]

Huang and collaborators[10b] have reported the CAN-mediated oxidative reactions of various vinylidenecyclopropanes (VCPs) with high regio- and stereoselectivity and developed a convenient approach to the synthesis of unsymmetrical divinyl ketone and functional enone derivatives. These novel compounds bearing unsaturated ketone skeletons would be useful in organic synthesis (Scheme 3B). On the basis of the oxygen- 18 tracer experiment, the authors clarified that the oxygen atom of the products arises from the H_2O in the reaction system, and a plausible reaction mechanism was proposed (Scheme 3C). Hence, the reaction may be of interest from both the mechanistic and synthetic standpoints. Further studies to expand the scope and synthetic utility of the method are reported to be underway by the authors.

The optimized conditions to afford divinyl ketones entail a refluxing mixture of methanol / water (5.5 equiv) for 5 h.

A possible reaction mechanism is depicted inScheme 3C.

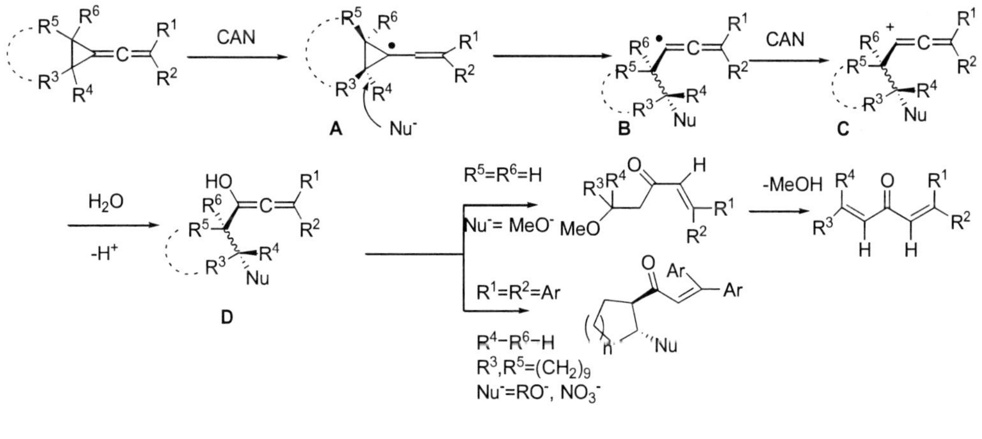

Scheme 3C. Mechanism proposed for the radical synthesis of divinylketones in water.

The initial event is likely to be a single-electron oxidation of VCPs by CAN to furnish radical cation A. The following nucleophilic attack of the solvent such as MeOH or water at the cyclopropane moiety may cause the rearrangement to produce the ring-opened radical intermediate B. B can be further oxidized by a second equivalent of CAN to give cation C, which is quenched by water in the solvent to produce D. For VCPs , a subsequent enol arrangement of the corresponding intermediate D and further elimination of MeOH may occur to furnish the product 2. For bicyclic VCPs 4, the key point is that the intermediate B is selectively attacked by the nucleophile from the less sterically hindered position to afford C. In this case, the *trans* isomer of functional enones 5 could not progress to the corresponding divinyl ketones (Scheme 3C). Moreover, when VCP (1h) with four methyl groups at the cyclopropane moiety was employed, the interesting product 2h was observed in 41% yield.[10b]

3. Main Group and Transition Metal Radicals in Water

3.1. Reformatzky Reactions in Water

The recent interest in aqueous medium metal-mediated carbon-carbon bond formation led to the continuing search for more reactive and selective metal species for such reactions.

The Reformatsky reaction (eq 1) between a 2-halo ester and a carbonyl compound in the presence of Zn was the first example of a large number of now commonly used carbon-carbon bond forming reactions: the addition of organometallic reagents to the carbonyl group. For nearly a century, these reactions were believed to require strictly anhydrous and oxygen-free conditions. Only during the past decade chemists have witnessed numerous examples of one-step reactions between organic halides, a reactive metal, and an electrophilic substrate commonly a carbonyl compound, which proceeded not only in wet solvents, but sometimes in water or salt solutions.[11] Many of these procedures gave addition products in preparative yields, comparable or superior to those obtained with preformed organometallic reagents under anhydrous conditions.

$$XC(R^3)(R^4)C(=O)OR^5 - R^1C(=O)R^2 \xrightarrow[2)\ H_2O]{1)\ Zn\ /\ C_6H_6} HOC(R^1)(R^2)C(R^3)(R^4)C(=O)OR^5 \quad (1)$$

Bieber et al.[12] demonstrated that the Reformatsky reaction can be carried out in water with a wide range of carbonyl substrates including saturated and unsaturated aldehydes and ketones where the previously indium-promoted reaction in water wasreported to be ineffective (*vide infra*).[13] Preparatively interesting yields comparable to those of the classical procedure in anhydrous solvents, can be obtained from substituted benzaldehydes with ethyl bromoacetate and from aromatic and unsaturated aldehydes with ethyl 2-bromo*iso*butyrate. From the mechanistic point of view, the reaction was inhibited by galvinoxyl and hydroquinone, which is consistent with a radical mechanism of two Single Electron Transfer (SET) processes proposed by Chan.[13] In Scheme 4, an alternative radical chain mechanism was postulated by the authors, which does not involve hydrogen abstraction.[12]When Zn reacts with 4 (Scheme 4), it produces directly the Reformatsky reagent5. This will react with water to form ethyl acetate or with a benzoyl radical to form benzoate and radical 6 which adds to the aldehyde 7, giving the oxyl radical9.Reduction of the intermediate 9 by another molecule of Reformatsky reagent 5 produces the final adduct 8 and a new radical 6 to continue the chain. Alternatively, the initial radical 6 may be produced by bromine abstraction from 4, either by a phenyl radical or on the zinc surface.

3.2. Alkylation of Carbonyl Compounds in Water

An efficient Barbier-Grignard-type alkylation of aldehydes in water in the presence of CuI, Zn, and catalytic InCl in dilute aqueous sodium oxalate affords alkylated alcohols in good yields.[14] According to eq 2, a series of alkyl halides can be used to afford alkylated alcohols in fairly good yields.

Hammond et al.[15] have recently synthesized *gem*-difluorohomopropargyl alcohols from *gem*-difluorohomopropargyl bromides 11 using indium and a catalytic amount of $Eu(Otf)_3$ (5 mol%) as a water tolerant Lewis acid.Later on, the same authors employed a combination of Zn and catalytic amounts of indium and iodine (eq 3).

Only fluorinated propargyl alcohols were observed as products under the reaction conditions.The reaction is highly regioselective as the corresponding fluoroallenylalcohols were not detected.[15]

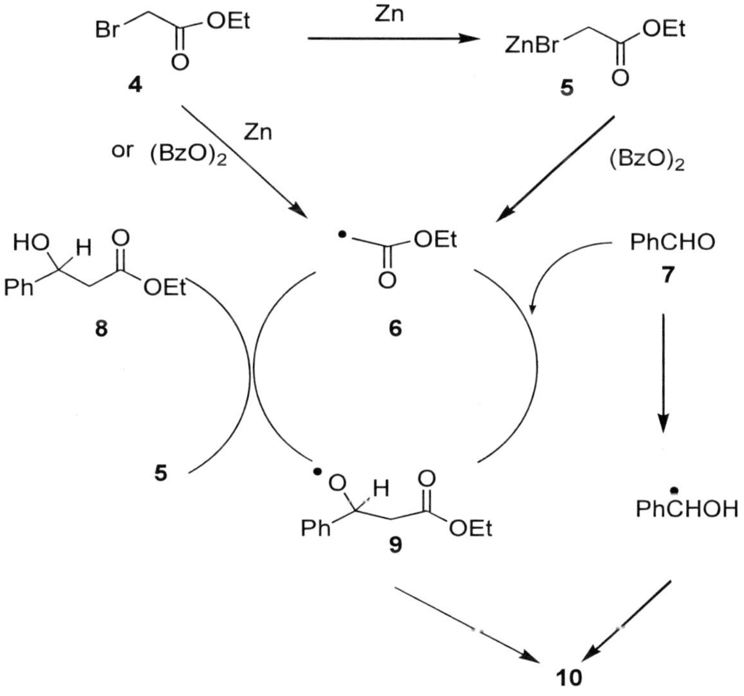

Br-CH2-C(=O)-OEt
4
Zn
ZnBr-CH2-C(=O)-OEt
5
Zn
or $(BzO)_2$
$(BzO)_2$
•CH2-C(=O)-OEt
6
PhCHO
7
HO H O
Ph OEt
8
5
•O H O
Ph OEt
9
PhĊHOH
10

Scheme 4.Proposedmechanism for the Reformatsky reaction in water.

MeO_2C–C(=N–NHBz) **27** $\xrightarrow{R^2I,\ Et_3B,\ H_2O\text{-}MeOH}$ MeO_2C–C(R^2)–NH–NHBz (4)

28a:R^2=*i*-Pr, 88%
28b:R^2=*c*-hexyl, 91%
28c:R^2=*t*-Bu, 51%

NNHBz (isatin hydrazone) $\xrightarrow{R^2I,\ Et_3B,\ H_2O\text{-}MeOH}$ R^2 NHNHBz (5)

29a:R^2=*i*-Pr
29b:R^2=*c*-hexyl
29c:R^2=*t*-Bu

30a, 73%
30b, 72%
30c, 40%

R^1COR^2 + $CH_2(CN)_2$ $\xrightarrow{H_2O\text{-}NH_4OAc}$ [$R^1R^2C{=}C(CN)_2$] — Et_3B, Et_2O-H_2O → $R^1R^2(Et)C$–$CH(CN)_2$ **34**; RI (**35**) / Et_3B, Et_2O-H_2O → $R^1R^2(R)C$–$CH(CN)_2$ **36,37** (6)

31 **32** **33**

31,33,34
a:R^1=Ph, R^2=H
b:R^1=C_4H_9, R^2=H
c: R^1+R^2=-$(CH_2)_6$-

35
a:R=*i*-Pr
b:R=*t*-Bu

36
a:R^1=Ph, R^2=H, R=*i*-Pr
b:R^1=C_4H_9, R^2=H, R=*i*-Pr
c:R^1+R^2=-$(CH_2)_6$-, R=*i*-Pr

37
a:R^1=Ph, R^2=H, R=*t*-Bu
b:R^1=C_4H_9, R^2=H, R=*t*-Bu
c:R^1+R^2=-$(CH_2)_6$-, R=*t*-Bu

An interesting alkylative amination of aldehydes has been reported by Porta and collaborators.[16] The strategy consists of an aqueous acidic $TiCl_3$ solution that promotes alkylative amination of aldehydes in a one-pot reaction involving up to four components, according to the stoichiometry of Scheme 5.

The reactions were carried out by adding the phenyldiazonium salt (2.5-3.75 mmol), as the fluoroborate (method I) or as the chloride (method II), portionwise over 2 h at 20 °C to a solution containing 12 (2.5 mmol), 13 (3.75 mmol), 14 (7.5 mmol), and $TiCl_3$ (5.0-6.5 mmol of the 15% commercially aqueous acidic solution) in 15 mL of glacial acetic acid under N_2 atmosphere (Scheme 5).

The mechanism proposed (Scheme 6A) involves generation of phenyl radical by a redox process (i), which abstracts an iodine atom from RI to generate the alkyl radicalwhich adds to the benzaldehyde-amine adduct, protonated imine derivative, to generate an aminyl radical cation (iv) which upon further reduction by Ti(III) generates the alkylative amination products 15.

Bieber and collaborators[16b] have made react benzylic chloridesin aqueous dibasic potassium phosphate under silver catalysis with aromatic aldehydes in the presence of zinc dust to give 1,2-diaryl alcohols in moderate to good yields (Scheme 3B). Dimerization to bibenzyls and reduction of the halide are important side reactions. A wide range of substituted aromatic and heteroaromatic aldehydes and of substituted benzylic chlorides can be used. Aliphatic aldehydes and ketones are unreactive. A mechanism of two SET on the metal surface is discussed.

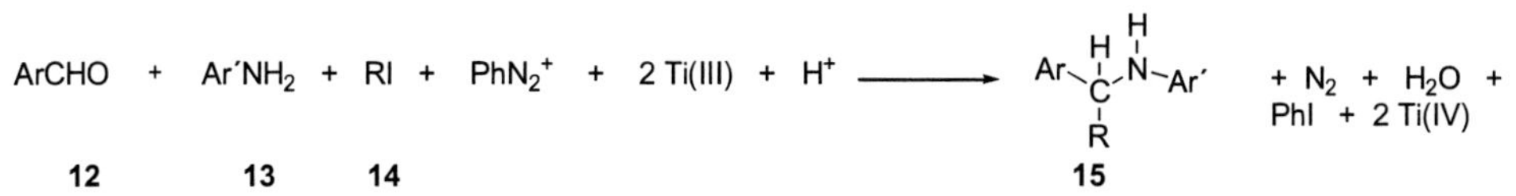

$$\underset{\mathbf{12}}{ArCHO} + \underset{\mathbf{13}}{Ar'NH_2} + \underset{\mathbf{14}}{RI} + PhN_2^+ + 2\,Ti(III) + H^+ \longrightarrow \underset{\mathbf{15}}{ArCH(R)NHAr'} + N_2 + H_2O + PhI + 2\,Ti(IV)$$

Scheme 5. Alkylative amination of aldehydes with Ti(III).

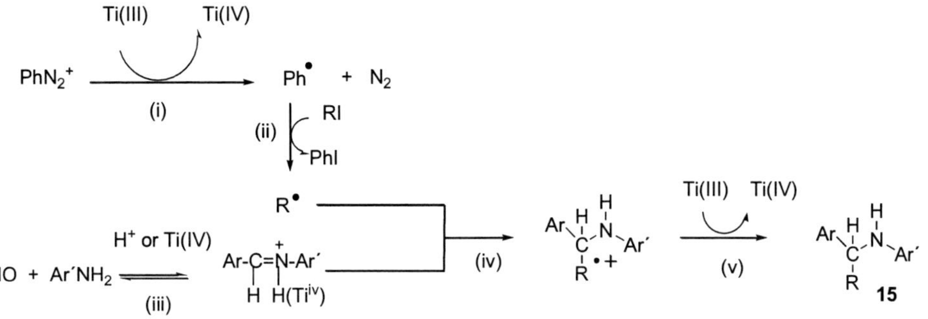

$PhN_2^+ \xrightarrow[\text{(i)}]{Ti(III)\ \to\ Ti(IV)} Ph^\bullet + N_2$

$Ph^\bullet \xrightarrow[\text{(ii)}]{RI\ \to\ PhI} R^\bullet$

$ArCHO + Ar'NH_2 \underset{\text{(iii)}}{\overset{H^+\ \text{or}\ Ti(IV)}{\rightleftharpoons}} Ar\text{-}\overset{+}{C}H\text{=}N(H(Ti^{IV}))\text{-}Ar'$

$R^\bullet$ + iminium $\xrightarrow{\text{(iv)}} ArCH(R)\text{-}NH\text{-}Ar'\ (\bullet +) \xrightarrow[\text{(v)}]{Ti(III)\ \to\ Ti(IV)} ArCH(R)NHAr'$ **15**

Scheme 6A.Proposed mechanism for the alkylative amination of aldehydes.

exo cyclization (k_{exo})

endo cyclization (k_{endo})

through

$k_{exo} / k_{endo} = 98{:}2$

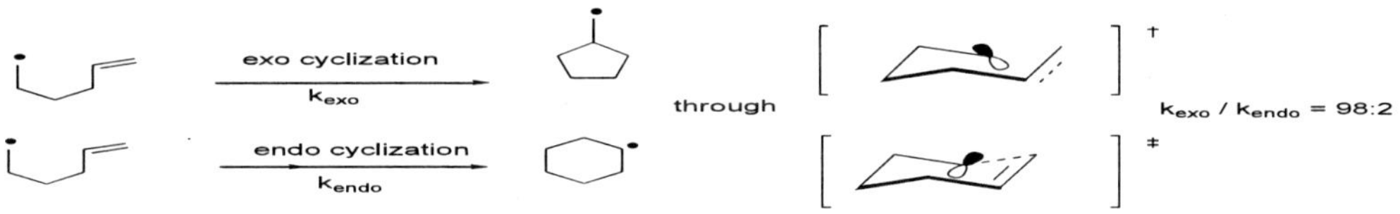

Scheme 6B. Alkylation of aldehydes with dibasic potassium phosphate under silver catalysis.

3.3. Alkylation of Imine Derivatives in Water

Among the many synthetic methods available for the synthesis of amines, the addition of organometallic reagents to imines provides one of the most straightforward methods to amines. Loh. et al.[17] reported on an efficient method for the alkylation of a wide variety of imines via a one-pot condensation of aldehyde, amine (including aliphatic and chiral amines), and alkyl iodides using indium-copper in aqueous media. These authors demonstrated that the combination of In / CuI/ $InCl_3$, was an efficient system for the activation of amine-alkylation in water, to generate the corresponding products in high yields (eq 4).

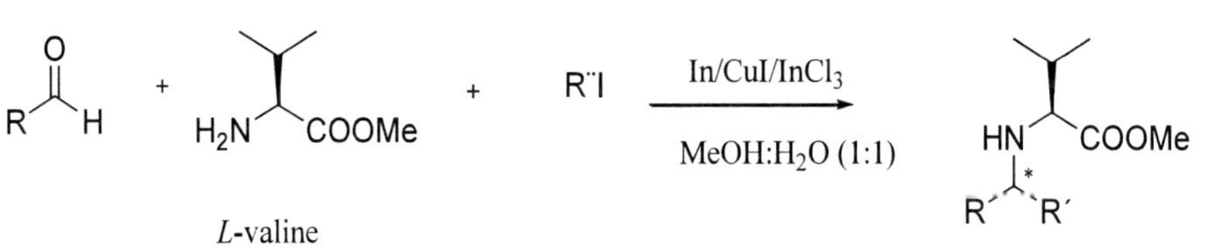

$$RCHO + \underset{L\text{-valine}}{H_2N\text{-}CH(iPr)\text{-}COOMe} + R'I \xrightarrow[MeOH:H_2O\ (1:1)]{In/CuI/InCl_3} RR'C^{*}H\text{-}NH\text{-}CH(iPr)\text{-}COOMe \quad (4)$$

Among the several metals screened, indium proved to be the best for this reaction, following the order for activation of the imine alkylation reaction:In > Zn > Al > Sn.[17]

It was worthwhile noting that the same reactions carried out in organic solvents such as MeOH, THF, CH_2Cl_2, DMF, DMSO, and hexane afforded the desired product in much lower yields. Even aliphatic amines, such as benzylamine could also react efficiently with different aldehydes and secondary alkyl iodides to furnish the desired products in fairly good yields. As shown in eq 4, enantiomerically-enriched amino compounds were also obtained. The one-pot reaction employing various aldehydes and alkyl iodides condensed efficiently with *L*-valinemethyl ester to generate the desired products in good yields and good diastereoselectivities. It was also worthwhile noting that even aliphatic aldehydes (cinnamaldehydes and nonyl aldehyde) were also good substrates for these reactions.A proposed reaction mechanism is shown in Scheme 7.

The reaction was initiated by a single electron transfer from indium-copper to alkyl iodide to generate an alkyl radical b (Scheme 7). This radical attacked the imine to furnish a radical intermediate c.Subsequent indium-promoted reduction of intermediate c and the quenching of the generated amino anion d in the presence of water, afforded the desired product e (Scheme 7).

As depicted above, the carbon-nitrogen double bond could be considered a radical acceptor, and therefore several radical addition reactions have been reported in organic solvents.[18] On the other hand, it has been shown that imine derivatives such as oxime ethers, hydrazones, and nitrones are excellent water-resistant radical acceptors for the aqueous-medium reactions using Et_3B as a radical initiator.[19]

The reaction of glyoxylic oxime ether 16, Scheme 8, with *i*-PrI (5 equiv) in H_2O-CH_2Cl_2 (4:1, v/v) and indium (7 equiv) afforded the *iso*propylated product 17 in 76% yield without formation of significant by-products.[20]

It is noteworthy that no reaction of 16 occurred in the absence of water.This result suggests that water would be important for the activation of indium and for the proton-donor to the resulting amide anion. In the presence of galvinoxyl free radical (radical scavenger) the reaction did not proceed, purporting that a free radical mechanism based on the single electron transfer (SET) process from indium is operative.

The indium-mediated alkyl radical addition to glyoxylic hydrazone 18afforded α-aminoacids 19(eq 5).[20]

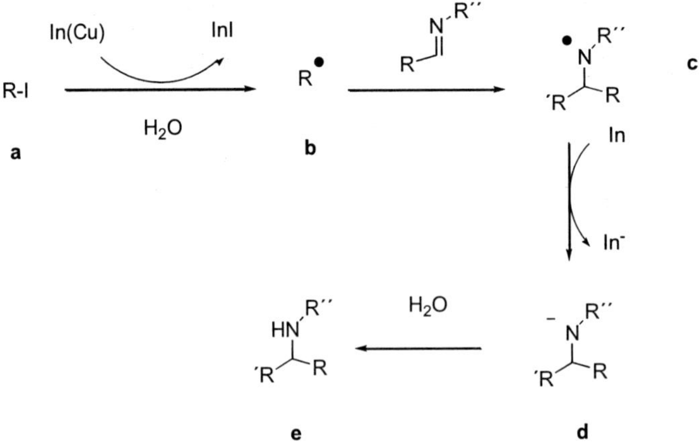

Scheme 7.Proposed reaction mechanism for the alkylation of imines in aqueous media.

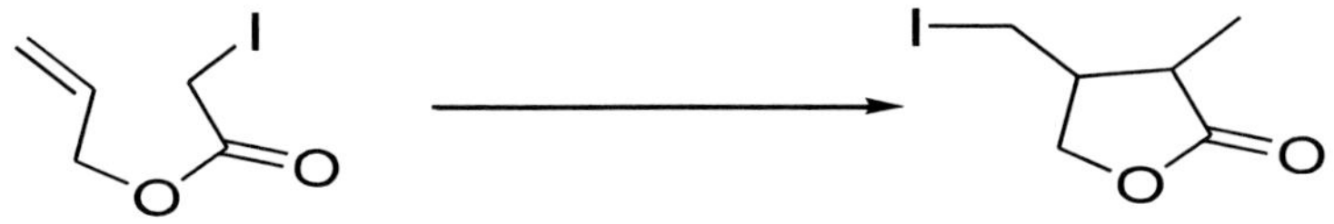

(7)

42a 43a

Scheme 8.In-Mediated reaction of oximes with alkyl halides in water.

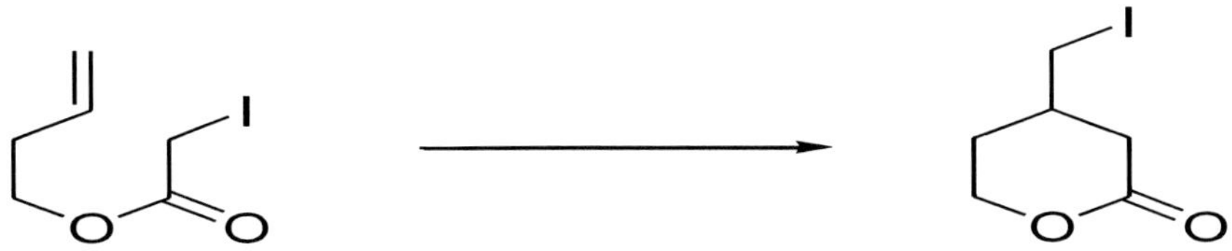

(8)

44 45

Integration of multi-step chemical reactions into one-pot reactions is of great significance as an environmentally benign method. The indium-mediated reaction leading to the one-pot synthesis of α-aminoacid derivatives was therefore considered (eq 6).

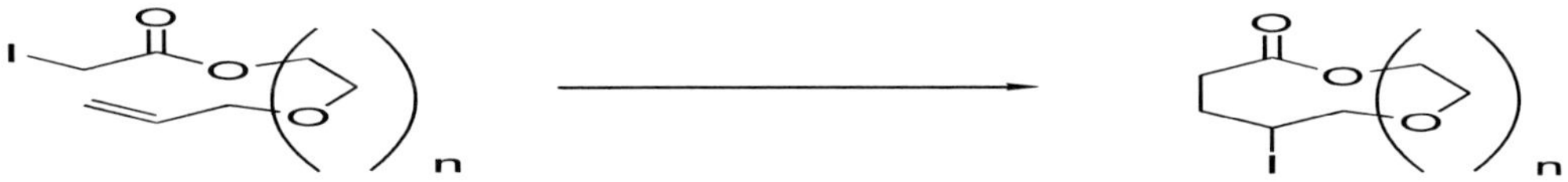

48 49

		benzene, %49	water, %49
48a	n=2	78	84
48b	n=3	43	86
48c	n=4	83	98

More recently, Loh and coworkers[21] have attempted the indium-copper-mediated Barbier-type alkylation of nitrones in water to furnish amines and hydroxylamines. Among the different metals investigated, indium and zinc were observed to be effective metals for the activation of the alkylation reaction in water to obtain the corresponding amines (eq 7).

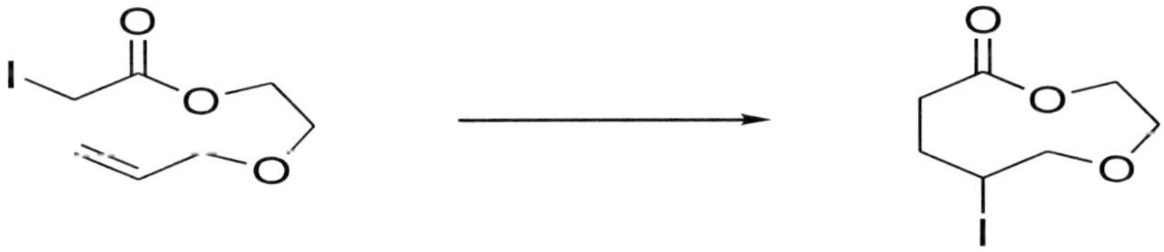

(9)

46 47

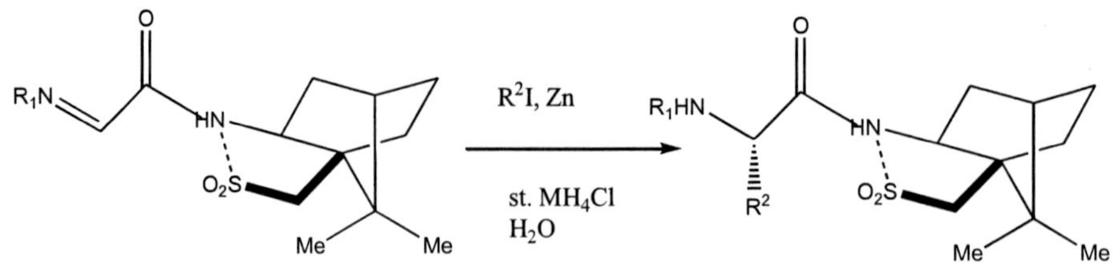

Scheme 9. Zn-mediated alkylation of imine derivatives.

Ueda proposed a zinc-mediated diastereoselective addition of alkyl iodides in water [22] (Scheme 9).

3.4. Alkylation of Electron-deficient Alkenes in Water

To test the utility of indium as a single-electron transfer radical initiator, the indium-mediated alkyl radical addition to electron-deficient C=C bonds (Scheme 10) was considered.

To a solution of phenylvinyl sulfone 20 and RI (5 equiv) in MeOH were added indium (7 equiv) and H_2O, and the reaction mixture was stirred at 20 °C for 30 min.As expected, 20 exhibits a good reactivity to render the desired alkylated products 21a-d in good yields with no detection of by-products such as reduced products. The reaction proceeded by SET process from indium as shown in Scheme 10.

The alkylation of sulfones can also be carried out in the absence of metals and in water. Jang reported the use of N-ethylpiperidine hypophosphite(EPHP) as a substitute for *n*-Bu_3SnH in radical conjugate additions (RCAs) to phenyl vinyl sulfone (Scheme 11). [23]

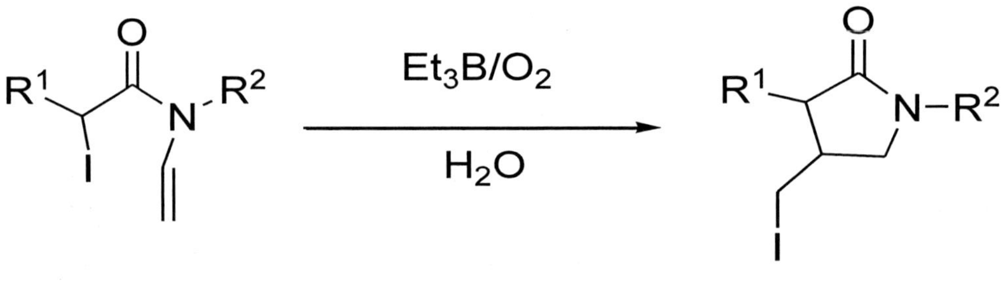

50 R^1=H, R^2=SO_2CH_3 77%
51 R^1=H, R^2=SO_2CF_3 61%
52 R^1=H, R^2=cyclohexyl 60%
53 R^1=CH_3, R^2=SO_2CH_3 78%
54 R^1=CH_3, R^2=CH_3 75%
55 R^1=H, R^2=CH_3 22%

Scheme 10. Indium-Mediated alkyl radical addition to electron-deficient C=C bond in water.

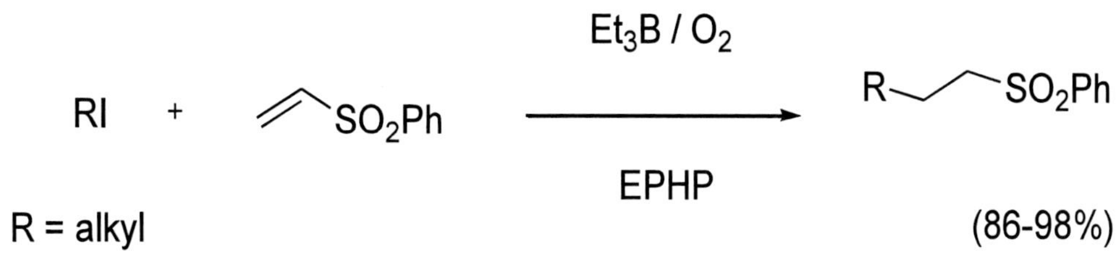

Scheme 11. Alkylation reactions of sulfones.

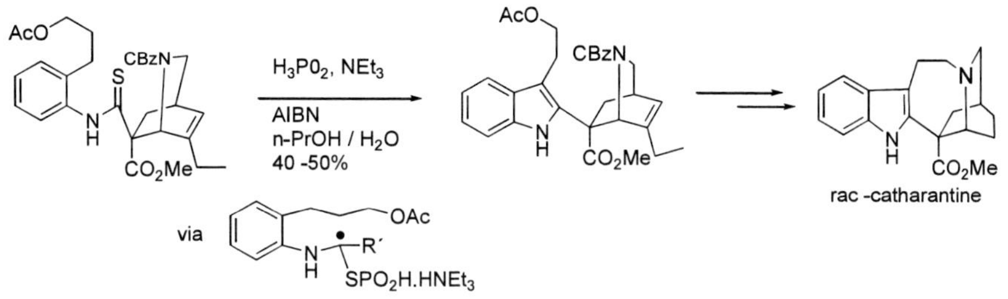

Scheme 12. Alkylation of sulfone derivatives.

α,β-Unsaturated esters and ketones could also be used as acceptors with reduced yields. Significantly, use of the initiator 4,4 -azobis(4-cyanovaleric acid) (ABCVA) in conjunction with cetyltrimethylammonium bromide (CTAB) allowed performance of the reaction in water (Scheme 12).[24]

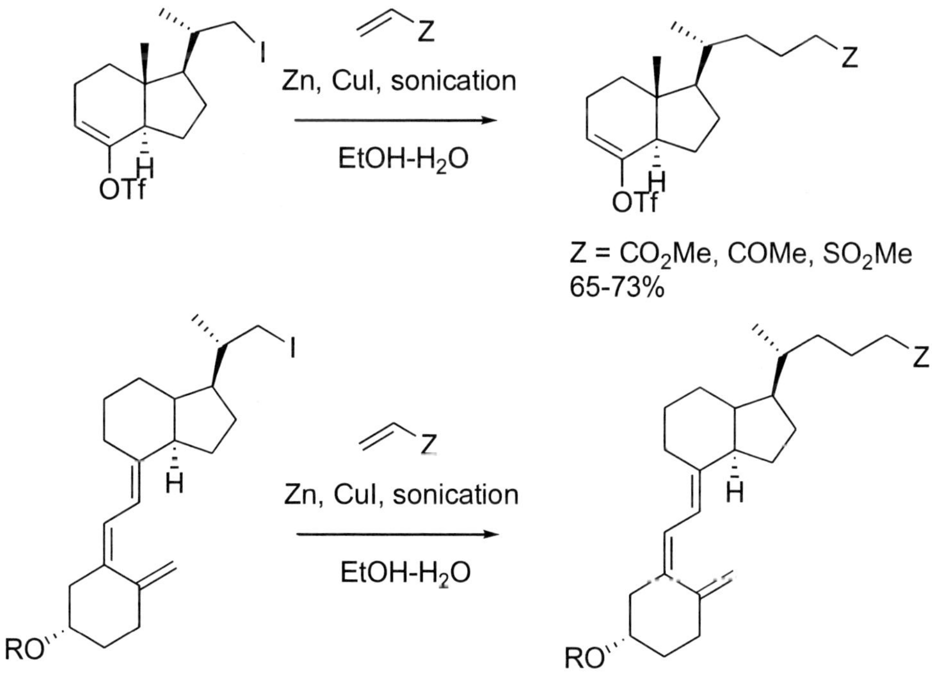

Scheme 13. Synthesis of vitamin D_3 analogues.

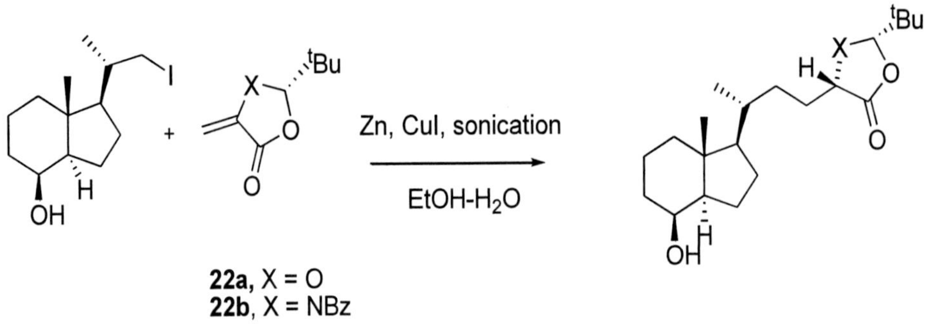

22a, X = O
22b, X = NBz

Zn, CuI, sonication
EtOH-H_2O

Scheme 14.

The surfactant could be omitted if tetraalkylammonium hypophosphites were employed instead of EPHP.[25]

Mourinño used the Luche conditions with acrylate, enolate, and vinyl sulfone acceptors to synthesize vitamin D_3 analogues (Scheme 13).[26, 27, 28]

In conjunction with studies of stereoselective Luche-type intermolecular RCAs,[29, 30, 31] Sarandeses and Perez Sestelo later employed chiral acceptors 22, allowing introduction of a stereocenter at C-24 (Scheme 14).[32, 33, 34] In a synthesis of C-18-modified vitamin D_3 analogues, Sarandeses performed an inter-molecular RCA of hindered neopentylic iodide 23 to methyl acrylate (Scheme 15).[35]

With a plethora of protective groups available for various types of functional groups, it is rather surprising that no practical protective groups have been developed for double bonds.Epoxidation can be used as a means of protecting double bonds, however, the successful implementation of this strategy would largely depend on the effective deoxygenation of epoxides back to alkenes.

Whereas indium metal has been used for so many reduction reactions and carbon-carbon bond forming reactions, it has not been exploited in the reduction of epoxides to form alkenes mediated by electron transfer from indium.

Miura, Hosomi *et al.* have recently developed a convenient In(III)-catalyzed intermolecular radical addition of organic iodides to electron-deficient alkenes.[36] In the presence of phenylsilane and catalytic amounts of indium(III) acetate, organic iodides add to electron deficient alkenes, according to Scheme 16.

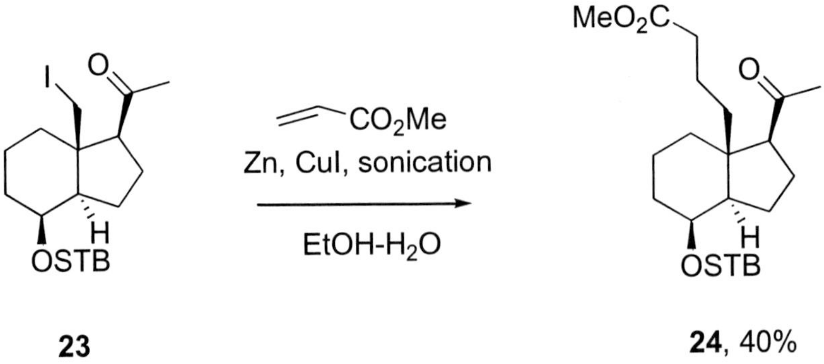

CO_2Me
Zn, CuI, sonication
EtOH-H_2O

23

24, 40%

Scheme 15.

In(OAc)_3 0.2 equiv
PhSiH_3

RI + E → R E

2,6-lutidine, 0.5 equiv
air, H_2O, EtOH, r.t.

R = alkyl, aryl
E = electron withdrawing group

Scheme 16. In-Silane-mediated addition of organic iodides to electron deficient alkenes.

The mechanism for this useful catalytic reaction is illustrated in Scheme 17A.

The first step is the formation of $(\text{AcO})_2\text{InH}$ by hydride transfer from PhSiH_3 to In(OAc)_3. The indium hydride undergoes H-abstraction by dioxygen from air to give $(\text{AcO})_2$-In• . The active species abstracts halogen from a halide 25 (R-X) to generate the corresponding carbon radical R• and $(\text{AcO})_2\text{InX}$. The addition of R• to an alkene 26 followed by H-abstraction from indium hydrides (In-H) gives the corresponding adduct 27 with regeneration of indium radicals (In•) . The indium salt formed$(\text{AcO})_2\text{InX}$ is converted into In-H by the reaction with PhSiH_3 in ethanol/water. The formation of 28 is the result of direct H-abstraction of R• from InH. The successive addition of R•to two molecules of 26 forms the adduct 29. The present system enables proper control of the concentration of InH to avoid these side reactions.

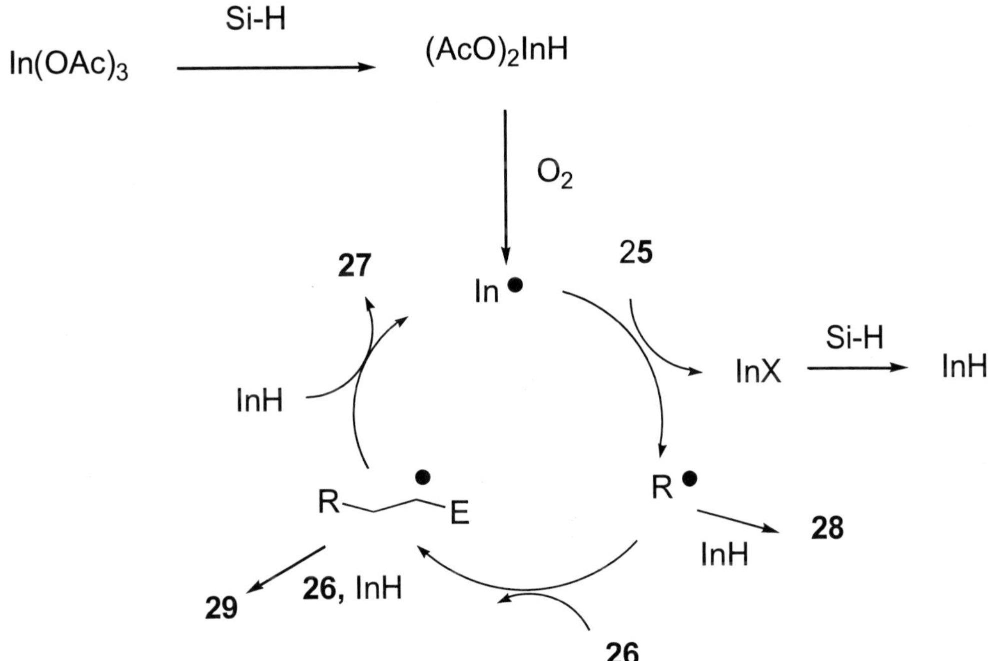

Scheme 17A. Proposed mechanism for the In-silane-mediated alkylation of alkenes.

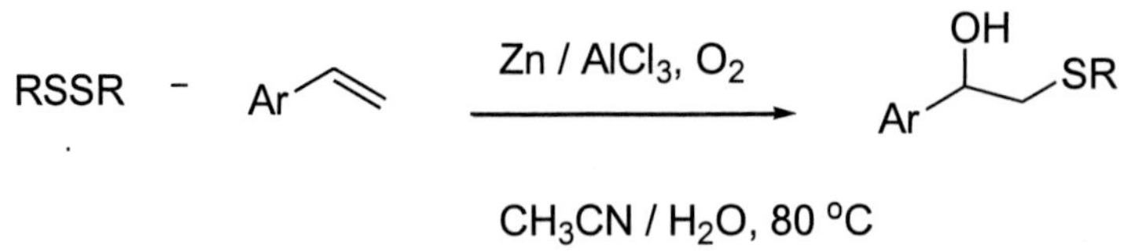

$$RSSR \; - \; Ar\text{-}CH{=}CH_2 \xrightarrow[CH_3CN\,/\,H_2O,\ 80\ ^{\circ}C]{Zn\,/\,AlCl_3,\ O_2} ArCH(OH)CH_2SR$$

Scheme 17B.

Movassagh and collaborators[36b] sought novel applications of zinc thiolates and zinc selenolates in chemical reactions. They investigated a convenient, catalyst-free method for the anti-Markovnikov addition of thiols to styrenes at room temperature in water. Theyhave examined a new methodology for the synthesis of β hydroxysulfides via the anti-Markovnikov addition of thiolate anions, generated in situ by reductive cleavage of diaryl disulfides in the presence of $Zn/AlCl_3$,to styrenes in aqueous acetonitrile in the presence of oxygen (Scheme 17B).

The experiments were initially conducted with styrene and diphenyl disulfide, as a model reaction by varying the molar ratios, solvents, and temperatures under ambient atmosphere. The authors found that the reactants were converted readily to the corresponding β hydroxysulfi.de using the $Zn/AlCl_3$ system with a molar ratio of disulfide/$AlCl_3$/Zn/styrene =0.5:1:3.5:1.2 in acetonitrile/water (4:1) at 80 °C. The formation of β hydroxysulfides may be explained as follows; :the oxygen may complex with the styrene assisted by hydro gen bonding with the water hydroxyls, and this would be followed by nucleophilic attack by zinc thiolate, $(RS)_2Zn$, prepared via reductive cleavage of the disulfide with $Zn/AlCl_3$.

3.5.Allylation of Carbonyl Compounds in Water

In the past two decades, the one-pot Barbier procedure for coupling allylhalides with carbonyl compounds has gained renewed interest.Contrary to the Grignard reaction, the Barbier procedure does not require strictly anhydrous solvents but can be performed very efficiently in aqueous media.In fact, the allylation of aldehydes and ketones under the Barbier conditions usually occursfaster and gives rise to higher yieldswhen water is used as a (co)solvent.[41] Allylation reactions of carbonyl compounds using Zn,[37] Bi,[38] Sn,[39] Mg,[40] Mn,[41] Sb,[42] Pb,[43] Hg,[44] and In [45] in aqueous media have been reported (Scheme 18).

Magnesium-mediated Barbier-Grignard-type alkylation of aldehydes with allyl halides was investigated by Li et al.[46] It was found that the magnesium-mediated allylation of aldehydes with allyl bromide and iodide proceeded effectively in aqueous 0.1 M HCl or 0.1 M NH_4Cl.Aromatic aldehydes reacted chemoselectively in the presence of aliphatic aldehydes. A variety of aldehydes were tested with this alkylation method, according to eq 8.

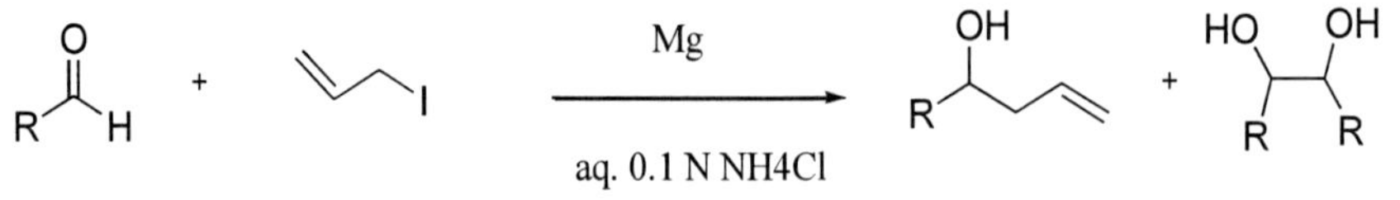

$$RCHO + CH_2{=}CHCH_2I \xrightarrow[aq.\ 0.1\ N\ NH4Cl]{Mg} RCH(OH)CH_2CH{=}CH_2 + RCH(OH)CH(OH)R \quad (8)$$

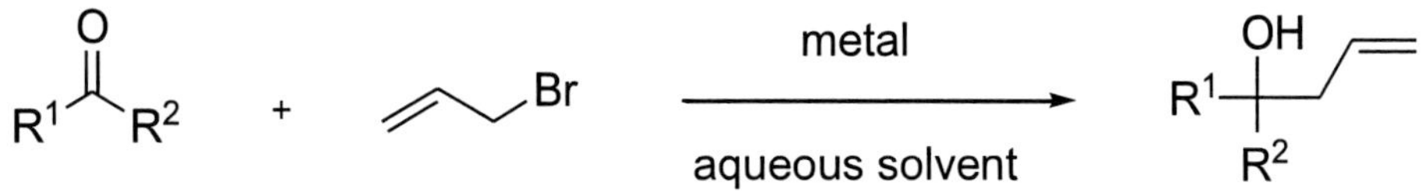

Scheme 18. General metal-mediated allylation of carbonyl compounds in aqueous systems.

The allylation of aromatic aldehydes bearing halogen atoms proceeded without any problems. The allylation of hydroxylated aldehydes also afforded the allylation products in good yields. Reaction of 4-hydroxybenzaldehyde under the standard conditions led to the formation of the allylation product.

The mechanism of the classical magnesium-mediated Barbier and Grignard reactions have been studied intensively by several groups.[47] It is generally believed that the radicals on the metal surface are involved in the organomagnesium reagent formation.For the Barbier allylation of carbonyl compounds with magnesium in anhydrous solvent, it is assumed that the reaction of allyl bromide on the metal surface generates an organometallic intermediate that is in equilibrium with the charge-separated form and the radical form, as proposed by Alexander,[48] as shown in Scheme 19.The two forms will also lead to either the protonation of the carbanion (overall reduction of the halide) or Wurtz-type coupling, whereas the intermediate reacts with aldehydes through the usual six- membered ring mechanism. The radical intermediate could lead to the formation of 1,6-hexadiene, pinacol product and benzyl alcohol.

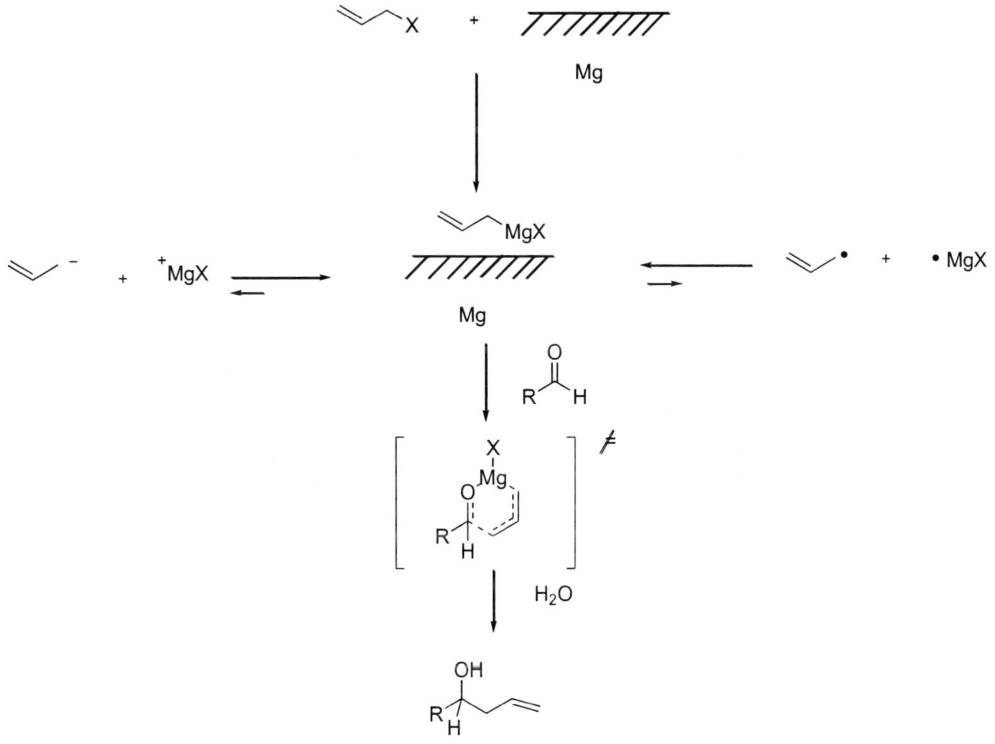

Scheme 19. Mg-surface mediated electron transfer mechanism of allylation of carbonyl compounds.

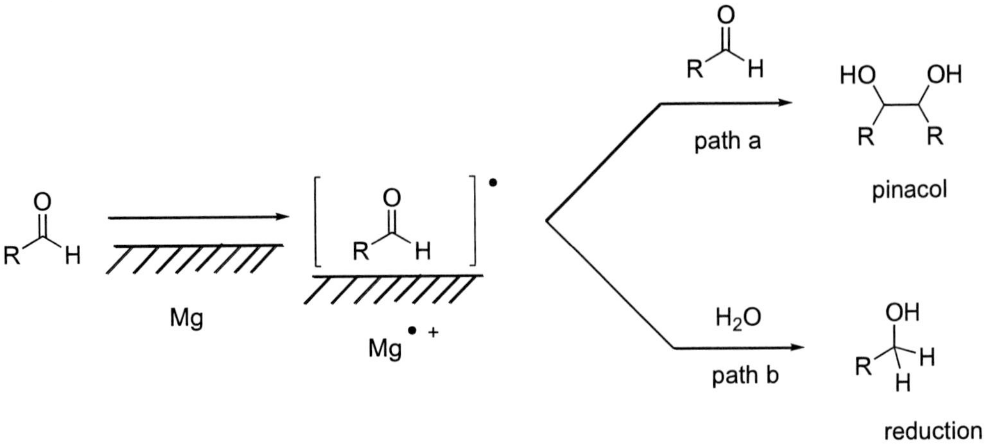

Scheme 20. Pinacol formation and reduction competing paths.

For the rationalization of the pinacol formation, the authors[46] postulate two potential pathways competing with each other generating either the pinacol-coupling product (path a, Scheme 20)or the benzyl alcohol product (path b, Scheme 20).The same authors observed that upon increasing the steric hindrance around the carbonyl group, a destabilization in the transition state responsible for the formation of the pinacol product (path a), would result in an increase in the formation of the benzyl product. Thus, they observed that no pinacol product was formed from the magnesium-mediated reaction with 2,6-dichlorobenzaldehyde, and a 74 % yield of the reduction product was encountered in this example (Scheme 19).

Madsen et al.[49] reported on a theoretical study of the Barbier-type allylation of aldehydes mediated by magnesium. They concluded that a radical anion was involved in the selectivity-determining event.

Other metals were used from time to time to mediate in the coupling reactions to construct new carbon-carbon bonds. Manganese was shown to be very effective for mediating aqueous medium carbonyl allylations and pinacol coupling reactions.

Li et al.[50] reported an unprecedented metal-mediated carbonyl addition between aromatic and aliphatic aldehydes. The allylation of aldehydes mediated by manganese in water in the presence of a catalytic amount of copper showed exclusive selectivity toward aromatic aldehydes. Manganese was found to be equally selective in promoting pinacol-coupling reactions of aryl aldehydes. When 3 equivalents of allyl chloride and manganese mediator were used, the isolated yield of the allylation product was 83 % (eq 9).

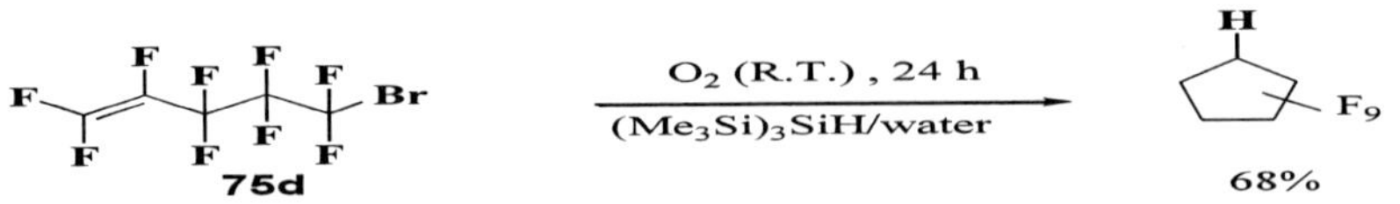

75d $\xrightarrow[(Me_3Si)_3SiH/water]{O_2\ (R.T.),\ 24\ h}$ 68%

It was found that[50] various aromatic aldehydes were allylated efficiently by allyl chloride and manganese in water. It is noteworthy to mention that aromatic aldehydes bearing halogen atoms were allylated without any problems. The allylation of hydroxylated aldehydes was equally successful. On the other hand aliphatic aldehydes were inert under the reaction conditions. Such an unusual reactivity difference between an aromatic aldehyde and an aliphatic aldehyde suggested the authors[50]the possibility of an unprecedented chemoselectivity. When competitive studies were carried out involving both aromatic and aliphatic aldehydes (eq 10), a single allylation of benzaldehyde was generated when a mixture of heptaldehyde and benzaldehyde was reacted with allyl chloride. Such a selectivity appeared unique when aqueous methodologies mediated by other metals such as Zn, Sn and In all generated a 1:1 mixture of allylation productsof both aldehydes (*vide infra*).

Mn/Cu (3.3:0.1)
H_2O
only product
(10)

Chang et al.[42] have recently reported that fluoride salts are equally effective in activating antimony in aqueous media to mediate in the coupling of allyl bromide with aldehydes to afford the corresponding homoallylic alcohols. 1 M Concentrations of NaF and KF were found to be equally effective as RbF and CsF or 2 M KF.The reaction proceeded well with either aromatic or aliphatic aldehydes. The allylation of α,β-unsaturated aldehydes as represented by *trans*-cinnamaldehyde occurred in a regiospecific manner and furnished solely the 1,2-addition product. Furthermore, electron donating or withdrawing groups on the aromatic ring did not seem to affect the reaction significantly either in the yield of the product or the rate of the reaction. With this metal, activated antimony, no alcohols or pinacols were detected as side products of the reactions, as has been shown previously for the Mn and Mg (*vide supra*) cases.Even the nitro substituent on the aromatic ring of the aldehyde was not reduced under the reaction conditions obtaining the corresponding allylated alcohol from *p*-nitrobenzaldehyde (usually the nitro group is sensitive to reduction by metals and cannot be allylated under Barbier conditions).[51] In this sense, the authors argued[42] that the use of a fluoride salt as an activating agent is superior to the use of Al, Fe, or $NaBH_4$, reported previously. Efforts to allylate ketones failed by this Sb-mediated methodology. From the mechanistic point of view,the reaction proceeds between the allylmetal species and the aldehyde.[49]

In 1983, Nokami et al. first reported on the successful coupling of allyl bromide (30) with carbonyl compound (31) mediated by tin to give the homoallylic alcohol (32) in water (Scheme 21).[53] However, the reaction requires a catalytic amount of hydrobromic acid.

Later on, the addition of metallic aluminum powder or foil was found to improve the yield of the product dramatically.[54] On the other hand, Wu and co-workers found that higher temperature can be used to replace aluminum.[55] Alternatively, Luche found that the reaction can be performed in the absence of aluminum or hydrobromic acid by the use of ultrasonic irradiation together with saturated aqueous NH_4Cl/THF solution.[56] Various mechanisms have been proposed for the aqueous Barbier reactions, including the intermediacy of a radical,[57] radical anion,[58] and an allylmetal species.[59] In the latter case, it has been presumed that diallyltin dibromide is the organometallic intermediate in the tin-mediated allylation reactions; however, no experimental proof has been offered.[60, 61]

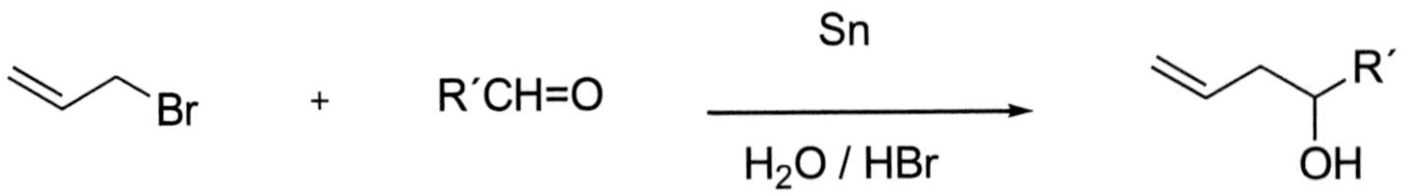

Br + R´CH=O $\xrightarrow[H_2O\ /\ HBr]{Sn}$ R´ OH

30 31 32

Scheme 21. Tin-mediated allylation of carbonyl compounds in water.

Chan and co-workers[62] found that allylation of carbonyl compounds can easily be accomplished in water using diallyl tin dibromide to obtain homoallyl alcohols in good yields. Table 1 summarizes some examples of carbonyl substrates that can be allylated through this latter reagent.

Table 1. Allylation reactions of carbonyl compounds with diallyltin dibromide in water

Entry	Carbonyl compound	Product	Yield
1	benzaldehyde	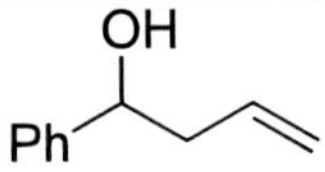 OH Ph	95
2	heptaldehyde	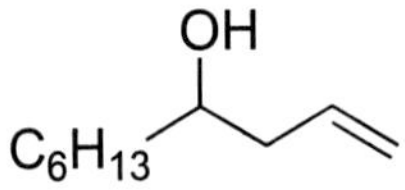 OH C_6H_{13}	99
3	4-nitrobenzaldehyde	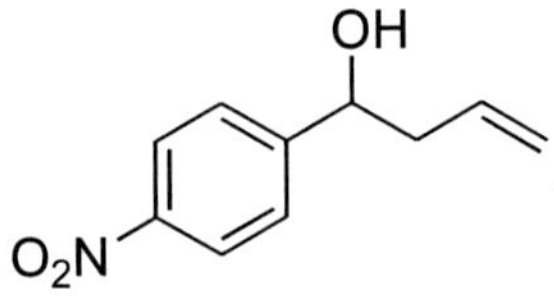 OH O_2N	97
4	Cyclohexylaldehyde	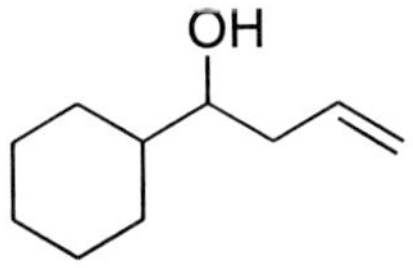 OH	95
5	OHCCHO	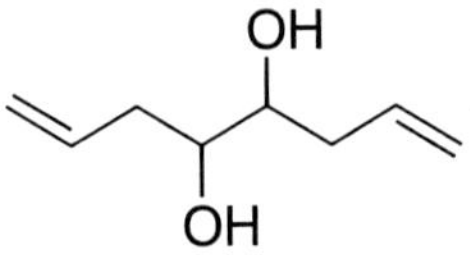 OH OH	95
6	Cinamaldehyde	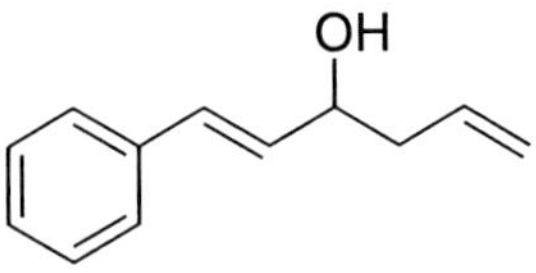 OH	99
7	4-chlorobenzaldehyde	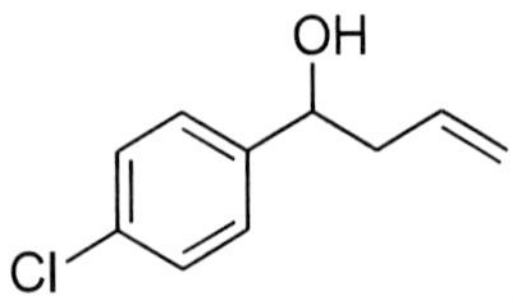 OH Cl	99

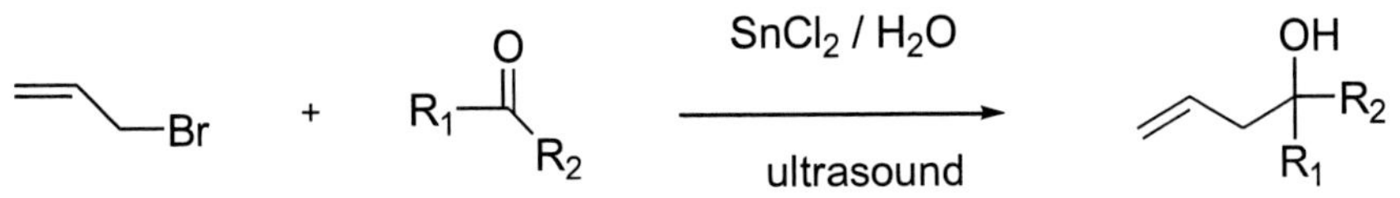

Scheme 22. Tin-mediated allylation of carbonyl compounds triggered by ultrasound.

Although an allyl tin bromide and diallyl tin bromide species have been postulated as intermediates in these reactions, these results however do not eliminate the possibility of a parallel process of metal surface-mediated radical or radical anion reactions. Nevertheless, the understanding that the reaction can proceed through an organotin intermediate has useful synthetic applications.

However, allylation reactions based on allylstannanes have serious drawbacks in the synthesis of biologically active compounds, because the inherent toxicity of organotin derivatives and the difficulty of removal of residual tin compounds often prove fatal.

Bian and collaborators[63] have accomplished the allylation reaction of aromatic aldehydes and ketones with tin dichloride in water.

The allylation reactions of aromatic aldehydes and ketones were carried out in 31 –86% yield using $SnCl_2$ –H_2O system under ultrasound irradiation at r.t. for 5 h. The reactions in the same system gave homoallyl alcohols in 21 –84%yield with stirring at r.t. for 24 h (Scheme 22). Compared with traditional stirring methods, ultrasonic irradiation is more convenient and efficient.

In Table 2, thearomatic aldehydes and ketones studied are summarized. It is observed that by ultrasound irradiation , the allylation products are obtained in yields ranging from73 to 86% , except for acetophenone, wich renders a low yield of the respective allylation product.

Chan et al[64] investigated the nature of the allylation reaction of carbonyl compounds with indium in water (Scheme 23).

The authors postulate a series of intermediates, among which those in Scheme 24 were considered.

Table 2. Allylation reactions of aldehydes with $SnCl_2$ / H_2O

Entry	Substrate	Yield (stirring)	Yield, % (ultrasound)
1	Benzaldehyde	84	86
2	4-chlorobenzaldehyde	77	76
3	Furfural	58	59
4	Cinnamaldehyde	71	73
5	3,4-$(CII_2O)C_6H_3CHO$	84	79
6	4-$CH_3OC_6H_5CHO$	-	-
7	$C_6H_5COCH_3$	21	31

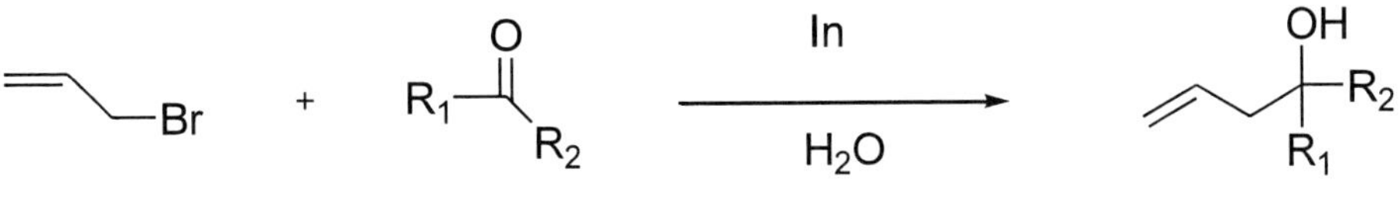

Scheme 23. Allylation of carbonyl compounds with indium metal in water.

Table 3. Indium-mediated allylation reaction of difluoroacetyltrialkylsilane in aqueous mixtures

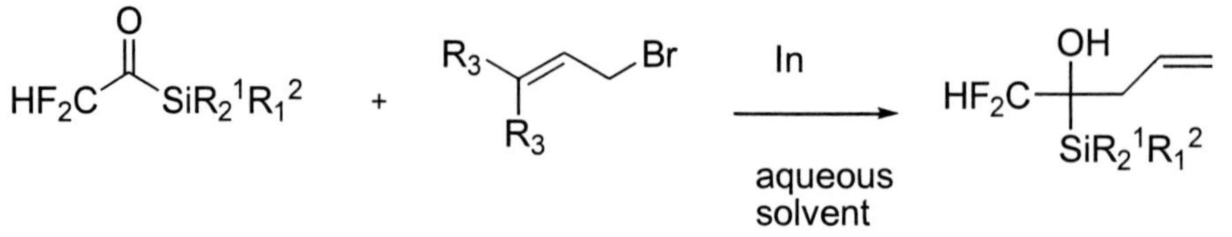

$$HF_2C\text{-}CO\text{-}SiR_2^1R_1^2 + (R_3)_2C{=}CH\text{-}CH_2Br \xrightarrow[\text{aqueous solvent}]{In} HF_2C\text{-}C(OH)(SiR_2^1R_1^2)\text{-}CH_2CH{=}CH_2 \ \mathbf{33}$$

Entry	R^1	R^2	R^3	Solvent	% 33
1	phenyl	t-butyl	H	THF-water (1:1)	97
2	Ethyl	Ethyl	H	THF-water (1:1)	83
3	Isopropyl	Isopropyl	H	THF-water (1:1)	85
4	Ethyl	Ethyl	CH_3	THF-water (1:1)	-
5	ethyl	Ethyl	CH_3	THF-water (1:1) NH_4Cl	-

However, through a series of experiments [64], it was concluded that indeed the allyl indium intermediate contained indium(I), and the structure of the intermediate supported by experiments was that shown in equation 11:

$$CH_2{=}CH\text{-}CH_2\text{-}In(I) \quad (11)$$

Welch and collaborators[65] explored the generality of the indium-mediated allylation reaction with various difluoroacetyltrialkylsilanes in water and THF mixture.

Desired homoallylic alcohols were synthesized in good yields without enol silyl ether formation. Substituents on silicon have no effect on the product formation. However, in reaction with a substituted allyl bromide such as 4-bromo-2-methyl-2-butene, the desired homoallylic alcohol was not formed,rather, the acylsilane was recovered (Table 3).

The regio- and stereoselectivity of metal-mediated allylation reactions in aqueous media is well understood.[66] In general, regioselectivity is governed by the substituent on the allyl halide. The formation of γ-adducts was exclusively observed under some conditions, however, allyl halides bearing bulky γ-substituents such as 1-bromo-4,4-dimethyl-2- pentene and (3-bromopropenyl)trimethylsilane resulted in the formation of α-adducts. In contrast, diastereoselectivity is affected by the steric and chelating effect of substituents. These reactions have also been attempted with zinc.

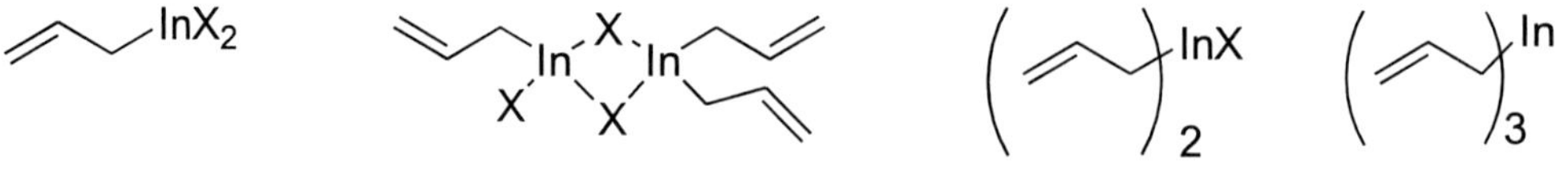

$CH_2{=}CHCH_2InX_2$ $[(CH_2{=}CHCH_2)(X)In(\mu\text{-}X)_2In(CH_2CH{=}CH_2)_2]$ $(CH_2{=}CHCH_2)_2InX$ $(CH_2{=}CHCH_2)_3In$

X = Cl, Br

Scheme 24. Intermediates proposed in the indium-mediated allylation of carbonyl compounds.

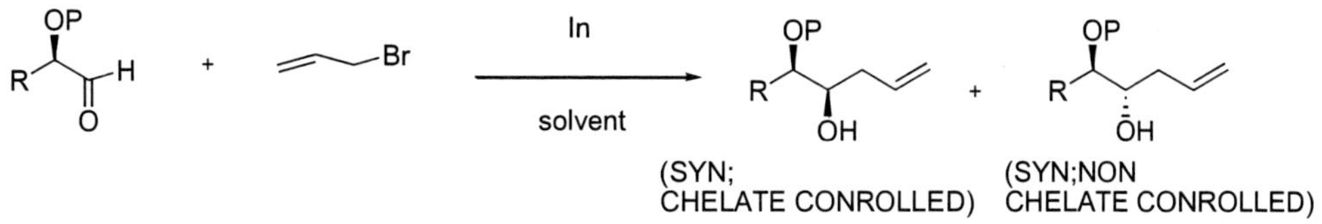

Scheme 25. In-mediated allylation of α-oxygenated aldehydes.

Paquette et al.[67] have investigated the stereochemical course of indium-promoted allylations to α- and β-oxy aldehydes in solvents ranging from anhydrous THF to pure H_2O. The free hydroxyl derivatives react with excellent diastereofacial control to give significantly heightened levels of *syn*-1,2-diols and *anti*-1,3-diols (Scheme 25). Relative reactivities were determined in the α-series, and the hydroxyl aldehydes proved to be the most reactive substrates. This reactivity ordering suggests that the selectivity stems from chelated intermediates. The rate acceleration observed in water can be heightened by initial acidification. Indeed, the indium-promoted allylation reaction mixtures become increasingly acidic on their own. Preliminary attention has been given to salt effects, and tetraethylammonium bromide was found to exhibit a positive synergistic effect on product distribution. Finally, mechanistic considerations are presented in order to allow for assessment of the status of these unprecedented developments at this stage of advancement of the field.

The results obtained with 34 (entry 1) and 35 (entry 2) provide important calibration points for non-chelate-controlled behavior. In both instances, the *anti*-product is favored. Presumably because the basicity of the *tert*-butyldimethylsiloxy substituent falls below that of the benzyloxy, the *anti*-percentages reach a maximum for 34.

It is noteworthy that in every example allylations performed in either H_2O or H_2O-THF (1:1) proceeded at appreciably more rapid rates than in THF alone. For 34, the diastereoselectivity realized is constant whether H_2O is present or not. Interestingly, the *anti*-preference in the case of 35 decreases by a factor of about 3 in pure H_2O. Product yields were found to be consistently high. Hemiacetal 36 must, of course, undergo ring opening prior to condensation with the allylindium reagent. Beyond that, it is not clear that carbon-carbon bond formation materializes prior to, or onlyafter, the loss of formaldehyde.

pH Considerations. The rate acceleration noted above for allylations promoted in water could, as for example with D-arabinose, be attributed to the improved solubility of the substrate in the aqueous medium. However, the phenomenon persists when the solubilities of the reagents are lower in H_2O than in THF. This behavior could be explained by attributing enhanced stability to the allylindium reagent in THF. Under these circumstances, reactivity toward an incoming aldehyde carbonyl would be reduced and condensation would proceed more slowly. Indeed, an increase in reaction rate has been observed by Araki et al.[68] when progressing from benzene to THF. In this latter study, it was noted that the pH of all allylations performed in water or aqueous THF dropped significantly as the reactions progressed. This aspect of indium-promoted condensations does not seem to have been previously recognized and was therefore explored more fully in order to elucidate the accompanying advantages or disadvantages. As indicated in Table 4, aldehydes 34-36 were closely scrutinized.

Table 4. Indium-mediated allylations of α-oxygenated aldehydes

entry	aldehyde	Reaction time	*Syn*	*Anti*	Yield, %
1	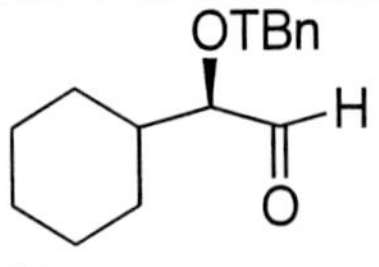 34	3.5h	1	3.9	90
2	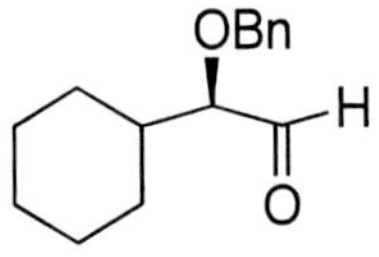 35	3h	1	1.2	92
3	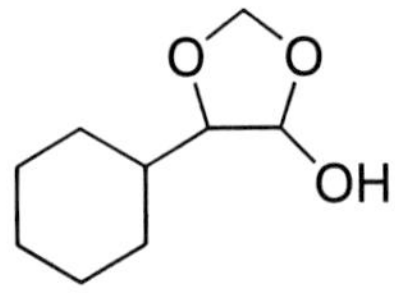 36	24-30h	2.3	1	90-95

When the pH was maintained at 7 by controlled infusion of sodium hydroxide solution, an increase in reaction times became necessary to achieve complete allylation (entries 1, and 3). This phenomenon could be due in part to increased dilution due to addition of the aqueous base and not constitute a manifestation of the pH itself. Reactions allowed to proceed without pH control were accompanied by a progressive development in acidity to a point below pH 4. When the allylations were initiated at a preset pH of 4 ,the transformations took place at notably accelerated rates. Note worthily, the benzyl- and silyl-protected substrates exhibit the same diastereoselectivities at all ranges of pH tested. These findings dispel any concerns that product distribution might be dependent upon pH to the point where the *syn/anti*-ratios would vary as the reaction progressed. This feature of indium catalysis in water requires that care be exercised when acid-sensitive reactants are involved. Consequently, these considerations must be taken into account when designing alternative applications of this chemistry.

β-Oxy-Substituted Aldehydes. A free hydroxyl substituent β- to a carbonyl group was considered to be exploitable for 1,3-asymmetric induction during condensation with allylindium reagents in water. Aldehydes 36-39 were selected because of their structural simplicity, similarity to 34-36, and varied basicity at the β-oxygen. If chelation were to gain importance and nucleophilic attack were to occur from the less hindered diastereotopic π-face of the aldehyde carbonyl, then anti-adduct (Scheme 26) 40 would result.

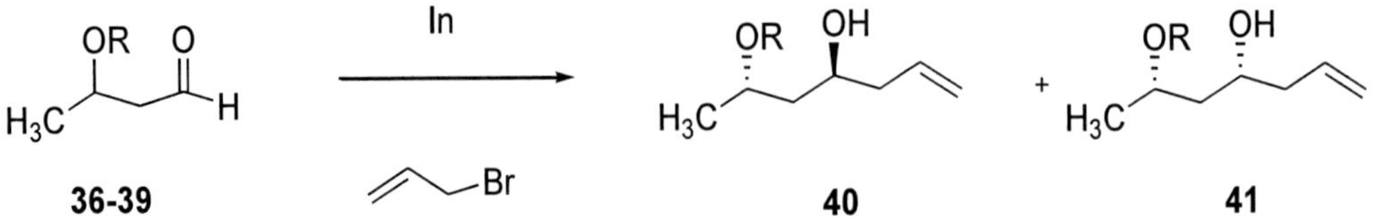

Scheme 26. In-mediated allylation of β-oxygenated aldehydes.

Table 5.Effect of pH on the rate and diastereoselectivity of allylindium additions in water at 25 °C

Entry	aldehyde	pH	Reaction time (h)	Syn product	Anti product	Yield, %
1	34	B	3.5	1	3.9	90
2	34	7	5.5	1	3.0	85
3	34	4	4	1	3.0	80
4	35	B	3	1	1.2	92
5	35	7	12.5	1	1.4	84
6	35	4	4	1	1.5	86
7	36	B	24-30	2.3	1	90-95
8	36	7	48	2.0	1	80-87
9	36	4	0.5	10.0	1	85-88

b. pH not controlled.

Homoallylic alcohols 40 are the Felkin-Anh (non-chelation-controlled) products. The diastereomeric ratios of 40 and 41 resulting from exposure of 36-39 to allyl bromide and indium in water are compiled in Table 6. The product distributions are seen to correlate closely with the β-alkoxy series. Thus, α,β-methoxyl substituent is capable of modest levels of chelation control, irrespective of whether the allylation is performed in an anhydrous or aqueous medium. Since a benzyloxy or a *tert*-butyldimethyl-siloxy group results in production of totally stereo random mixtures of 40 and 41, there is no evidence for transient structural rigidification prior to nucleophilic attack in these examples. This crossover could reflect the operation of a steric effect. The unprotected hydroxyl derivative 36 exhibits the most pronounced face selectivity as anticipated. In fact, the 8.5:1 ratio of 40 to 41 compares quite favorably with the product distributions exhibited by the β-hydroxy aldehydes 37 and 38. Clearly the free β-OH group is capable of chelation control in water, finding it possible to coordinate to the indium ion despite its preexisting solvation by water molecules.

Mechanistic Considerations. Additions of the allylindium reagent to α- and β-hydroxy aldehydes in water have been demonstrated to be highly stereoselective and synthetically useful operations. The corresponding methoxy and MOM derivatives exhibit comparable properties, although to a demonstrably lessened degree. The free hydroxyl derivatives represent the more reactive substrates in either series, this reactivity ordering conforming expectedly to chelation-controlled addition. This ability of the indium cation to lock the carbonyl substrate conformationally prior to nucleophilic attack is indicative that coordination to the substrate can indeed overcome the H_2O solvation forces, especially when the neighboring functionality is an unprotected hydroxyl substituent.

The sense of asymmetric induction in the α-series, *viz.* a strong kinetic preference for formation of the syn diol, is consistent with operation of the classic Cram model as in A (Figure 1). Once complexation occurs, the allyl group is transferred to the carbonyl carbon from the less hindered π-surface opposite to that occupied from the R group. In B (Figure 1), the chelation pathway is seen to be capable of adoption of a chair conformation which concisely accommodates favored formation of the syn diol. The reversal in stereoselectivity in going from 36 to 39 in the same aqueous environment is the classical test for chelation. For the β-chelate reactions, the factors which influence product formation appear to be the same. When C forms (Figure 1), intramolecular attack is guided to occur syn to the preexisting hydroxyl. This reaction trajectory leads preferentially to the *anti* diol, provided that a chair like transition state approximatingD (Figure 1) is followed. Importantly, it is one single allylindium that chelates and reacts. Although similar working models have been advanced in explanation of the mode of addition of titanium[69] or borane reagents,[70] this behavior is distinct from other chelation-controlled reactions where the reacting reagent is different from the chelating agent. This may well be an argument that the indium-mediated reaction takes place on the metal surface.

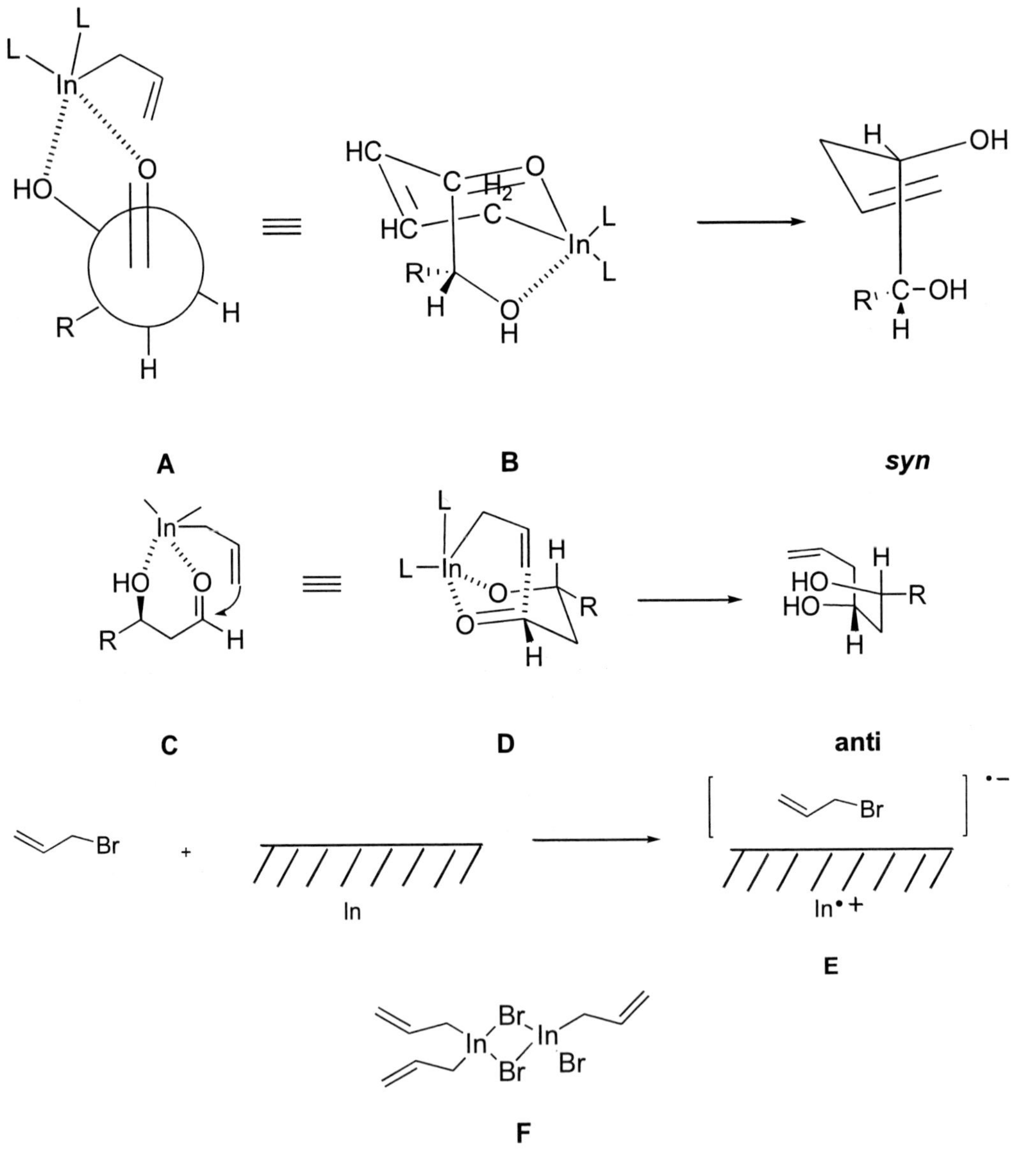

Figure 1.

Table 6. Indium-promoted C-allylation (with allyl bromide) of β-oxygenated aldehydes in water

Entry	Aldehyde	Reaction time (h)	Syn	Anti	Yield, %
1	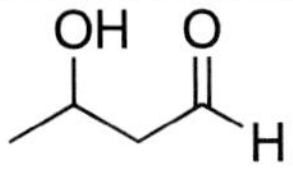 OH O H **36**	2	1	8.2	77
2	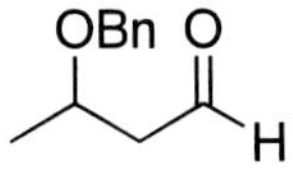 OBn O H **37**	2.5	1	1	80
3	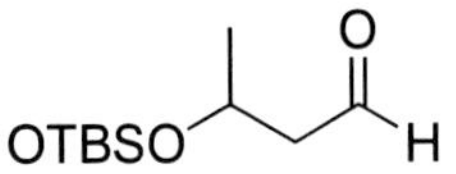 OTBSO O H **38**	3.5	1	1	84
4	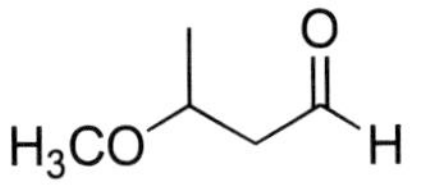 H_3CO O H **39**	2.7	1	4	78

The present studies have demonstrated a direct kinetic link between stereoselectivity and the presence of a neighboring hydroxyl group. While this relationship has been extensively discussed,[71, 72] the support of this concept is not universal. Several experimental and theoretical reports have appeared supporting the notion that π-complexation is not a kinetically important event.[73, 74] Clearly, additional studies of this entire question would be welcomed.

The precise mechanism of indium-promoted reactions remains unclear. In the mid1980s, the involvement of radical pairs was advanced in explanation of tin-promoted allylations.[75] Subsequent recourse to radical clock experiments demonstrated unambiguously that radicals could not be involved.[76] Single electron transfer process similar to that advanced by Chanhave been proposed.[77] According to this reaction profile, the allyl bromide approaches the surface of the indium metal where the SET process generates the reactive radical anion/indium radical cation pair E. These conditions operate, of course, only when indium metal is present as a reactant. Acyclic diastereo-facial control is presently recognized to occur in a wide range of reactions.[78, 79] Suffice it to indicate at this point that the preformation of allylindium reagents may well bypass the involvement of E, suggesting an alternative pathway involving the more conventional species F can also operate.[80, 81] Proper selection of reaction conditions could alter the precise pathway at work.

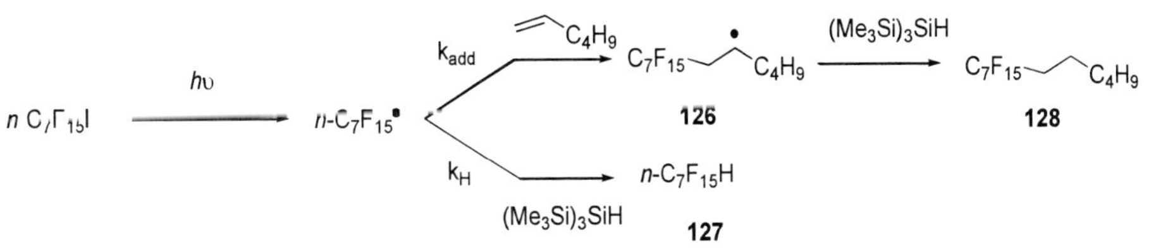

n-$C_7F_{15}I$ —hν→ n-$C_7F_{15}^{\bullet}$ —k_{add} (=\\$C_4H_9$)→ C_7F_{15}\\•/\\$C_4H_9$ (**126**) —$(Me_3Si)_3SiH$→ C_7F_{15}\\/\\$C_4H_9$ (**128**) (21)

n-$C_7F_{15}^{\bullet}$ —k_H, $(Me_3Si)_3SiH$→ n-$C_7F_{15}H$ (**127**)

Scheme 27.In-mediated 1,3-butadien-2-ylation of carbonyl compounds in water.

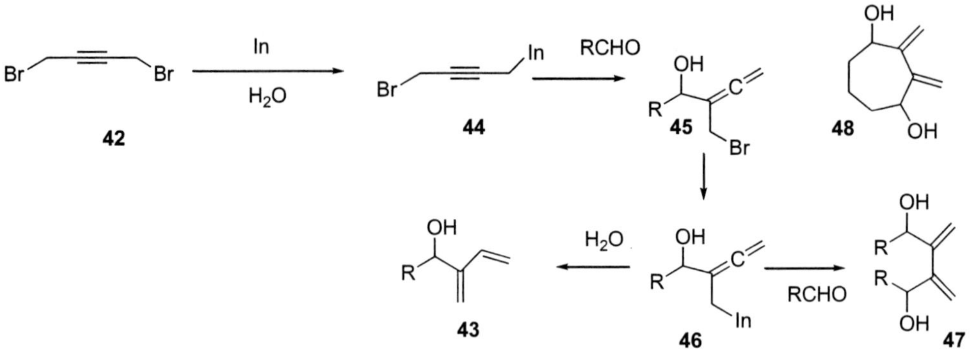

Scheme 28. Proposed mechanism for the In-mediated 1,3-butadien-2-ylation of carbonyl compounds in water.

Chan al. reported[82] thatindium can effectively mediate the coupling between 1,4-dibromobut-2-yne (42) and carbonyl compounds in aqueous media to give regioselective 1,3-butadien-2-ylmethanols 43 in good yields (Scheme 27).

The reaction is likely to proceed via an organoindium intermediate 44 which reacts with the aldehyde to give adduct 45 (Scheme 28). Further reaction of the bromide with indium can lead to another organoindium intermediate, 46, which quenched by water to give 1,3-butadienyl-2-methanol 47. Reaction of 47 with another molecule of aldehyde to give di-adduct 48 was not observed, presumably because of steric hindrance. However, the authors were able to show that with glutaric dialdehyde intramolecular trapping of intermediate 46 was possible and the cyclic di-adduct 48 was obtained in 40% isolated yield.[82]

With this methodology, a series of dienes was synthesized, and the yields are reported in Table 7.

When benzaldehyde is treated with 3-bromo-2-chloro-l-propene 49 and indium in water gave the corresponding allylation product 50 (Scheme 29).[83] ^{1}H NMR measurement of the crude product indicated, virtually, a quantitative reaction. Isolation with chromatography provided 78% of the allylation product 50. Upon ozonolysis, compound 50 was converted to the corresponding β-hydroxyl methyl ester51 in 82% yield (Scheme 29).

Other aldehydes reacted similarly, and the results are summarized in Table 8. Attachment of various functionalities on the aromatic ring provides an equivalent or better overall yields of the product (Entries 2,4,5 and 7, Table 8). Aliphatic aldehydes were similarly transformed to the corresponding γ-hydroxyl esters (entries 3 and 6, Table 8). The use of a mixture of water and THF as solvent for the indium mediated allylation did not affect the reaction result (entry 4, Table 8). It is noteworthy to mention that compounds with a free hydroxyl group (entry 8, Table 8) can be converted directly to the corresponding γ~-hydroxyl ester. In entry 9, Table 8, even though the aldehyde existed in its cyclized hemiacetal form, the compound was transformed to the desired compound without any difficulty.

Oshima and collaborators have [84] performed the radical allylation of α-halo carbonyl compounds with allygallium in water (Scheme 30).

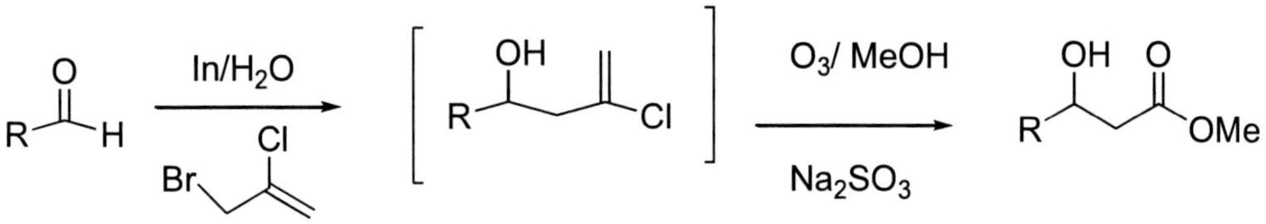

In/H_2O — O_3/ MeOH, Na_2SO_3

49 **50** **51**

Scheme 29. In-mediated Allylation of aldehydes with 3-bromo-2-chloro-l-propene.

Table 7. Indium-mediated 1,3-butadien-2-ylation of carbonyl compounds in water

Entry	Aldehyde	Product	Yield, %
1	Benzaldehyde	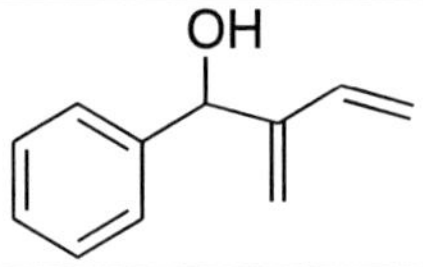	53
2	4-methoxybenzaldehyde	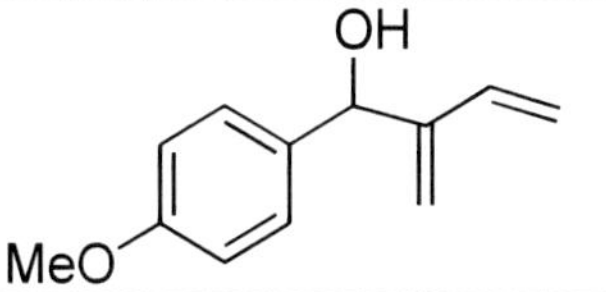	55
3	cinalmaldehyde	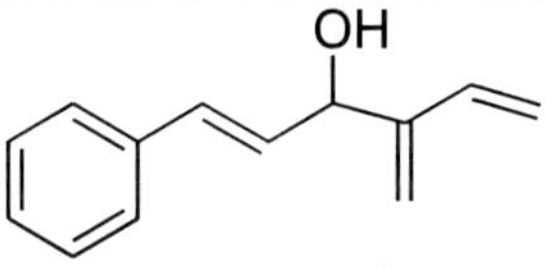	60
4	heptaldehyde	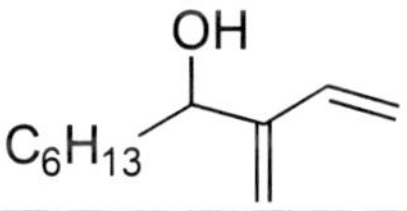	
5	cyclohexylaldehyde	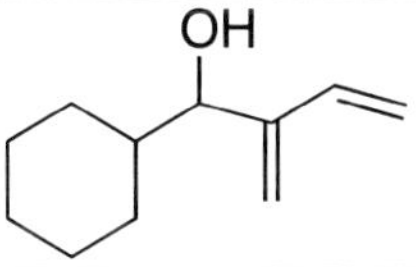	64
6	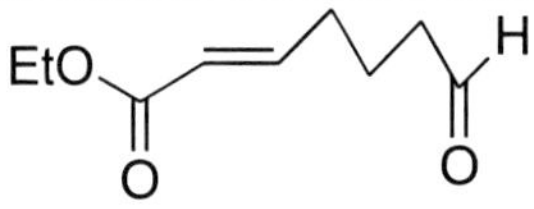	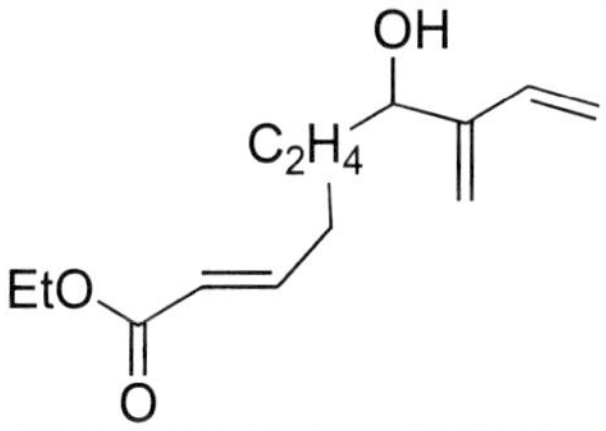	67
7	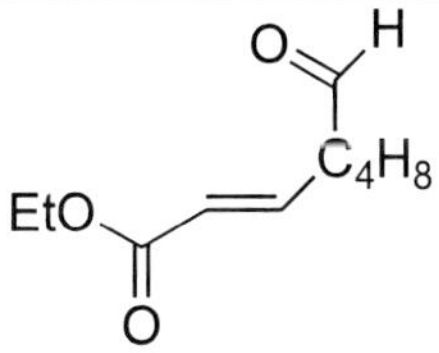	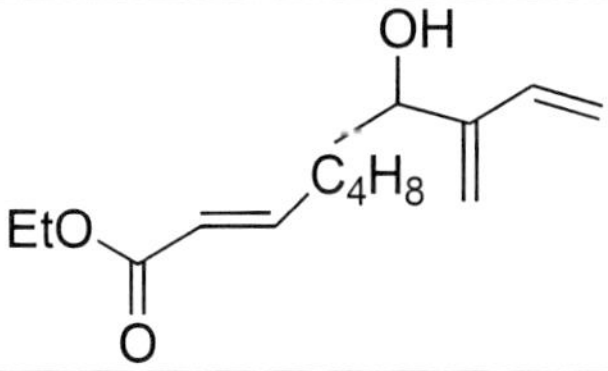	68

Table 8. In-mediated Allylation of aldehydes with 3-bromo-2-chloro-l-propene 49

Entry	Substrate	Product, %
1	Benzaldehyde	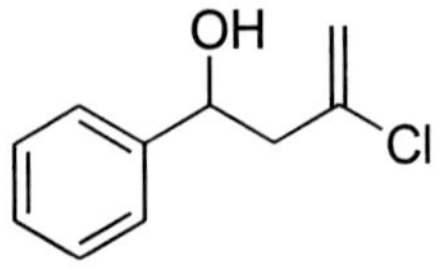 OH Cl 78
2	3-bromobenzaldehyde	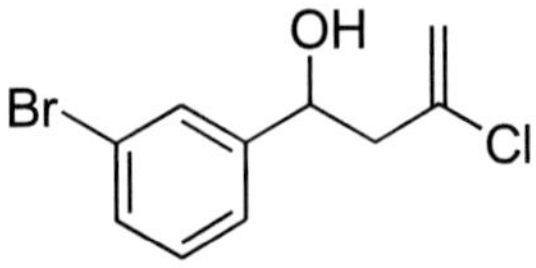 Br OH Cl 91
3	Heptaldehyde	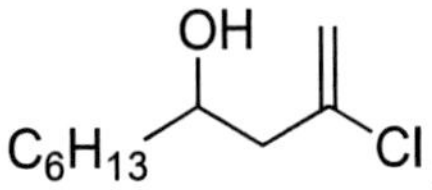 C_6H_{13} OH Cl 58
4	4-chlorobenzaldehyde	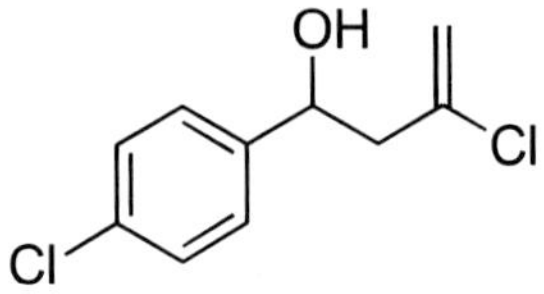 Cl OH Cl 82
5	4-methylbenzaldehyde	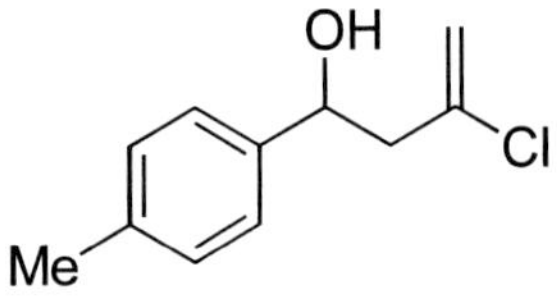 Me OH Cl 91
6	Cyclohexylaldehyde	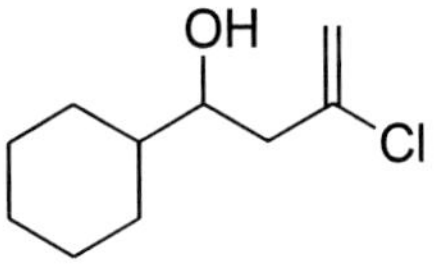 OH Cl
7	4-methoxybenzaldehyde	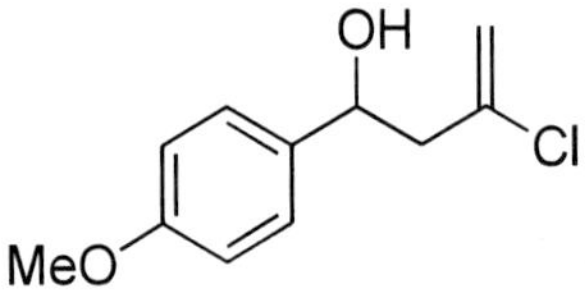 MeO OH Cl 77
8	4-benzylic carbaldehyde	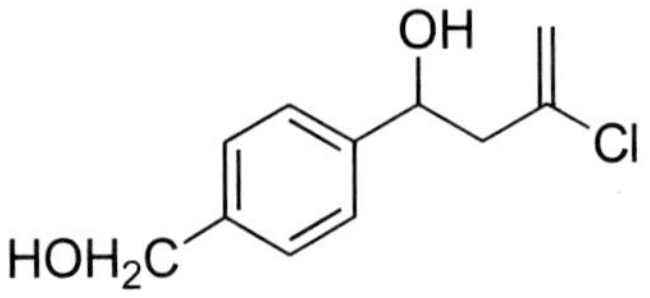 HOH_2C OH Cl
9	2-hydroxy-tetrahydropyrene	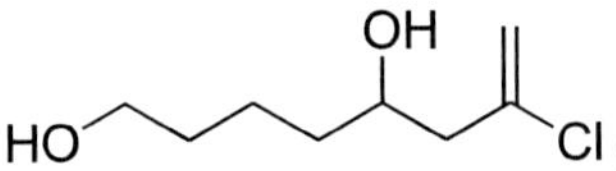 HO OH Cl 84

The origin of the favorable solvent effect is not clear at this stage. Similar phenomena were reported on atom- transfer radical reaction of α-iodo carbonyl compounds in aqueous media (see this Chapter), where the high cohesive energy density of water causes reduction of the volume of an organic molecule. In the present case, the addition step could be accelerated because the addition necessarily accompanies the decrease of the total volume of the reactants. It is also probable that the structure of the allylgallium species would change and that the addition of water could increase the reactivity of allylgallium. Allylgallium dichloride is likely to be transformed into allylgallium hydroxide that is possibly more reactive for radical allylation.

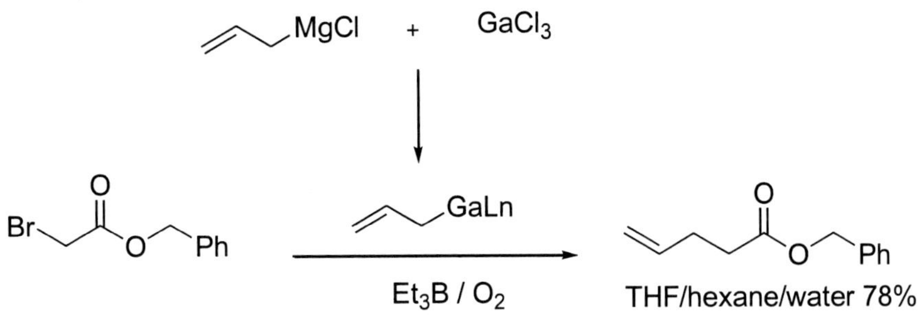

MgCl + $GaCl_3$

GaLn

Br O Ph → Et_3B / O_2 → O Ph

THF/hexane/water 78%

Scheme 30. Radical allylation of α-halo carbonyl compounds with allygallium in water.

Various combinations of α-carbonyl compounds and allylic gallium reagents were examined (Table 9). More reactive α-iodo carbonyl compounds gave better results compared with their bromo analogs. 2-Halopropanoate or 2-halopropanamide also reacted with the allylgallium reagent to give 2-methyl-4-pentenoate or 2-methyl-4-pentenamide (entries 2, 3, and 5, Table 9). In contrast, 2-bromo-2-methylpropanoate did not give the anticipated product, and the starting material was recovered unchanged, probably due to the steric hindrance around the carbon-centered radical. Interestingly, allylation was effective for the substrates having a terminal carbon-carbon double bond (entries 9 and 10, Table 9).

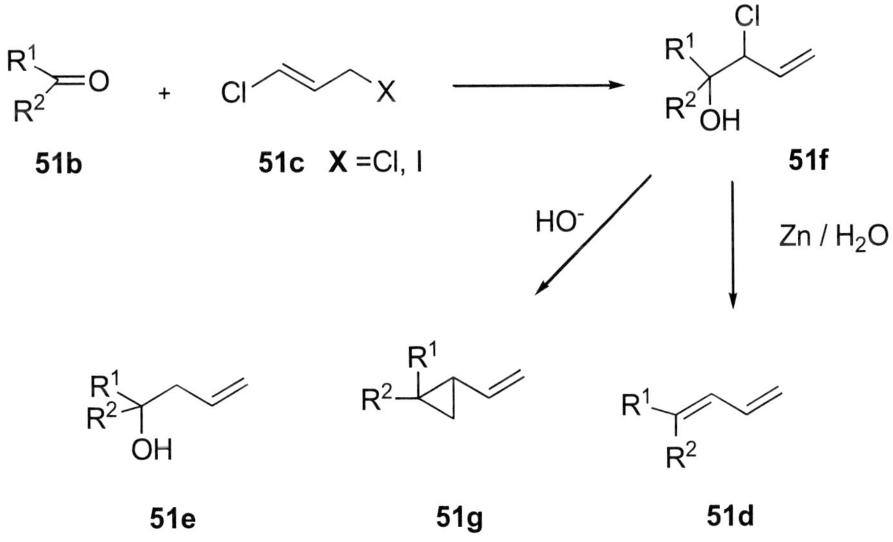

R^1 R^2 =O + Cl X → Cl R^1 R^2 OH

51b **51c X** =Cl, I **51f**

HO^- Zn / H_2O

R^1 R^2 OH R^1 R^2 R^1 R^2

51e **51g** **51d**

Figure 2a.

Table 9. Radical allylation of α-halocarbonyl compounds with allyl gallium in water

R_2-CH=CH-CH$_2$GaLn, Et_3B / O_2 THF / water

entry	X	Y	R^1	R^2	Time	Yield, %
1	Br	OCH_2Ph	H	H	2	78
2	Br	OCH_2Ph	Me	H	2	63
3	Br	NMe_2	Me	H	2	64
4	I	OCH_2Ph	H	H	0.5	89
5	I	OCH_2Ph	Me	H	0.5	81
6	I	$NHCH_2CHCHC_{10}H_{21}$	H	H	1	87
7	I	$OCH_2CHCHC_3H_7$	H	H	0.5	95
8	I	$O(CH_2)_6Cl$	H	H	1	85
9	I	$O(CH_2)_2$ OCH_2CHCH_2	H	H	0.5	64
10	I	$OCHPhCH_2CHCH_2$	H	H	2	71
11	I	$OCH(C_3H_7)COC_3H_7$	H	H	1	84
12	I	OCH_2Ph	H	Me	0.5	60
13	I	$NHCH_2CHCHC_{10}H_{21}$	H	Me	2	85
14	I	$O(CH_2)_6Cl$	H	Me	2	46
15	I	OCH_2Ph	H	Me	1	50
16	Br	NMe_2	Me	Me	7	65

An electron-deficient (alkoxycarbonyl)methyl radical reacted faster with the highly electron-rich alkene moiety of the allylgallium species than with the olefinic parts of the substrate and of the product. Allylation of the ketone moiety was not observed.

Chan and collaborators[84b] reported on a simple synthesis of 1,3- butadienes from carbonyl compounds in aqueous medium according to Figure 2. Atypical experimental procedure consists of amixture of the carbonyl com pound 51b (1 mmol), 1,3-dichloropropene (51c, X = C1; 1 mmol), and zinc powder (2 mmol) in 10 mL of water was heated to 35 "C with vigorous stirring for 3-4 h. The reaction mixture was cooled and quenched with ether. The organic product was isolated from the ether phase and purified by flash column chromatography to give the corresponding 1,3-butadiene51d. The reaction has a number of interesting features. First, it is important to note that, in the reaction of benzaldehyde, the yield of 1-phenylbutadiene was quite satisfactory under these conditions in aqueous medium but failed to proceed at all in diethyl ether or other organic solvents normally used for organometallic reactions. Second, the reaction seems to proceed with both aldehydes and ketones. With cinnamaldehyde, the corresponding triene was obtained. Third, the butadienes were formed stereoselectively and, in the case of aldehydes, exclusively as the *E* isomers. Furthermore, unprotected hydroxyl compounds such as glyceraldehyde and 5-hydroxypentanal underwent the diene conversion without difficulty. On the other hand, the yield of 51d was modest at best in all cases, in spite of efforts to vary the reaction temperature, time, amount of metal, etc. The poor yield was traced to the formation of the homoallylic alcohol 51e, which must have been formed by the zinc-mediated reduction of the intermediate chlorohydrin 51f.

3.6. Allylation of Imine Derivatives in Water

The reaction of allylmetals with electrophiles has been developed as an important carbon-carbon bond forming method. In general, γ-adducts (branched products) are obtained in the allylation of aldehydes and imines using allylmetals. The selective synthesis of α-adducts (linear products) has been a subject of current interest (Figure 2b).

Recently, the preparation of allylindium reagents via transient organopalladium intermediates has been studied by Araki et al.[85]

Allylation reactions of electron-deficient imines with allylic alcohol derivatives in the presence of a catalytic amount of palladium(0) complex and indium(I) iodide was studied by Takemoto et. al.al.al.[86]. The reversibility of allylation was observed in the reaction of glyoxylic oxime ether having camphorsultam. γ-Adducts were observed with high regioselectivity in water.

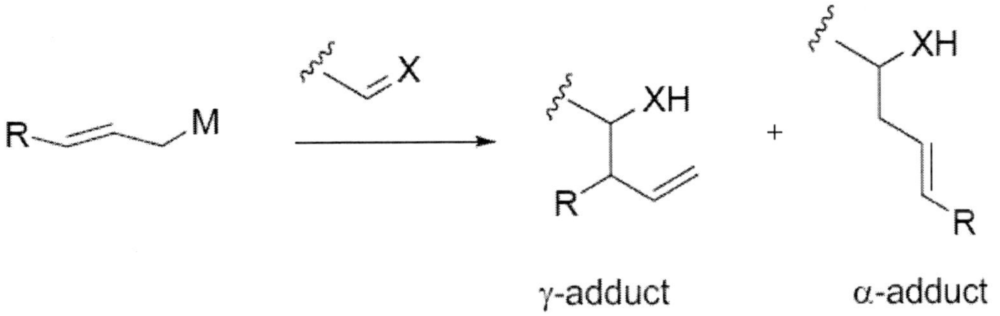
X
M
R
XH
+
XH
R
R
γ-adduct
α-adduct

Figure 2b. Regioselectivity in allylation reactions.

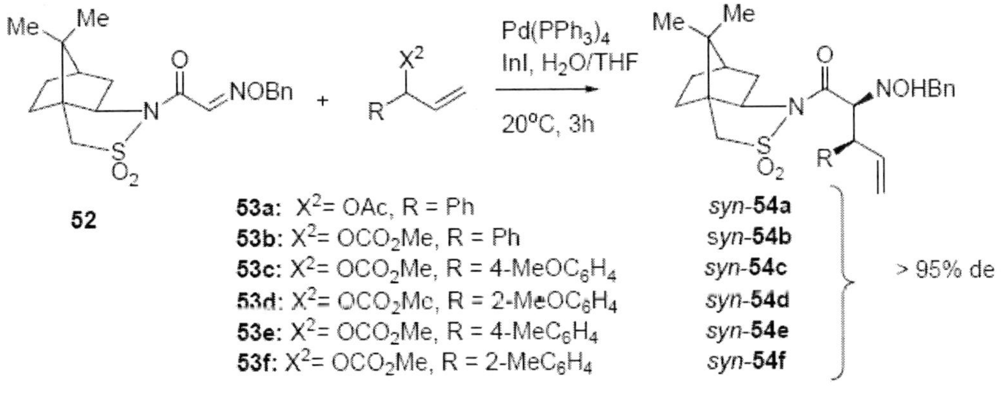
Me
Me
O
N
NOBn
S
O2
+
X2
R
Pd(PPh3)4
InI, H2O/THF
20°C, 3h
Me
Me
O
N
NOHBn
S
O2
R
52
53a: X2= OAc, R = Ph
53b: X2= OCO2Me, R = Ph
53c: X2= OCO2Me, R = 4-MeOC6H4
53d: X2= OCO2Me, R = 2-MeOC6H4
53e: X2= OCO2Me, R = 4-MeC6H4
53f: X2= OCO2Me, R = 2-MeC6H4
syn-54a
syn-54b
syn-54c
syn-54d
syn-54e
syn-54f
> 95% de

Scheme 31. The reaction of imine derivatives with allylic acetates in the presence of $Pd(PPh_3)_4$ and indium(I) iodide in water, to afford the γ-adduct *syn*-54a.

Table 10. Reaction of 52 with allyl alcohol derivatives 53a-f

Entry	Reagent	Solvent	Time(h)	Product	Yield (%)
	53c	H_2O-THF	3	*Syn*-54c	90
2	53d	H_2O-THF	3	*Syn*-54d	81
3	53e	H_2O-THF	3	*Syn*-54e	71
4	53f	H_2O-THF	3	*Syn*-54f	72
5	53g	H_2O-THF	3	*Syn*-54g	66

The reaction of 52 with allylic acetate 53a in the presence of $Pd(PPh_3)_4$ and indium(I) iodide in the presence of water, affords the γ-adduct *syn-5*4a in 90% yield as a single diastereomer (Scheme 31, Table 10).

In comparison with the reaction of glyoxylic oxime ether, the authors investigated the allylation of N-sulfonylimine under similar reaction conditions (Scheme 32).[86]

Although the reaction of 55 proceeded smoothly, the formation of α-adducts were obtained even under anhydrous THF. These observations indicate that the allylation of N-sulfonylimine 55 was not a reversible process due to extra stabilization of indium-bondingadduct by electron withdrawing N-sulfonyl group. The bulky γ,γ-dimethylallyl acetate58c and carbonate 58d were less effective for the allylation reaction of N-sulfonylimine 55. The reaction of 55 with α,α,-dimethylallyl acetate 58c gave the γ-adduct 56d in 36% yield.

In another account, Takemoto et al. successfully attempted the propargylation reaction of hydrazones and glyoxylic oxime ethers in water mediated by Pd-indium iodide, demonstrating the role of water in directing the diastereoselectivity.[87]

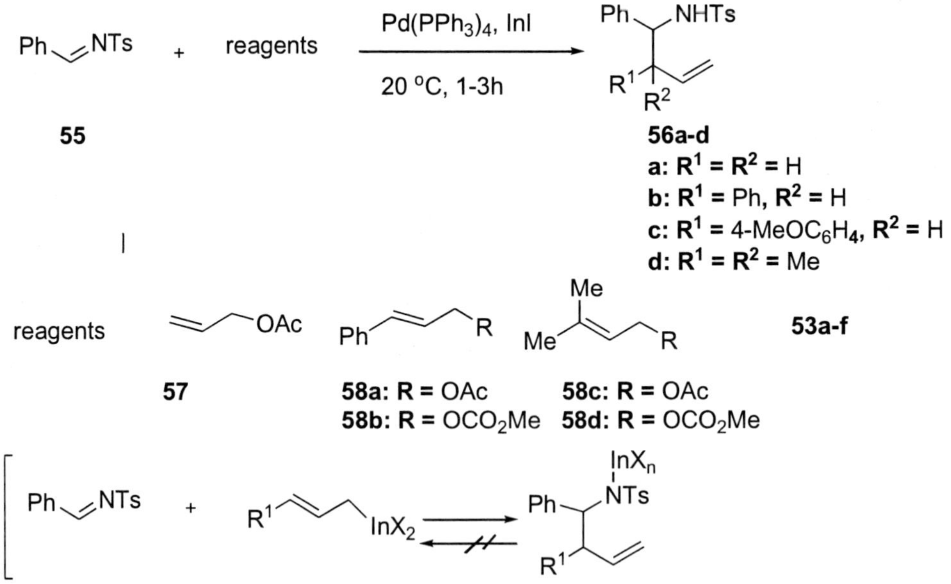

Ph NTs + reagents → [$Pd(PPh_3)_4$, InI; 20 °C, 1-3h] Ph NHTs, R^1 R^2

55 **56a-d**

a: R^1 = R^2 = H
b: R^1 = Ph, **R^2** = H
c: R^1 = 4-$MeOC_6H_4$, **R^2** = H
d: R^1 = R^2 = Me

reagents OAc (**57**); Ph R (**58a: R** = OAc; **58b: R** = OCO_2Me); Me Me R (**58c: R** = OAc; **58d: R** = OCO_2Me) **53a-f**

Ph NTs + R^1 InX_2 ⇄ Ph NTs InX_n R^1

Scheme 32. Allylation of N-sulfonylimine with allylic acetate derivatives in the presence of $Pd(PPh_3)_4$ and indium(I) iodide in water.

3.7. Radical Conjugate Additions to α,β-unsaturated Carbonyl Compounds in Water

Conjugate addition of alkyl groups to α,β-unsaturated carbonyl compounds is a versatile synthetic method for the construction of C –C bonds. Among the various methods available, the most commonly employed strategies involve the use of organometallic species such as Grignard reagents (RMgX) or organolithium (RLi) reagents. However, the use of these highly reactive organometallic reagents can lead to undesired side reactions such as hydrolysis, Wurtz coupling, β elimination of the organometallic reagent, and the reduction of carbonyl compounds. Also,1,2-addition of the alkyl group to the carbonyl group can compete with the 1,4-conjugate addition reaction. If this reaction could be developed to take place in water without the above-mentioned side reactions, it would greatly aid organic chemists.

Luche found that the combination of Zn–Cu couple and sonication mediates in the intermolecular addition of alkyl radicals to α,β-unsaturated aldehydes, ketones, esters, and amides in aqueous EtOH (Scheme 33).[88. 89.90]

These reactions were typically most efficient with tertiary and secondary radicals. Mechanistic studies suggested that the intermediate a-carbonyl radical is reduced to an enolate and subsequently protonated.[91]

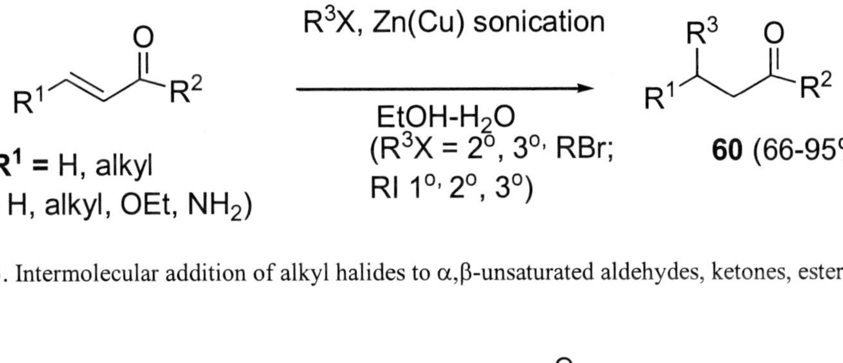

R^3X, Zn(Cu) sonication

EtOH-H_2O

(R^3X = 2°, 3°, RBr; RI 1°, 2°, 3°)

59 (**R^1** = H, alkyl **R^2** = H, alkyl, OEt, NH_2)

60 (66-95%)

Scheme 33. Intermolecular addition of alkyl halides to α,β-unsaturated aldehydes, ketones, esters.

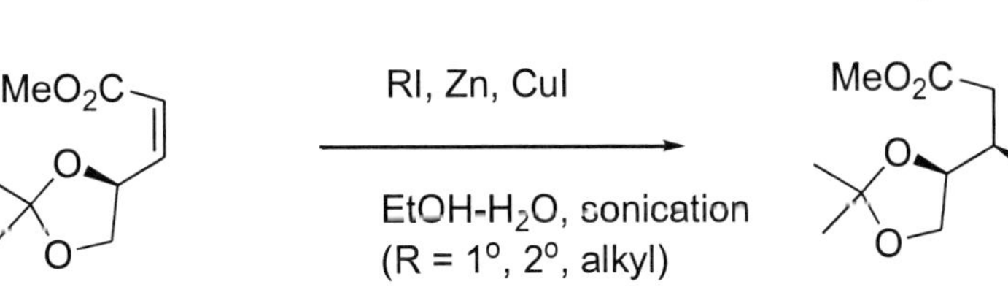

RI, Zn, CuI

EtOH-H_2O, sonication (R = 1°, 2°, alkyl)

61

62 (67-96% > 82:8 *cis:trans*)

RI, Zn, CuI

EtOH-H_2O, sonication (R = 1°, 2°, alkyl)

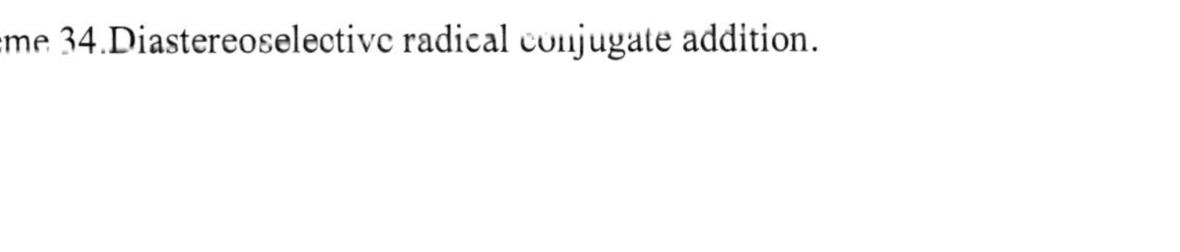

63

64 (51-82%)

Scheme 34.Diastereoselective radical conjugate addition.

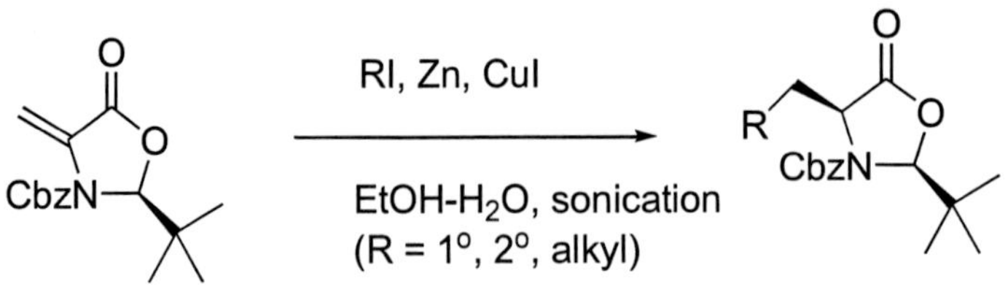

CbzN O O
RI, Zn, CuI
EtOH-H_2O, sonication
(R = 1°, 2°, alkyl)
R CbzN O O

65 **66** (76-96% >98:2 *cis:trans*)

Scheme 35.Diastereoselective radical conjugate addition of N-Cbz methyleneoxazolidinone with alkyl iodides in aqueous media.

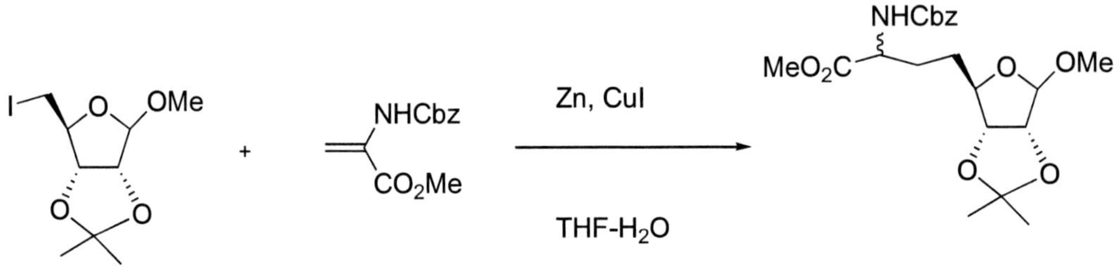

I O OMe O O + NHCbz CO_2Me
Zn, CuI
THF-H_2O
NHCbz MeO_2C O OMe O O

67 **68** **69 (87%)**

Scheme 36. Radical conjugate addition with a ribose derivative.

Sarandeses and Perez Sestelo have employed the Luche protocol in diastereoselective radical conjugate addition (RCA) reactions. Methylene-dioxolanone 61 and γ,δ-dioxolanyl-α,β -unsaturated ester 63 serve as effective chiral acceptors for the stereoselective synthesis of α- and γ-hydroxy acid derivatives 62 and 64, respectively, (Scheme 34).[29, 30]

Similar RCAs conducted with N-Cbz methyleneoxazolidinone 65 provided α-amino acid derivatives 66 with very good yields and dr's (Scheme 35).[31]

In the course of a synthesis of sinefungin analogues, Fourrey discovered that sonication was not required to promote RCA of ribose derivative 67 to dehydroalanine 69 under modified Luche conditions (Scheme 36).[92, 93, 94]

Rather, vigorous stirring was sufficient. Similar reactions could also be induced by a Zn–Fe couple.[95]

Crich demonstrated that alkyl radicals generated via the reductive mercury method underwent intermolecular conjugate addition to dehydroalanine 71 (Scheme 37).[96]

Primary, secondary, and tertiary alkylmercury bromides and chlorides could all be used in this reaction.

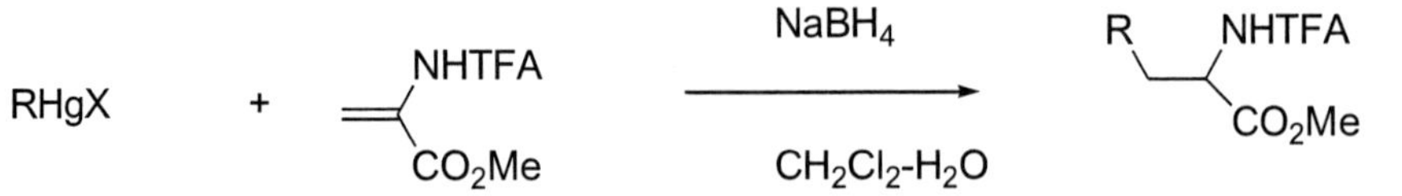

RHgX + NHTFA CO_2Me
$NaBH_4$
CH_2Cl_2-H_2O
R NHTFA CO_2Me

(R = alkyl) **70** **71** (62-85%)

Scheme 37. Synthesis of dehydroalanine derivatives via RCA.

Table 11. Alkylation of peptides in aqueous mixtures

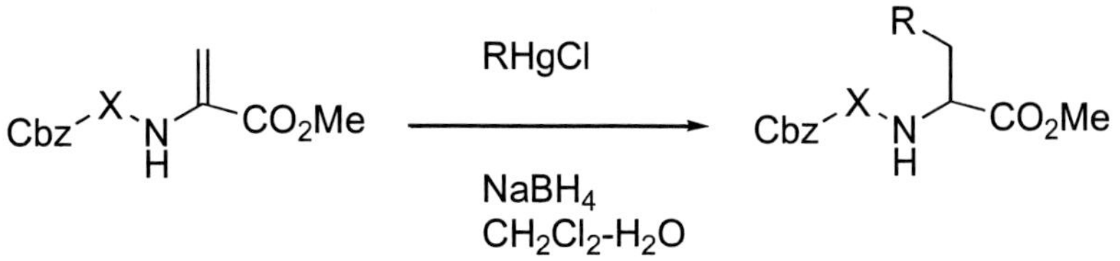

Cbz–X–NH–C(=CH$_2$)–CO$_2$Me (**72**) → [RHgCl, NaBH$_4$, CH_2Cl_2-H_2O] → Cbz–X–NH–CH(CH$_2$R)–CO$_2$Me (**73**)

Entry	X	R	Yield (%)	de (%)
1	L-Val	*c*-C_6H_{11}	71	11
2	L-Phe	*c*-C_6H_{11}	64	5
3	L-Cys(Z)	*c*-C_6H_{11}	70	6
4	L-Ser	*c*-C_6H_{11}	42	1
5	L-Pro	i-Pr	98	1

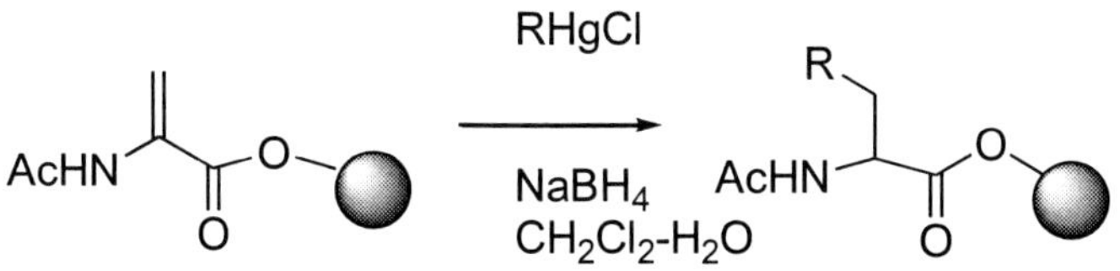

AcHN–C(=CH$_2$)–C(=O)–O–(resin) (**74**) → [RHgCl, NaBH$_4$, CH_2Cl_2-H_2O] → AcHN–CH(CH$_2$R)–C(=O)–O–(resin) (**75** (49-60%))

Scheme 38. Solid state synthesis of α-amino acids.

Table 12. Reactions of a α,β-unsaturated ester with cyclohexyliodide under different conditions

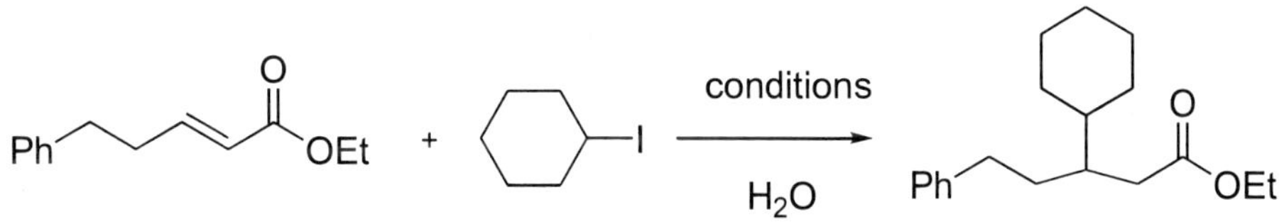

Ph–CH$_2$CH$_2$–CH=CH–C(=O)OEt (**75b**) + cyclohexyl–I → [conditions, H_2O] → Ph–CH$_2$CH$_2$–CH(c-C_6H_{11})–CH$_2$–C(=O)OEt (**76**)

entry	Conditions	Yield (%)
1	In / CuI / $InCl_3$	80
2	In / $InCl_3$	<20
3	In / CuI	54
4	CuI / $InCl_3$	0
5	In / CuBr / $InCl_3$	68
6	In / CuCl / $InCl_3$	<50
7	In / AgI / $InCl_3$	48

Table 13. Radical conjugate additions of alkyl iodides to different α,β-unsaturated esters in water, employing In / CuI / $InCl_3$

R, O, R´ + R¨I —In/CuI/$InCl_3$, H_2O→ R, R¨, O, R´

Entry	α,β-unsaturated ester	Alkyl iodide	Yield (%)
1	Ph, O, OEt	Cyclohexyl iodide	80
2	Ph, O, OEt	Cyclopentyl iodide	84
3	Ph, O, OEt	Isopropyl iodide	70
4	Ph, O, OEt	2-iodobutane	73
5	Ph, O, OEt	Cyclohexyl iodide	70
6	Ph, O, OEt	Cyclohexyl iodide	75
7	O, OEt	Cyclohexyl iodide	61
8	C_5H_{13}, O, OEt	Cyclohexyl iodide	84
9	O, OEt	Cyclohexyl iodide	76
10	Ph, O, O, OEt	Cyclohexyl iodide	46

Crich extended this process to the intermolecular RCA of alkyl radicals to dehydroalanine-containing dipeptides 72 (Table 11).[97]

The stereocenter present in each substrate exerted little influence over the hydrogen atom abstraction, as the products 73 were obtained in low de. Tripeptides were also viable substrates in this reaction, producing RCA adducts in good yield (87–88%) and poor de (3–15%).

Yim and Vidal showed that the Crich method could be employed in the solid-phase synthesis of α-amino acids by anchoring the dehydroalanine radical acceptor to Wang resin (Scheme 38).[98] Cleavage of the N-acetyl amino acid from the resin was accomplished by acid treatment.

Table 14. Radical conjugate additions of alkyl iodides to different α,β-unsaturated enones in water, employing In / CuI / $InCl_3$

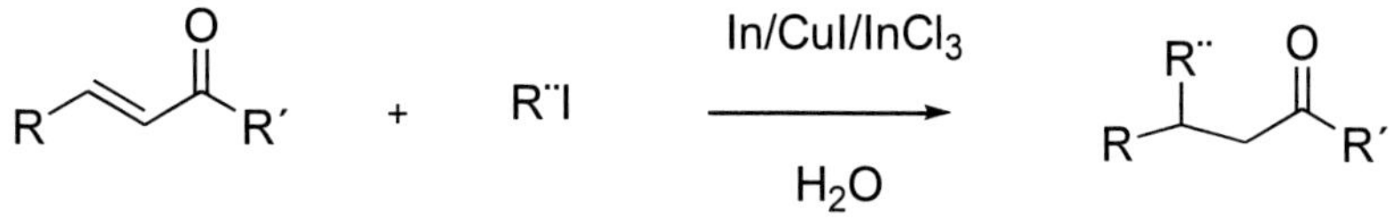

entry	Enone	Alkyl iodide	Yield (%)
1	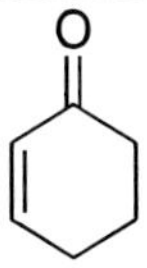	Cyclohexyl iodide	85
2	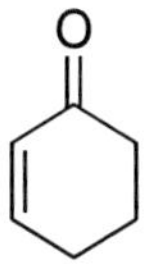	Cyclopentyl iodide	78
3	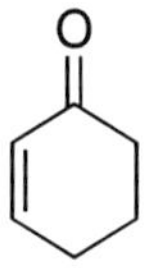	Isopropyl iodide	73
4	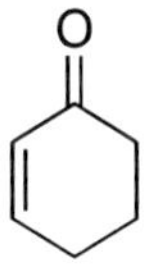	Hexyl iodide	45
5	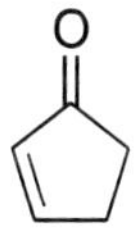	Cyclohexyl iodide	53
6	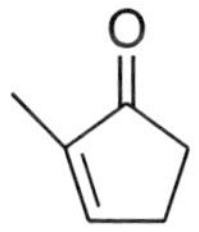	Cyclohexyl iodide	81
7	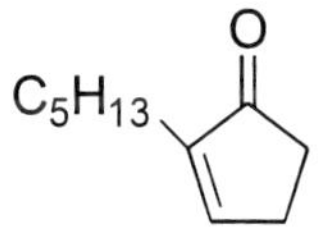 C_5H_{13}	Cyclohexyl iodide	83
8	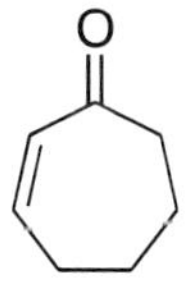	Cyclohexyl iodide	74
9	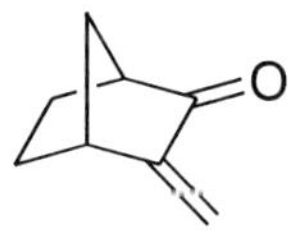	Cyclohexyl iodide	65

The pioneering works by Luche, Li, Naito and others have shown that it is possible to carry out alkyl additions to conjugated systems in water. Unfortunately, in most cases, the use of harsh reaction conditions such as ultrasonication, inert atmospheres, cosolvent systems, and the narrow substrate scope limit their applicability to complex molecule synthesis. Therefore, the development of more general and practical methods for alkyl addition to α,β-unsaturated carbonyl compounds under mild conditions is highly desirable.

More recently, Loh and collaborators[99] have attempted the alkylation reaction of carbonyl compounds in water using unactivated alkyl halides and In / CuI/I_2 or In / AgI / I_2 system.From their results, it became apparent that the use of organic solvents inhibited the occurrence of the Barbier-Grignard-type alkylation reaction. In contrast to the work reported by Li and coworkers[52] it was noteworthy that even aliphatic aldehydes could also react efficiently with alkyl iodides to furnish the alkylated products in good yields.

Lately, Loh and coworkers developed alkylation reactions of unactivated alkyl iodides to α,β-unsaturated carbonyl compounds in water(including a chiral version) using indium/copper in water.[67] In addition, the formation of symmetrical *vic* -diarylalkanes was observed when aryl-substituted alkenes were used as the substrate.

Initial studies focused on the reaction of α,β-unsaturated ester and cyclohexyl iodide under different reaction conditions.

As shown in Table 12, it was found that the combination of In / CuI / $InCl_3$ (6:3:0.1) was an efficient system for activation of the conjugate addition reaction of 75b in water (see Table 12). The reaction proceeded smoothly at room temperature to generate the corresponding adduct 76 in 80%yield (entry 1). It is important to note that, without the use of CuI, the reaction proceeded sluggishly to give the desired product in poor yield (entry 2). Without the addition of $InCl_3$, the yield of the product decreased to 54% (entry 3). In addition, it was found that the use of the metal (*i.e.*, indium) was also indispensable (entry 4). Among the several metals screened, indium proved to be the best for this reaction. The following order was apparent for activation of the conjugate alkylation: In >Zn >Al >Sn. Other copper and silver compounds such as CuBr, CuCl, and AgI were also investigated, but all gave the products in lower yields in comparison to CuI (entries 5 –7). It was worthwhile to note that the reactions proceeded more efficiently in water than in organic solvents such as MeOH,THF,CH_2Cl_2,DMF,DMSO,and hexane. Furthermore, the reactions were carried out without an inert atmosphere and ultra-sonication was unnecessary.

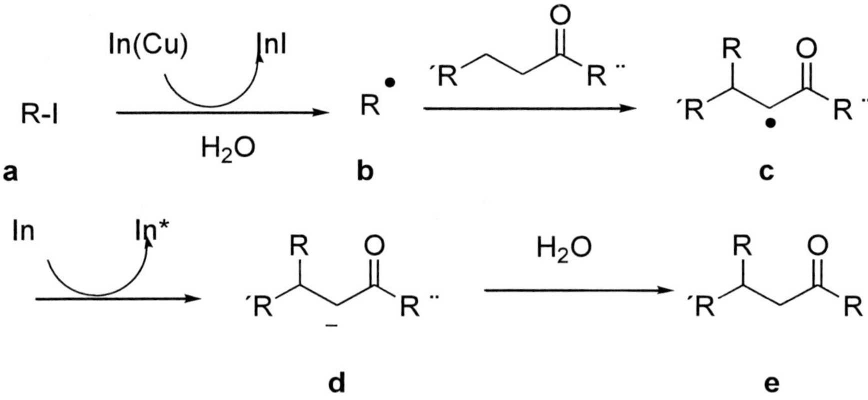

Scheme 39A. Proposed mechanism for the RCA of α,β-unsaturated compounds with alkyl iodides employing In / CuI / $InCl_3$.

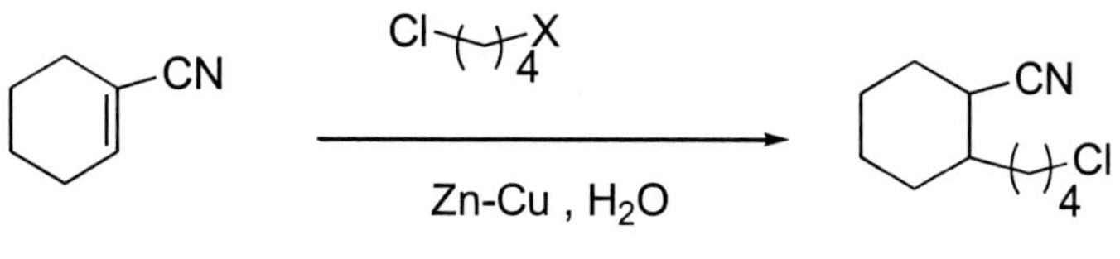

Cl–(CH₂)₄–X

Zn-Cu , H_2O

Scheme 39B. Conjugate additions of alkyl halides to alkenenitriles in water.

The reaction was extended to various α,β-unsaturated esters and enones, as shown in Tables 13 and 14.

A plausible reaction mechanism is proposed (Scheme 39A). The reaction is possibly initiated by a single-electron transfer from indium/copper to alkyl iodide a (Scheme 39) to generate an alkyl radical b. This radical can attack the α,β-unsaturated carbonyl compound via 1,4-conjugate addition to furnish a radical intermediate c .Subsequent indium-promoted reduction of intermediate c and quenching of the generated anion d in the presence of water affords the expected product e . This method works with a wide variety of α,β-unsaturated carbonyl compounds. The mild reaction conditions, moderate to good yields, and the simplicity of the reaction procedure make this method an attractive alternative to conventional methods using highly reactive organometallic reagents in anhydrous conditions.

Numerous and useful indium-mediated allylation reactions of carbonyl compounds have been reported.[100] However, the corresponding reaction of imine derivatives has not been widely studied because of the lower electrophilicity of carbon-nitrogen double bonds.Therefore, the development of indium-mediated reactions of imines in aqueous media has been a subject of recent interest. Chan et al.[56][101] reported on the first studies of indium-mediated allylation of N-sulfonylimines in aqueous media. This is likely an area for further development and new discoveries.

Then, it was demonstrated that trialkylboranes are excellent reagents for conjugate addition to vinyl ketones (eq 12), acrolein, α-bromoacrolein and quinones.

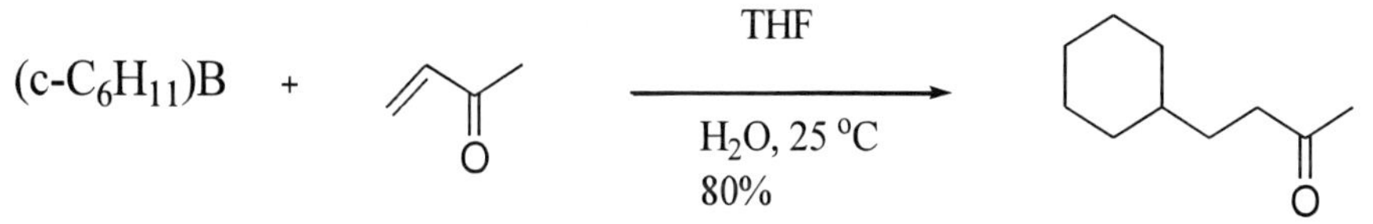

$$(c\text{-}C_6H_{11})B \quad + \quad \xrightarrow[\text{H}_2\text{O, 25 °C; 80\%}]{\text{THF}} \qquad (12)$$

Various attempts to extend this reactionto β-substituted-α,β-unsaturatedcarbonyl compounds such as *trans*-3-penten-2-one, mesityl oxide, 2-cyclohen-1-one, and *trans*-crotonaldehyde were unsuccessful unless radical initiators were used.

Fleming and collaborators[101b] have utilized a silica-supported zinc-copper matrix for promoting conjugate additions of alkyl iodides to alkenenitriles in water. Acyclic and cyclic nitriles react with functionalized alkyl iodides, overcoming the previous difficulty of performing conjugate additions to disubstituted alkenenitriles with nonstabilized carbon nucleophiles. Conjugate additions with ω-chloroalkyl iodides generate cyclic nitriles primed for cyclization, collectively providing one of the few annulation methods for cyclic alkenenitriles (Scheme 39B).

3.8. Synthesis of α,β-unsaturated Ketones

Indium- and zinc-mediated carbon-carbon bond forming reactions in aqueous media have been of great importance from both economic and environmental points of view.Indium and zinc are stable under air, and it is much easier to run their reactions than those with SmI_2, and the toxicity is quite low. Moreover, the metal species can be easily removed from the reaction mixture by simple filtration and washing with water, unlike tributyltin hydride.

Electron-transfer reactions mediated by indium have attracted the attention of synthetic chemists due to its low first ionization potential at 5.79 eV, which is lower than many reducing metals such as aluminum(5.98 eV), tin (7.34 eV), magnesium (7.65 eV), zinc (9.39 eV), and close to that of alkali metals such as sodium (5.12 eV) and lithium (5.39 eV).The second ionization potential for indium is much higher (18.86 eV). Having such low first ionization potential makes indium attractive for conducting reduction reactions.This is particularly so because it is so much easier to handle than alkali metals, for example, the metal remains unaffected by air or oxygen at ordinary temperatures and is practically unaffected by water even at high temperatures, and very resistant to alkaline conditions. As indium has a low toxicity, has found considerable utility in dental alloys. It has also to be pointed out that a favorable experimental feature of indium-mediated radical reactions is that the reactions proceed in the absence of toxic tin hydride, providing the carbon-carbon bond forming method in aqueous media.

Indium has shown great potential for a number of carbon-carbon bond forming reactions such as Reformatsky, Barbier type alkylation, allylation, and propargylation of carbonyl compounds. This is largely due to the fact that a highly reactive metal, such as indium, is required to break the non-activated carbon-halogen bond (as well as to react with the carbonyl once the organometallic intermediate is formed). However, even if the desired intermediate is successfully generated, various competing side reactions may occur when utilizing a highly reactive metal, for example, the reduction of water, the reduction of starting materials, the hydrolysis of the organometallic intermediate, and pinacol-coupling (*vide infra*).[52]

The indium-mediated reaction of benzaldehydes and methyl vinyl ketones proceeded smoothly in the presence of $InCl_3$ in aqueous media to form β,γ-unsaturated ketones.[102] Thus benzaldehydes were reacted with methyl vinyl ketone in the presence of indium powderand $InCl_3$ in a solvent mixture of THF and H_2O at ambient temperature for 6-8 h.Addition of NH_4Cl to the reaction mixture afforded the desired β,γ-unsaturated ketones in good to moderate yields (eq 13).

R + O O → (1.-In / $InCl_3$; H_2O / THF; 2.- H^+) R O (13)

Generally, the yields of products are not affected by the nature of the substituents on the phenyl ring. The reaction also proceeded with heteroaromatic aldehydes. With the absence of In, or $InCl_3$ the reaction did not occur. When other Lewis acids such as $SnCl_4$, $FeCl_3$, and $CuCl_2$ were used instead of $InCl_3$, low yields of β,γ-unsaturated ketones resulted (20-

38%).When an aldehyde reacted with ethyl vinyl ketone instead of methyl vinyl ketone as a Michael acceptor, a coupling product was produced in 62 % yield. The reaction conditions were extended to other Michael acceptors such as acrolein, acrylonitrile, ethyl acrylate, and acrylic acid; however, the reactions didnot proceed.[102]

The reaction mechanism was postulated to be a radical mechanism involving the radical anion intermediate of methyl vinyl ketone formed from indium (Scheme 40).

The reaction intermediate undergoes radical cyclopropanation and addition to benzaldehyde. Upon addition of butylated hydroxytoluene (BHT) a rate retardation effect was observed.[102]

Upon addition of BHT (butylated hydroxy-toluene) to the reaction, a rate retardation effect was detected. When the reaction was followed by ^{1} H NMR in a 1:1 mixture of THF-d_8 and D_2O, the cyclopropanyl proton signals were observed at ä 1.2-0.5 as multiplet. Quenching the reaction mixture with DCl in D_2O after an appropriate reaction time and examination of the $CDCl_3$-extracted products by ^{1} H NMR showed the signal of the 5-phenyl-4-penten-2-onetogether with peaks of some MVK decomposed compounds.

Murphy and collaborators[103], however, developed a facile and an environment-friendly protocol for the deoxygenation of epoxides with good radical-stabilizing groups adjacent to the oxirane ring, using indium metal and indium (I) chlorideor ammonium chloride in alcohol / water mixtures (eq 14).

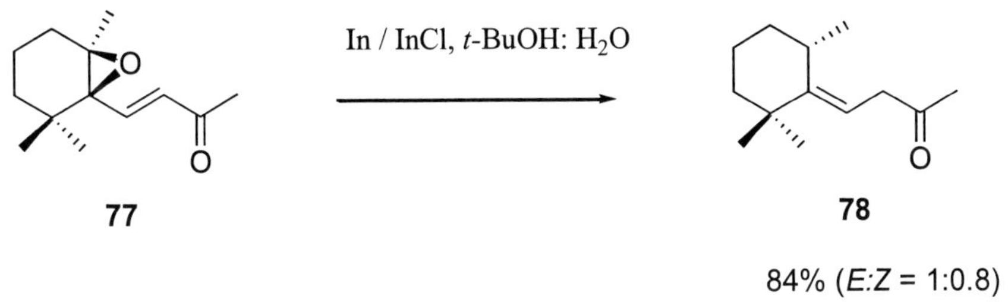

In / InCl, *t*-BuOH: H_2O (14)

77 **78**

84% (*E*:*Z* = 1:0.8)

Oxirane 77 underwent smooth deoxygenation to afford the alkene 78 in an excellent 81% yield.The reaction sequence is depicted in Scheme 41.

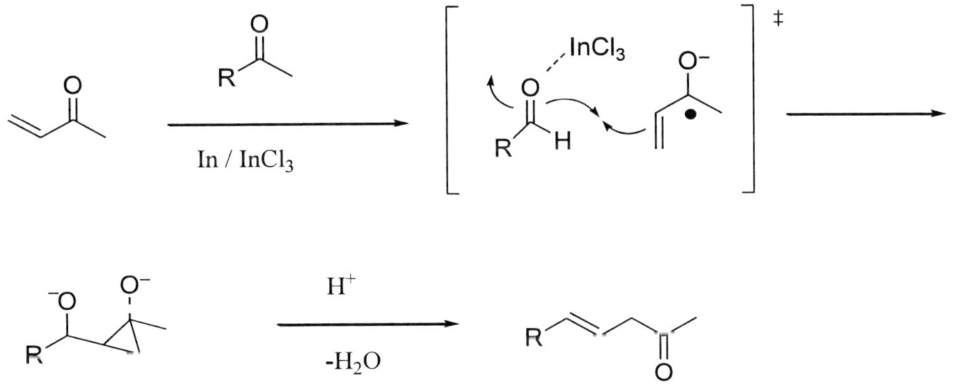

In / $InCl_3$ ‡ $InCl_3$ O⁻ R H

H^+ $-H_2O$ R

Scheme 40. Possible reaction mechanism for the In-mediated synthesis of β,γ-unsaturated ketones.

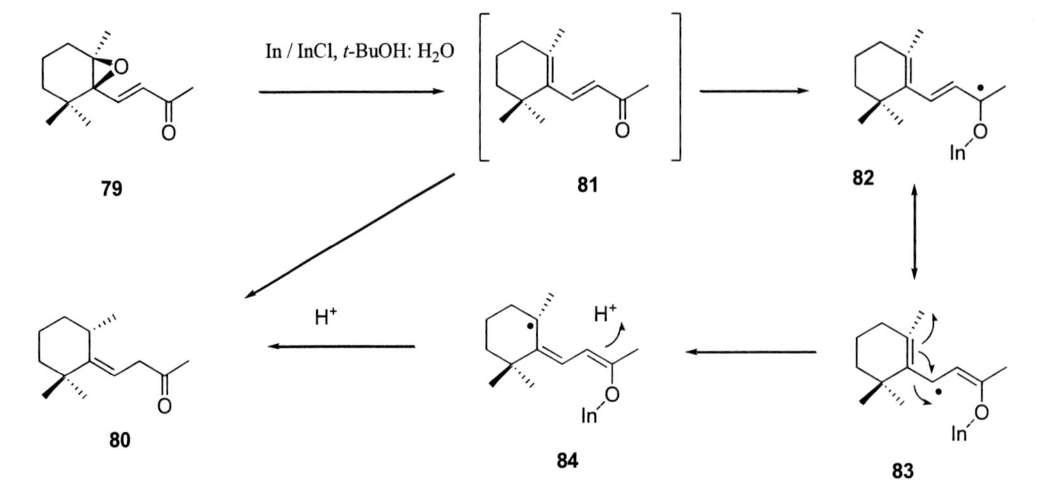

In / InCl, *t*-BuOH: H_2O

79 81 82

H^+ H^+

80 84 83

84% (*E*:*Z* = 1:0.8)

Scheme 41. Plausible mechanism for the Indium-mediated deoxygenation of epoxides in aqueous media.

The formation of alkene 80 can be explained via the reduction of the expected dienone intermediate 81 by indium metal (Scheme 41).[103]

3.9. Metal-mediated Mannich Type Reactions in Water

A convenient method that allows the easy, mild and efficient synthesis of a large number of differently substituted propargylamines characterized by high atom economy, low environmental impact and use of non-toxic solvents and reagents has been developed by Bieber et. al.[104] The possibility to react unprotected primary amines and alkynols allows further reactions on these sites. As shown in Scheme 42, the reaction involves a three component procedure between terminal alkynes 85, formaldehyde and secondary amines.

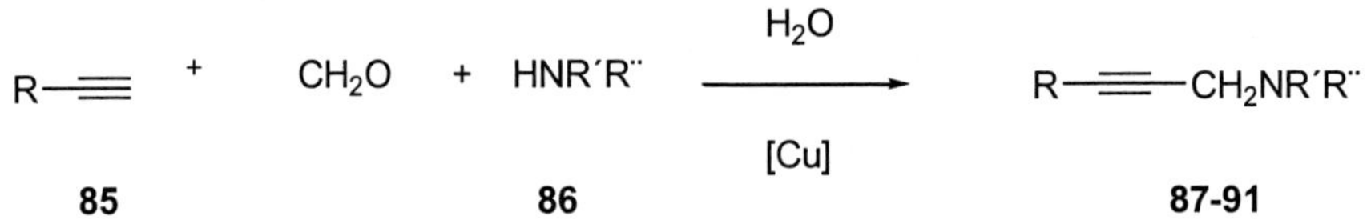

$$R{-}C{\equiv}CH + CH_2O + HNR'R'' \xrightarrow[\text{[Cu]}]{H_2O} R{-}C{\equiv}C{-}CH_2NR'R''$$

85 **86** **87-91**

Reactants	Product	Yield
R= CH_2OH R′, R′′=$(CH_2)_4$	87	70%
R= CH_2OH R′,R′′ =$(CH_2)_5$	88	96%
R= CMe_2OH R′, R′′ =Et, Et	89	87%
R= CMe_2OH R′,R′′= $(CH_2)_4$	90	80%
R=CMe_2OH R′,R′′= $(CH_2)_5$	91	82%

Scheme 42. Mannich-type reaction for the synthesis of propargylamines in water, mediated by CuI.

Thus, it was possible to obtain propargylamines 87-91 in pure water. Other propargylamines can also be obtained in DMSO-water mixtures, in yields ranging from 70 to quantitative.

In view of the need for a copper catalyst and the exceptionally mild reaction conditions, a radical intermediate should also be considered. A radical addition of phenyl, acetyl and alkyl radicals to iminium ions has been postulated before in the Zn-Barbier reaction[105] and in the $TiCl_3$ -promoted reaction with diazonium salts or tert -butyl hydroperoxide.[106]

3.10.Metal-Mediated Radical Cyclizations in Water

Strategies involving tandem radical reactions or radical annulations offer the advantage of multiple carbon-carbon bond formations in a single operation. Thus a number of extensive investigations to this effect were reported in recent years.[107] However, the aqueous-medium tandem construction of carbon-carbon bonds has not been widely explored, and therefore, tandem radical reactions in aqueous media have been a subject of recent interest.[108]

Naito et al.[109] investigated the indium-mediated reaction of substrates having two different radical acceptors. At first, the tandem addition-cyclization-trap reaction (ACTR) of substrate 92 having acrylate and olefin moieties was examined (eq 15).To a suspension of 92 in water were added *i*-PrI (2 x 5 equiv) and indium (2 equiv), and then the reaction mixture was stirred at 20 °C for 2 h.The reaction proceeded smoothly affording the desired cyclic product 93a in 63% yield as a *trans/cis* mixture in 3:2.1 ratio, along with 13% yield of the addition product 94a.

RI, In (15)

$CHPh_2$ $CHPh_2$ $CHPh_2$

92 **93a-c** **94a-c**

a: R =*i*-Pr, **b:** R =*c*-Pentyl, **c**: R=*t-Bu*

The preferential formation of cyclic products 93 a-c could be explained by a radical mechanism (Scheme 43).

The indium-mediated reaction was initiated by SET to RI with generation of an alkyl radical which then attacked the electrophilic acrylate moiety of 92 to form the carbonyl-stabilized radical A (path A, Scheme 43). The cyclic products 93a-c were obtained via intramolecular reaction of radical A with the olefin moiety followed by iodine atom-transfer reaction from RI to the intermediate primary radical B. Although there are many examples of anions adding to isolated double bonds, these reactions have been limited to lithium-mediated reactions.[110]

Sulfonamides (electron-deficient alkenes) such as 95 (eq 16) have also been examined in indium-mediated tandem radical reactions.[109] As expected, sulfonamide 95 exhibited good reactivity to afford moderate and good yields of the desired cyclic products 96a-c without the formation of other by-products.

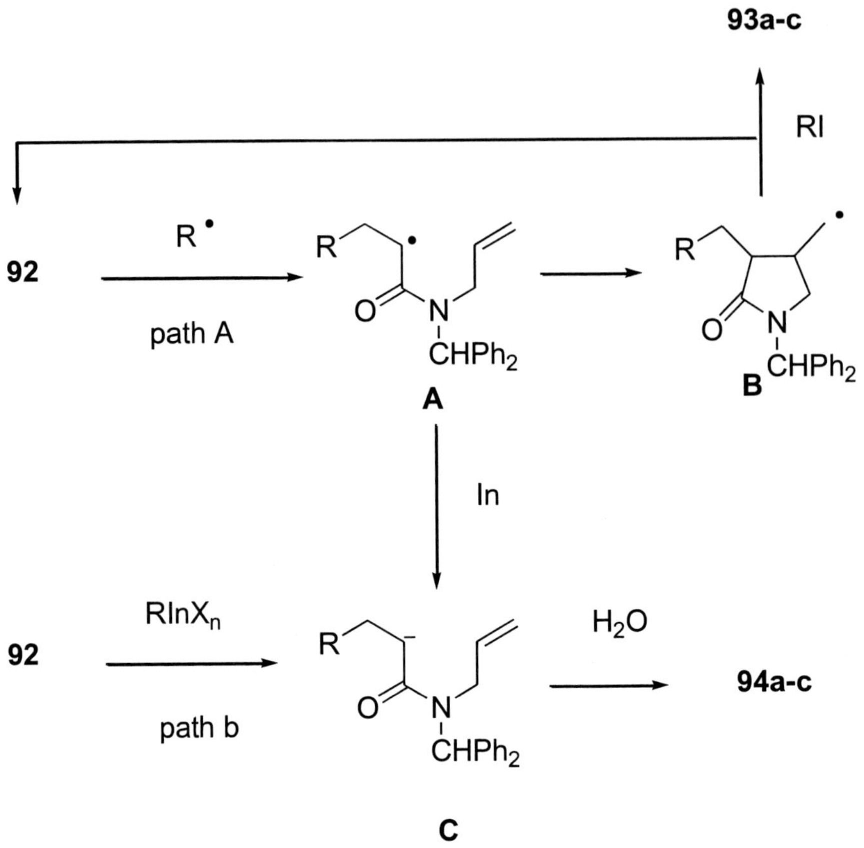

Scheme 43. Indium-mediated tandem addition-cyclization-trap reaction.

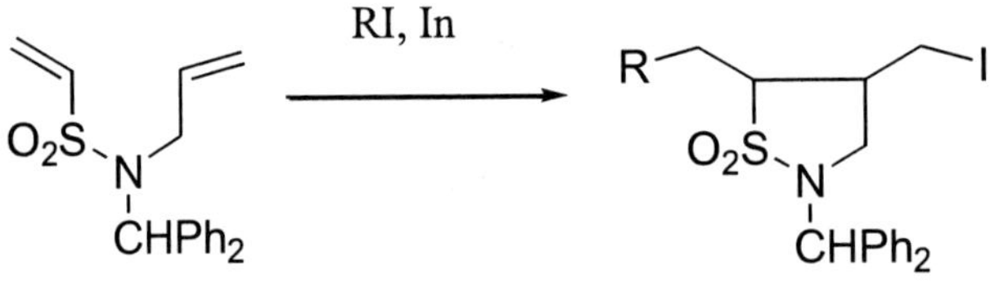

(16)

a: R = *i*-Pr, **b :** R = *c*-pentyl, **c** : R = *t*-Bu

The indium-mediated tandem reaction of 95 with *i*-PrI in water afforded selectively the cyclic product 96 a in 81% yield as a *trans/cis* mixture in 1:1.4 ratio, with no detection of simple addition products. Thus indium was found to be a highly promising radical initiator in aqueous media. Hydrazones connected with a vinyl sulfonamide group such as 97 have also been investigated in tandem-radical addition-cyclization reaction of imines (eq 17).

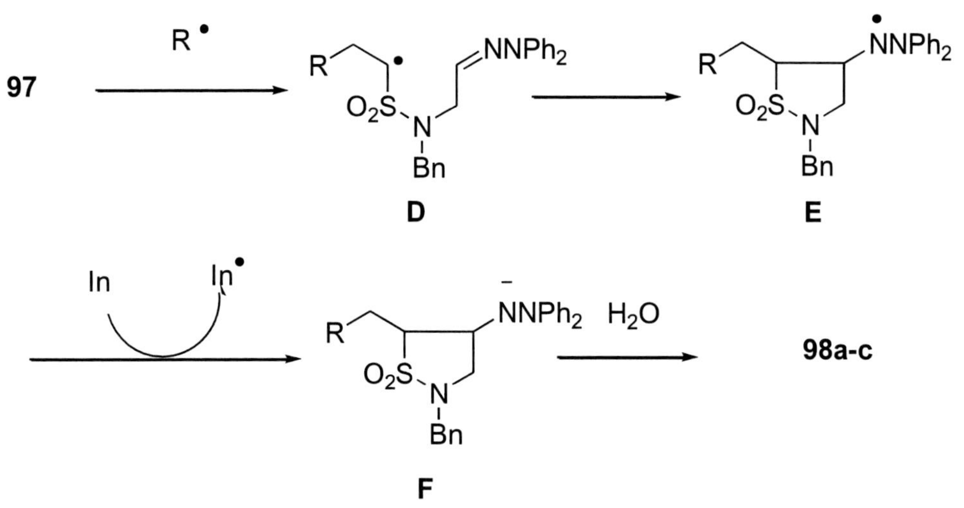

Scheme 44. Proposed reaction pathway for the indium-mediated radical-addition-cyclization of hydrazones in water.

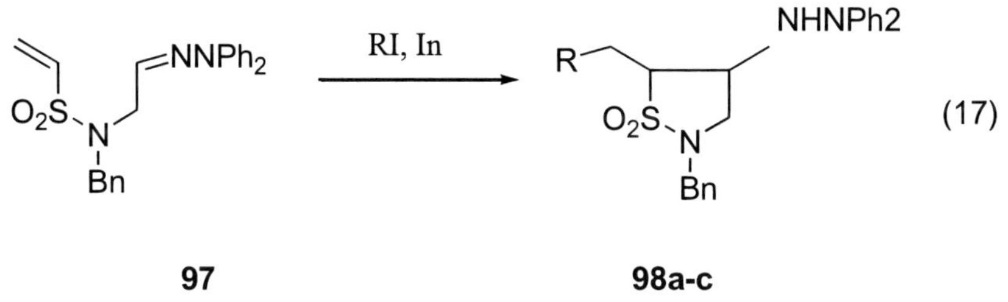

(17)

97 **98a-c**

a: R = *i*-Pr, **b :** R = *c*-pentyl, **c** : R = *t*-Bu

The radical reaction of 97 does not proceed via a catalytic radical cycle such as iodine atom-transfer; thus a large amount of indium was required for a successful reaction to take place (Scheme44).

The tandem reaction of hydrazone 97 with isopropyl radical was carried out in water-methanol for 5 h by using *i*-PrI (2 x 5 equiv) and indium (10 equiv). As expected, the reaction proceeded smoothly to render the *iso*propylated product 98 a in 93% yield as a *trans/cis* mixture in a 1:1.2 ratio, without the formation of the simple addition product. The biphasic reaction of 97 in water-CH_2Cl_2 also proceeded effectively to afford 94% yield of 98a. A cyclopentyl radical and a bulky *tert*-butyl radical worked well to give the cyclic product 98b and 98c in 86% and 42% yields, respectively. The stereochemical outcome for the cyclization of hydrazone 97 is almost the same as that in the case of olefin 95 (eq 16) in which *cis* products were the major products.

An intrinsic drawback of indium is the need for almost stoichiometric amounts of this relatively expensive metal, or as seen above, a large excess. In response to the cost factor, various combinations containing catalytic amounts of indium and a secondary cheaper metal (such as Al, Zn, Sn, or Mn) have been developed, but these protocols are limited to allylation of carbonyl compounds

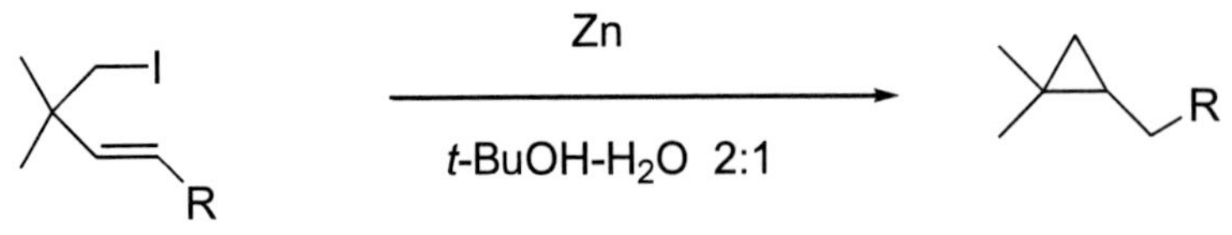

Zn

t-BuOH-H_2O 2:1

99 (R = CO_2Bn, COPh
CONR′$_2$, SO_2Ph)

100 (62-84%)

Scheme 45. Formation of cyclopropanes via rare 3-exo-trig cyclizations of β-iodo-alkenyl substrates.

Togo reported the Zn-mediated formation of cyclopropanes 100 via rare 3-exo-trig cyclizations of substrates 99 (Scheme 45).[111]

Geminal dialkyl substitution was required. The zinc presumably functions as a single-electron reductant both in forming the initial alkyl radical and in reducing the incipient α-carbonyl or sulfonyl radical faster than the potential fragmentation can occur. This method compares well to similar reactions promoted by SmI_2, as the latter reagent is air-sensitive.

Mangeney performed a regio- and stereoselective 6-exo intramolecular RCA of 1,4-dihydropyridine 101 under Luche conditions [88-91] (Scheme 46).[112, 113]

The product was transformed into both (K)-lupinine and (C)-epi-lupinine 102.

In 1990, Marshall and co-workers published that, upon treatment with $AgNO_3$ or $AgBF_4$, allenals (103, R^1 = H) and allenones (103, R^1 = CH_3) alkyl) afford furans (104) (Scheme 47).[114|] These authors have developed this methodology and published many applications, always trying to improve the experimental conditions. The best set of conditions can be $AgNO_3$ / $CaCO_3$/ acetone / water[115] or 10% $AgNO_3$ on silica gel and hexane.[116] The following reaction pathway has been proposed: the process is initiated by coordination of Ag(I) with the allenyl π-system. Attack by the carbonyl oxygen would lead to the oxo-cation 105. Ensuing proton loss from cation 105 would result in the Ag(I)-furan intermediate 106. This could undergo direct protonolysis with loss of Ag(I) to afford furan product 104. Deuterium incorporation experiments support this mechanism.[117] It is important to point out that the choice of the transition-metal catalyst is crucial to form this kind of substituted furans, because, under similar conditions, allenic ketones delivered different products when catalyzed by Pd(II) or Hg(II).[118]

Among various types of radical reactions, radical cyclizations in the *5-exo-trig* and *6-exo-trig* manners are the most powerful and versatile methods for the construction of five- and six-membered ring systems. Recently, two-atom carbocyclic enlargement based on an indium-mediated Barbier-type reaction in water was reported.[119] A series of different ring-sized α-iodomethyl cyclic β-keto esters in a mixture of *tert*-amyl alcohol (TAA) and water was examined (eq 18).[120]

The same ring expansions of α-iodomethyl cyclic β-keto esters with zinc powder, instead of indium powder were examined in a mixture of TAA (*tert*-amyl alcohol) and water (1:1).In these cases, the yields were surprisingly much increased. The presence of water in these reactions was found to be essential for an effective and high yielding of the ring-expanded products. Moreover, both bromomethyl and iodomethyl cyclicβ-keto esters can be used for the ring-expansion reaction to provide 6-membered, 8-membered, 9-membered, 13-membered, and 16-membered products in yields ranging from 60 % up to 87%.[120] The reactions were extremely clean and operationally simple for isolation of products.A plausible reaction mechanism is depicted in Scheme 48.

The reaction is initiated by the first single electron transfer from metal (indium or zinc) to a α-halomethyl cyclic β-keto ester to form the corresponding methyl radical derivative, followed by *3-exo-trig* cyclization and its β-cleavage.In this mechanism, only ring-expansion products are formed since there is no hydrogen donor such as a tin hydride or silicon hydride.[120]

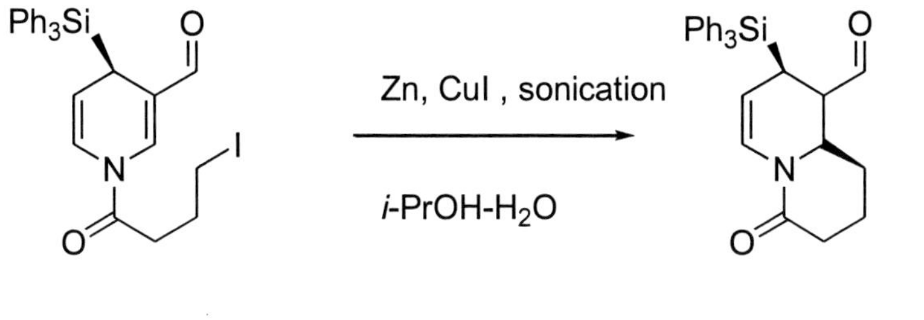

101 **102** (50%)

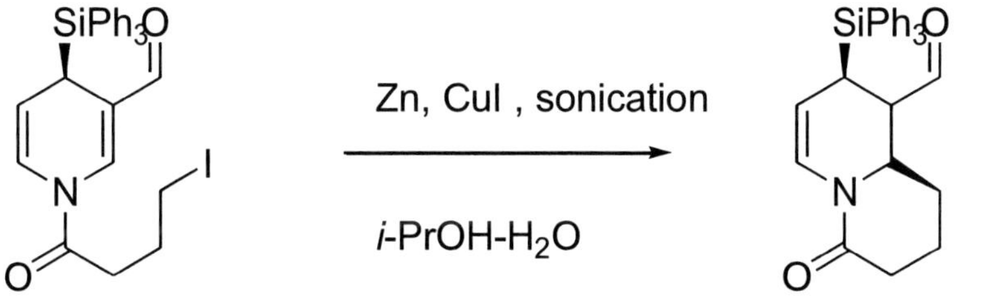

101 **102** (50%)

Scheme 46.6-exo intramolecular RCA of 1,4-dihydropyridine derivatives under Luche conditions.

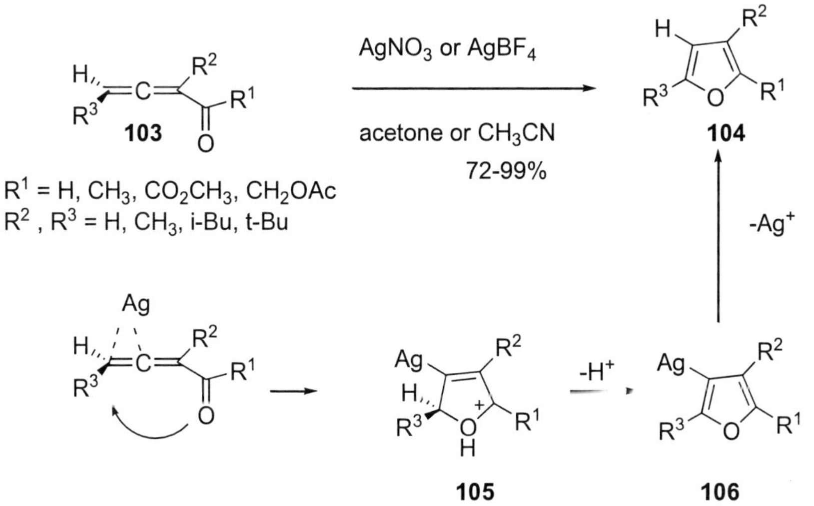

Scheme 47. Proposed reaction mechanism for the Ag-mediated synthesis of furans.

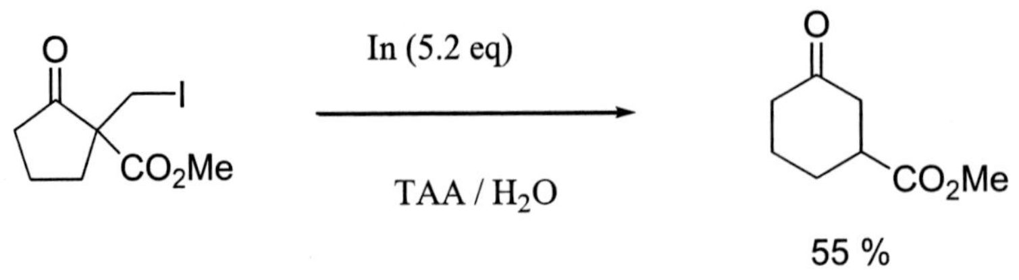
O
I
CO_2Me
In (5.2 eq)
TAA / H_2O
O
CO_2Me
55 %

(18)

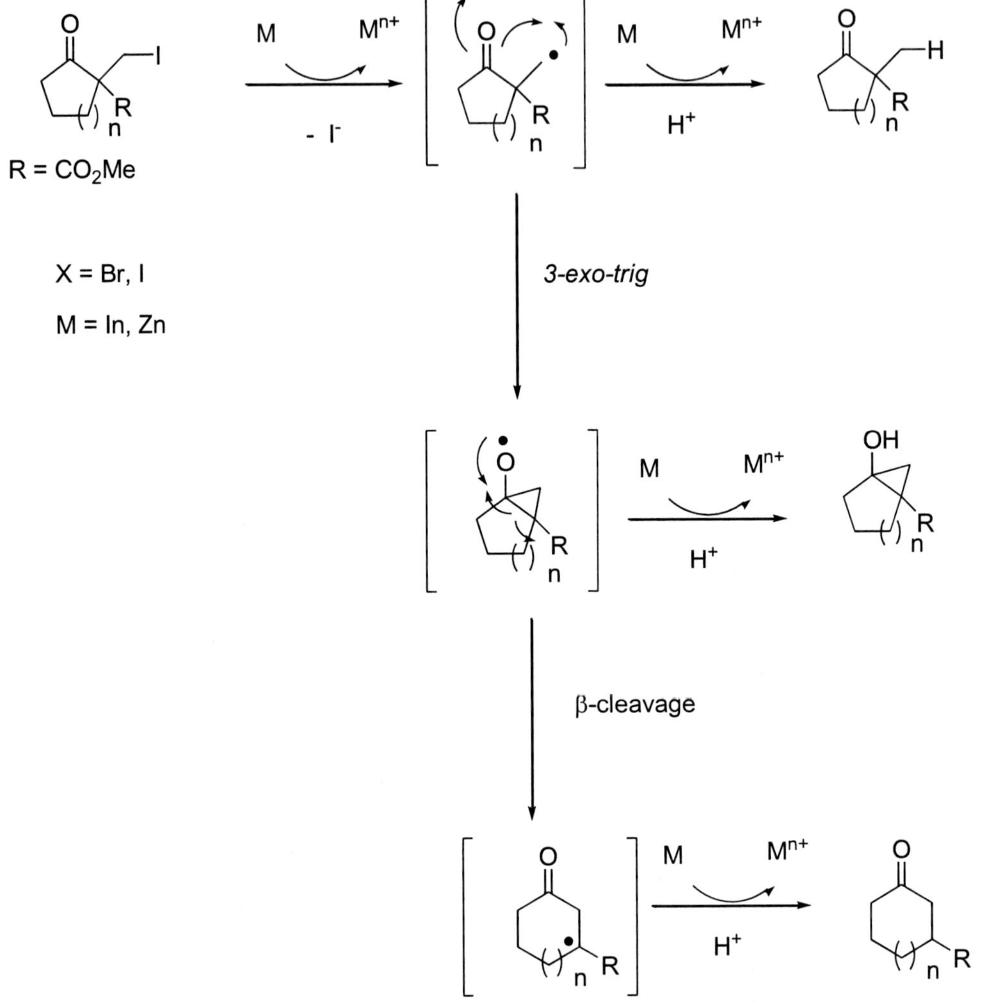
O
I
R
n
R = CO_2Me
X = Br, I
M = In, Zn
M
M^{n+}
- I^-
O
R
n
M
M^{n+}
H^+
O
H
R
n
3-exo-trig
O
R
n
M
M^{n+}
H^+
OH
R
n
β-cleavage
O
n
R
M
M^{n+}
H^+
O
n
R

Scheme 48. Proposed reaction mechanism for the metal-mediated ring expansion of α-halomethyl cyclic β-keto esters.

Recently, Li and Cao[121] demonstrated the efficiency of *p*-methoxybenzenediazonium tetrafluoroborate-$TiCl_3$ couple in promoting / initiating the halogen atom-transfer radical addition (ATRA) reaction and the iodine atom-transferradical cyclization(ATRC) reactionas an entry to heterocycles such as lactones and lactams, as shown in Table 15.

The active species in the *p*-methoxybenzenediazonium tetrafluoroborate / $TiCl_3$ –chain process are the aryl radicals.Initiation relies on the fact that the aryl radical is generated selectively, and it abstracts an iodine atom from the substrate rather than adding to the C=C bond.This is because the rate constant for the iodine atom abstraction ofa phenyl radical from an alkyl iodide is close to the diffusion-controlled limit (>10^9 $M^{-1}s^{-1}$) which is about 100 times faster than the rate of phenyl radical addition to a monosubstituted alkene (*ca.* 3 x 10^7 $M^{-1}s^{-1}$).More importantly, the rate constant for the iodine atom-transfer from the substrate to the adduct radicalis around 2.7 x 10^7 $M^{-1}s^{-1}$, at least one order of magnitudehigher than that for the trapping of the adduct radical by the diazonium ion *p*-methoxybenzenediazonium tetrafluoroborate.This allows the iodine atom-transfer chain process to evolve smoothly without the intervention of a termination step.[121]

The reaction of trialkylboranes with 1,4-benzoquinones to give 2-alkylhydroquinones in quantitative yields was the first reaction of this type occurring without the assistance of a metal mediator.[122] The reaction is inhibited by a radical scavenger such as galvinoxyl and iodine.[123]

Togo et al [124] undertook the In-mediated cyclopropanation of 2,2-disubstituted 1,3-diiodopropanes and 1,3-dibromopropanes in dioxane solution of 20% water and THF solution of 20% water. However, the cyclopropanation of 2,2-disubstituted 1,3-dichloropropanes with indium powder could only be accomplished in ionic liquids.

As regards the reaction mechanism, when 2,2-disubstituted-1,3-diiodopropanes were treated with In powder (2.4 equiv) in THF solution of 20% H_2O and dioxane solution of 20% H_2O,only 1,1-disubstituted cyclopropanes were obtained in high yields without the formation of 2,2-disubstituted 1-iodopropanes and 2,2-disubstituted propanes, which could be formed through the reactions of the corresponding carbanions with H_2O. This result suggests that the present cyclization reaction of 2,2-disubstituted 1,3-dihalopropane with In powder may proceed in the radical *3-exo -tet* manner, as shown in Scheme 49.

A rapid stereoselective route to the *trans* hydrindane ring system was achieved by Khan et al using tin-, indium-, and ruthenium-based reagents starting from tetrabromonorbornyl derivatives.[125]

Table 15. *p*-Methoxybenzenediazonium Tetrafluoroborate/ $TiCl_3$ Mediated Iodine Atom-transfer Radical Cyclization

107 → **108**

p-$AnN_2^+BF_4^-$ / $TiCl_3$

EtOH/H_2O, rt , 3 h

R^1	R^2	Product (yield, %)
H	allyl	99
II	Ts	96
H	Ms	86
H	Me	40
Me	allyl	96
Me	Ts	91

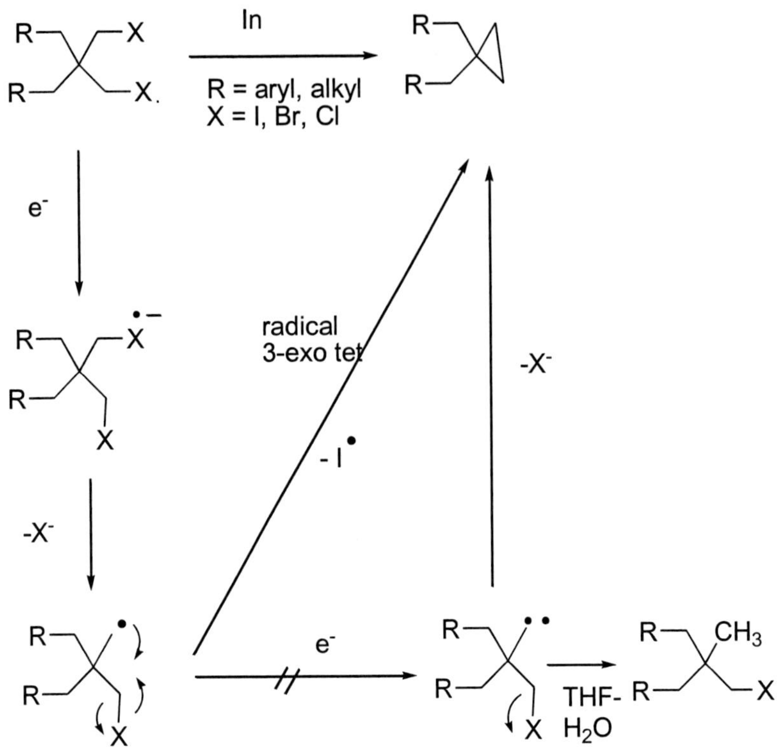

Scheme 49. Proposed reaction mechanism for the cyclpropanation of 2,2-disubstituted-1,3-dihalopropanes.

Kim and collaborators[126] reported on the syntheses of 2,1-benzisoxazol derivatives in aqueous media employing indium and 2-nitrobenzaldehydes, in the presence of 2-bromo-2-nitro-propane (BNP) in a methanol:water (v/v : 1:2) mixture (Scheme 50).

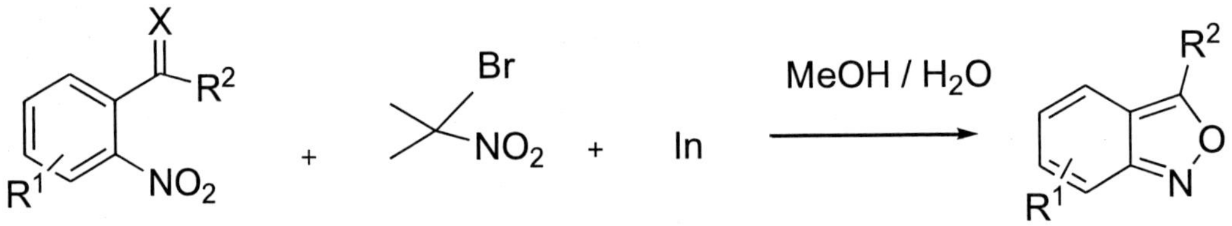

Scheme 50. Reactions of 2-bromo-2-nitropropane with nitrobenzaldehydes in water.

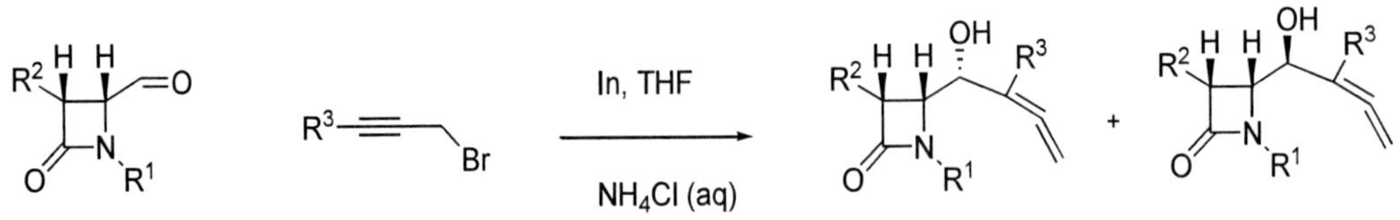

109 **110** ***anti*-110**

Scheme 51. Synthesis of 2-azetidinone-tethered intermediates.

Table 16. Regio and stereoselective allenylation of 4-oxaazetidine-2-carbaldehydes 109 in water

Aldehyde	R[1]	R[2]	R[3]	Product	*Syn/anti*	Yield (%)
(+)-109a	Allyl	MeO	Me	(+)-110a	95:5	75
(+)-109a	allyl	MeO	Ph	(+)-110b	100:0	64
(+)-109b	3-methyl-but-2-enyl	MeO	Me	(+)-110c	85:15	68
(+)-109b	3-methyl-but-2-enyl	MeO	Ph	(+)-110d	100:0	61
(+)-109c	3-methyl-but-2-enyl	MeO	Ph	(+)-110e	100:0	51
(+)-109d	methallyl	PhO	Me	(+)-110f	90:10	79
(+)-109d	Methallyl	PhO	Ph	(+)-110g	100:0	58
(+)-109e	methallyl	Vinyl	Me	(+)-110h	10:90	60
(+)-109e	PMP	Vinyl	Ph	(+)-110i	70:30	89
(+)-109f	PMP	Isopropyl	Me	(+)-110j	10:90	73
(+)-109f	PMP	Isopropyl	Ph	(+)-110k	65:35	63

Usually, the reaction in an aqueous solution completed much faster than in methanol. The optimum condition was obtained when 2 equiv. of BNP and 5 equiv. of indium were applied in methanol/water (v:v=1:2) with 2-nitrobenzaldehyde at 50°Cand the reaction time was diminished dramatically compared to the previous zinc-mediated reaction. It produced almost a quantitative yield of desired 2,1-benzisoxazole within 10 min. The role of BNP is to be an electron acceptor due to its low-lying antibonding π-orbital and the utility of BNP has been described by Russell et al.[127]. Furthermore, addition of di-*tert*-butyl nitroxide or *m*-dinitrobenzene has shown strong inhibitory effects. Di-*tert*-butyl nitroxide is a known radical scavenger and*m*-dinitrobenzene is known to quench radical anion intermediates. The reactions of nitrobenzaldehyde /BNP / indium in MeOH in the presence of 10 mol% of di-*tert*-butyl nitroxide or *m*-dinitrobenzene had shown about 1 h initial retardation for each and completion time.

Alcaides, Almendros and collaborators have performed the synthesis of 2-azetidinone allenol derivatives in high yields through the indium-mediated reaction in water.[128]

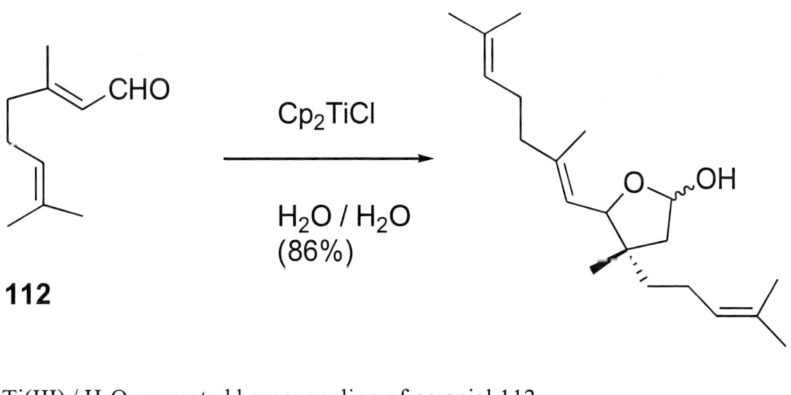

Scheme 52. Ti(III) / H_2O-promoted homocoupling of geranial 112.

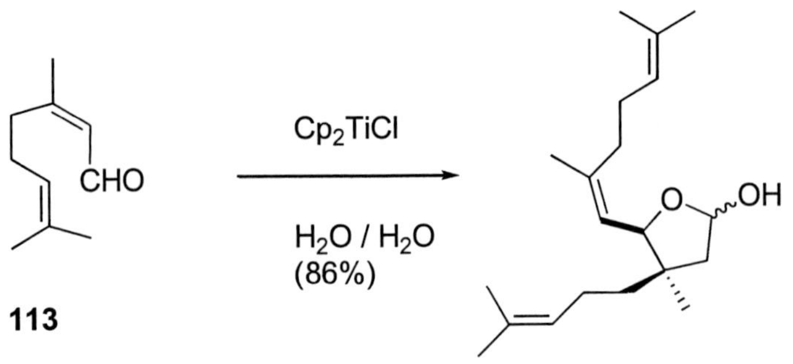

$$\xrightarrow[\substack{H_2O / H_2O \\ (86\%)}]{Cp_2TiCl}$$

113

Scheme 53. Ti(III) / H_2O-promoted homocoupling of neral 113.

2-Azetidinone-tethered allenols 110 a–k (Table 16) were obtained by a metal-mediated Barbier-type carbonylallenylation reaction of β-lactam aldehydes 109 a-f in aqueous media by using our previously described methodologies (Scheme 51, Table 16).

It is observed that the yields of lactam 110 range from 50 to 89% yields, with the prevalence of the *syn* isomer. In several cases, the *syn* isomer is obtained exclusively.

In the presence of water, titanocene(III) complexes (111) were reported[129] to promote a stereoselective carbon-carbon bond-forming reaction that provides γ-lactols by radical coupling between aldehydes and conjugated alkenals. The method is useful for both intermolecular reactions and cyclizations. The relative stereochemistry of the products can be predicted with confidence with the aid of model Ti-coordinated intermediates. The procedure can be carried out enantioselectively using chiral titanocene catalysts.

Thus, the Ti(III) homocoupling of geranial was achieved (Scheme 52), and that of neral 113 was obtained according to these techniques (Scheme 53).

3.11. Pinacol and other Coupling Reactions in Water

Since the first report of the reaction of acetone with sodium in 1858,[130] various low-valent metals such as Al,[131] Sm,[132] V,[133] Mg,[134] Zn,[135] Mn,[136] Sn,[137] Ti,[138], Ce,[139] Te,[140] U,[141] Cr,[142] Ga,][143] and In[144] have been used to promote this reductive coupling reaction. Among these methods, some require absolutely anhydrous system under inert atmosphere, and some reagents and solvents are costly, moisture-sensitive, and toxic. In order to find environmental friendly condition, it is very attractive to develop a new convenient method for the pinacol coupling by utilizing less toxic reagents and solvents. During past decades, great efforts have been devoted by chemists to explore environmentally benign systems for pinacol reaction. Different catalysts / co-catalysts in aqueous media including $TiCl_3$, VCl_3 / Al, Mn / HOAc, Al / MF, etc. have been reported with promising results.

As is well known the reductive (pinacol) coupling of carbonyl compounds is a useful method for the creation of carbon-carbon bonds with 1,2-difunctionality (eq 19). Although detailed mechanistic studies of pinacol coupling are lacking, the reactions are generally considered to involve the generation and reaction of the substrate ketyl (radical anion) with either the neutral substrate or another ketyl species.

$$2\ RC(O)R' + 2e^- + 2H^+ \longrightarrow \text{pinacol} \quad (19)$$

The pinacol-coupling reaction (*vide supra*) is a fundamental reaction in organic chemistry. The pinacol coupling reaction in water mediated by Ti(III) and other metals such as Zn-Cu have also been found to promote pinacol formation under ultrasonic radiation conditions in aqueous acetone. When benzaldehyde was reacted with manganese in the presence of a catalytic amount of acetic acid in water, the corresponding pinacol coupling product was obtained smoothly. Other aryl aldehydes were coupled similarly.On the other hand, aryl and aliphatic ketones appeared to be inert under the same reaction conditions, and only the reduced product was obtained with aliphatic aldehydes.[50]

Pan, Wu, and collaborators reported on the pinacol coupling of aromatic aldehydes and ketones using $InCl_3$ / Al in aqueous media.[145]

Under the optimized conditions (Table 17, entry 1), various substituted benzophenones were investigated to give a series of 1,1,2,2 -tetraaryl substituted -diols (115 a-h) (Table 17).

From these results, the authors found that 115 a-h were obtained in excellent yields without formation of by-product due to reduction of the carbonyls to the corresponding alcohols (Table 17, entries a–c). Benzophenones bearing electron-donating groups (114 d,e) *para* to ketones were reduced to the corresponding pinacols in the moderate yields even after 11 h (Table 17, entries d and e). Unfortunately, the reactions of benzophenones bearing electron-withdrawing groups (114 f,g) *para* to ketones with $InCl_3$ / Al reagent at 80 °C for 3 h gave miscellaneous products, in which most starting materials had been consumed (Table 17, entries f and g).

Table 17. Pinacol coupling of substituted benzophenones by using $InCl_3$ / Al

$$Ar_1C(O)Ar_2\ (\textbf{114a-h}) \xrightarrow[\text{EtOH-}H_2O,\ 80\ ^\circ C]{InCl_3 / Al,\ NH_4Cl} \text{HO(}Ar_1\text{)(}Ar_2\text{)C–C(}Ar_2\text{)(}Ar_1\text{)OH}\ (\textbf{115a-g})$$

Entry	Ar_1	Ar_2	Time (h)	Yield of 115 (%)
A	Ph	Ph	5	115a
B	$4\text{-}CH_3C_6H_4$	Ph	5	115b
C	4-F C_6H_4	4-F C_6H_4	5	115c
D	4-MeO C_6H_4	4-F C_6H_4	11	115d
E	4-MeO C_6H_4	Ph	11	115e
F	4-Cl C_6H_4	Ph	3	Mixture
G	$4\text{-}C_6H_5$ C_6H_4	Ph	3	Mixture
F	4-Cl C_6H_4	Ph	5	115f
G	$4\text{-}C_6H_5$ C_6H_4	Ph	5	115g
H	4-Cl C_6H_4	Ph	5	-

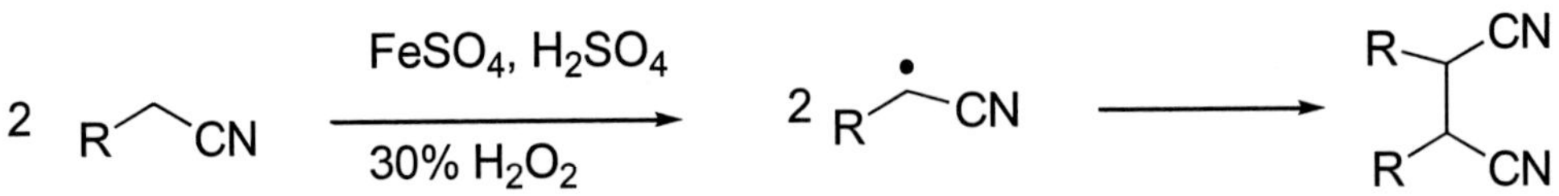

$$2\ \mathrm{R}\frown\mathrm{CN} \xrightarrow[30\%\ H_2O_2]{FeSO_4,\ H_2SO_4} 2\ \mathrm{R}\overset{\bullet}{\frown}\mathrm{CN} \longrightarrow \mathrm{R(CN)CH{-}CH(CN)R}$$

Scheme 54. Pinacol coupling by Fenton reagent..

On the other hand, lower temperature gave pinacols in moderate yields (Table 17, entries f and g) along with a small quantity of the corresponding alcohols. Unexpectedly, benzophenone bearing a chlorine group (114h) ortho to ketone only gave the corresponding alcohol as the main product.

Coupling of water-soluble acetonitrile derivatives has been developed by Holtz, Pinhas, and coworkers using a Fenton′s reagent (Scheme 54).[146]

For this reaction, it does not matter if a free radical or an iron oxo complex is formed. What matters is that the Fenton chemistry generates a radical[147, 148] or radical-equivalent[149] that can remove a hydrogen atom from the alkyl chain of the alkyl nitrile, and then two of these "alkyl radicals" can couple. This type of coupling reaction was first mentioned about 50 years ago,[150] but unfortunately, the yields were low and the regiochemistry was not investigated in detail. In this paper, the authors discuss the coupling of acetonitrile and other water-soluble alkyl nitriles.[151] Reaction yields are improved, and in addition, the authors have investigated the regiochemistry of the coupling reaction. This regiochemistry not only is important from a synthetic perspective, but it tells the energetics of hydrogen-atom removal from various positions on the alkyl chain.

Since the cyanomethyl radical is the only radical that can be obtained from acetonitrile, succinonitrile is the only dinitrile product. However, propionitrile can form two radicals: a resonance-stabilized secondary radical (116) formed by abstraction of an α-hydrogen atom or a primary radical (117) by abstraction of α-hydrogen atom. The products formed by all possible combinations of these two radicals are illustrated in Scheme 55.

Stereoisomers of 2,3-dimethylsuccinonitrile (119) (*d,l*-pair and a *meso* compound) are formed by the coupling of two molecules of the secondary radical (116), while two primary radicals (117) combine to form adiponitrile (120). Cross-coupling of these two radicals produces 2-methylglutamonitrile (118). The observed ratio for dinitrile isomers 118:119:120 is 50:45:5. After correcting for the number of abstractable hydrogens at each position, it was determined that the resonance-stabilized secondary radical (116) and the primary radical (117) form in a 3.5:1 ratio rather than the statistical ratio of 2:3.

*Iso*butyronitrile can also form two radicals: a resonance-stabilized tertiary radical (121) formed by abstraction of an α-hydrogen atom or a primary radical (122) by abstraction of α-hydrogen atom. The products formed by all possible combinations of these two radicals are illustrated in Scheme 56.

Stereoisomers of 2,5-dimethyladiponitrile (125) (*d,l*-pair and a *meso* compound) are formed by the coupling of two molecules of the primary radical (122), while two tertiary radicals (121) combine to form 2,2,3,3-tetramethysuccinonitrile (124). Cross-coupling of these two radicals produces 2,2,4-trimethylglutamonitrile (123).

The observed ratio for dinitrile isomers 123:124:125 is 26:29:45. After correcting for the number of abstractable hydrogens at each position, it was determined that the resonance-stabilized tertiary radical (122) and the primary radical (121) form in a 4.3:1 ratio rather than the statistical ratio of 1:6.

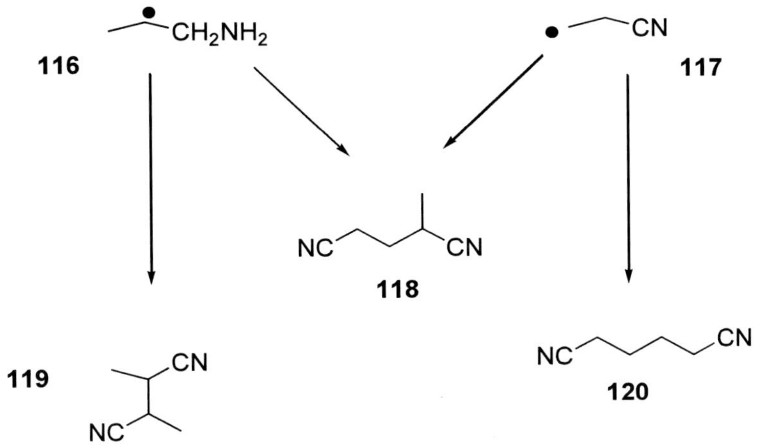

Scheme 55. Possible combination products from pinacol coupling of propionitrile.

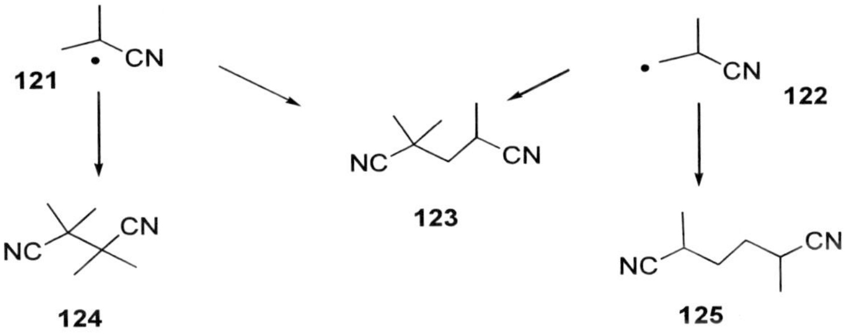

Scheme 56. Possible combination products from pinacol coupling of *iso*butyronitrile.

Since the cyanomethyl radical is easily reduced to the cyano-methyl anion by iron(II), with a stoichiometric amount of iron, the lower yield is not surprising. To improve the yield, the concentration of iron(II) should be kept low. Thus for large scale preparation of succinonitrile via Fenton's reagent, the reaction should be catalytic in iron(II), keeping its concentration as low as possible. Iron(0) is an attractive candidate as a reducing agent because iron (II) is the only product of the oxidation-reduction reaction.

Although the use of iron(0) creates a heterogeneous reaction which leads to greater variability in the product yield, the increased production of succinonitrile indicates iron(0) has an effect on the coupling reaction by reducing iron(III) to iron(II).

Loh and collaborators[152] have very recently reported an efficient pinacol cross-coupling reaction of aldehydes and α,β-unsaturated ketones using Zn / $InCl_3$ in aqueous media. The 1,2-diols were thus obtained in moderate to good yields, with up to 93.7 % diastereoselectivity (eq 20).

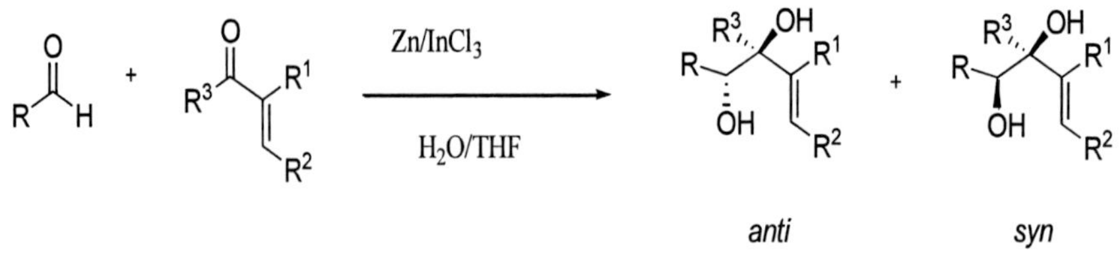

$RCHO$ + $R^3COC(R^1)=CHR^2$ $\xrightarrow[H_2O/THF]{Zn/InCl_3}$ *anti* + *syn* (20)

R = Ph $R^1 = R^2 = H, R^3 = CH_3$ 61:39 55%

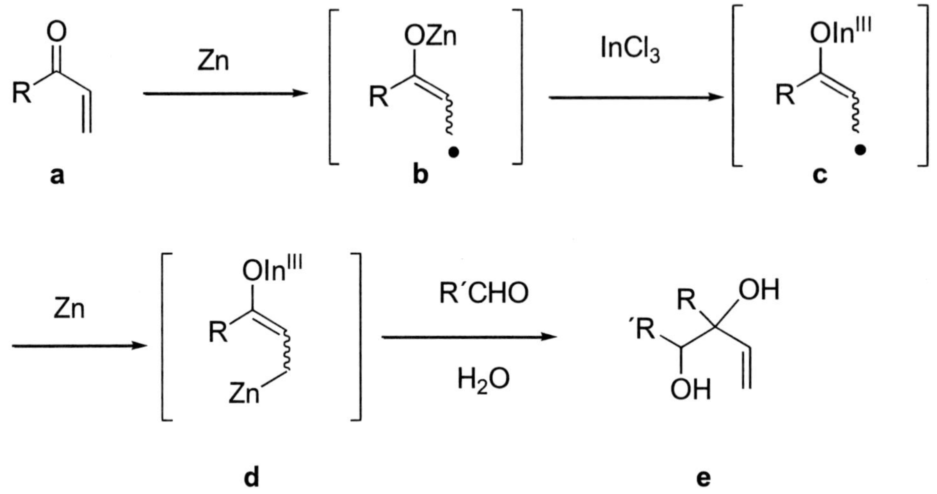

a $\xrightarrow{Zn}$ [**b**] $\xrightarrow{InCl_3}$ [**c**] $\xrightarrow{Zn}$ [**d**] $\xrightarrow[H_2O]{R'CHO}$ **e**

Scheme 57. Proposed mechanism for the pinacol cross coupling of enones and aldehydes.

A possible reaction mechanism is shown in Scheme 57.The reaction is initiated by a single electron transfer from zinc to the α,β-unsaturated ketone to form a radical enolate anion b.Fast trapping of the oxygen-metal bond in the radical enolate anion b by $InCl_3$ affords the γ-In(III)-substituted allylic radical c. The radical c is further reduced by zinc to furnish the corresponding allylic zinc species d. Finally, coupling of the γ-In(III)-substituted allylic zinc species d with an aldehyde followed by quenching of the resulting 1,2-diolate with water generates the desired product e.

Kalyanaman[153] reported a novel reductive coupling of aldimines brought about by indium to vicinal diamines as described in eq 21. The reaction occurs in aqueous ethanol. While NH_4I is not essential for the reaction, the reaction is accelerated by its presence. The reaction fails completely in CH_3CN, DMF and DMF containing small quantities of water. Indium used in the reaction was in the form of small rods made from a sheet of indium of about 1 mm thickness. It may be mentioned that optically pure derivatives of 127 have considerable potential in asymmetric synthesis as evidenced by recent publications.'

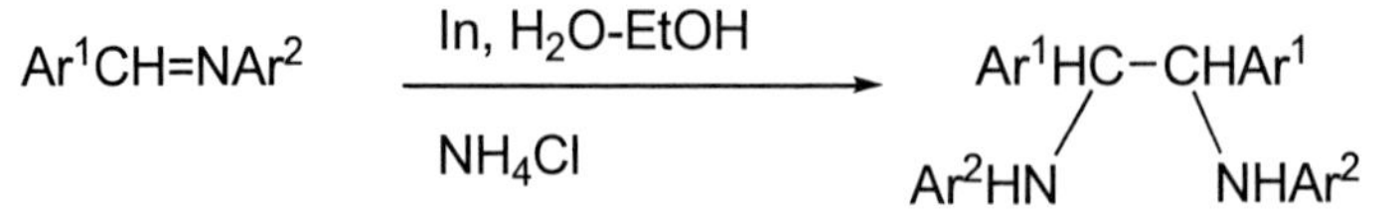

$$Ar^1CH{=}NAr^2 \xrightarrow[NH_4Cl]{In,\ H_2O\text{-}EtOH} Ar^1HC(NHAr^2){-}CH(NHAr^2)Ar^1 \quad (21)$$

126 **127** (meso + dl)

Some important aspects of the reaction are to be noted. The reaction occurs in aqueous medium and does not require exclusion of oxygen or anhydrous conditions as required by other reagents mentioned in the literature for effecting the same transformation as in eq 21. The complete absence of the side product in the crude reaction mixture, Ar'CH_2-NHAr (eq 21) (a resultant of unimolecular reduction as happens in some of the methods employed for this transformation') is noteworthy. Further, the reaction is brought about by indium rods and does not require fine indium powder. The product 127 is invariably a 1:1 mixture of *dl* and *meso* isomers indicating fast coupling of any putative radical intermediates. The reaction fails in the case of substrates, Ph(CH_3)C=NPh and PhCH=NCH_2Ph, showing selectivity for the coupling of aldimines obtained from aromatic aldehydes and aromatic amines.

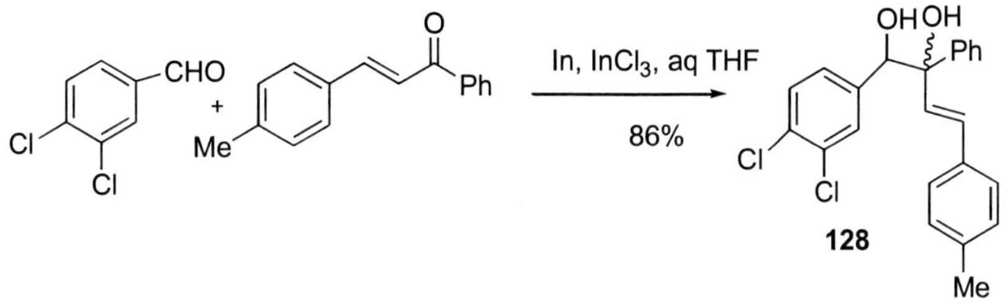

Scheme 58. Indium / indium trichloride –mediated pinacol cross coupling reaction of aldehydes and enones in aqueous media.

Nair et al. have reported on the indium / indium trichloride –mediated pinacol cross coupling of reaction of aldehydes and chalcones in aqueous media to obtain substituted but-3-ene-1,2-diols.[154]In an initial experiment, 3,4-dichlorobenzaldehyde was treated with 4-methylbenzylidene acetophenone in the presence of indium and indium trichloride in aqueous THF at room temperature to yield an isomeric mixture of 1-(3,4-dichlorophenyl)-2-phenyl-4-*p*-tolylbut- 3-ene-1,2-diols in 66% yield (Scheme 58).

Table 18. Synthesis of substituted but-3-ene-1,2-diols

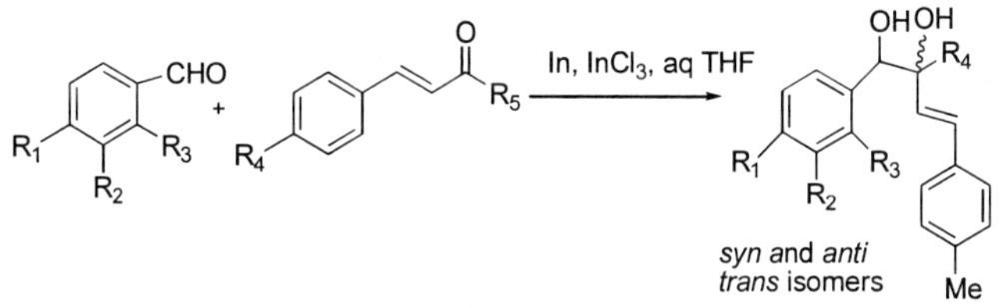

Entry	Substituents	products	Yield (%)	Ratio
1	R^1= R^2 =Cl, R^3= R^4= H, R^5=Ph	129	56	1:2
2	R^1= R^2 =Cl, R^3= R^4= H, R^5=Me	130	42	0:1
3	R^1 =Cl, R^2= R^3=H, R^4=Me, R^5=Ph	131	60	1:2
4	R^1=Cl, R^2 = R^3 = R^4=H, R^5=Ph	132	61	0:6.2
5	R^1=Cl, R^2 = R^3 = R^4=H, R^5=Me	133	46	0:1
6	R^1 = R^2 = H,R^3 = Cl, R^4=Me, R^5=Ph	134	56	0:6.1
7	R^1 = R^2 = R^4=H,R^3 =Cl, R^5=Ph	135	42	0:4.1
8	R^1 = R^2 = R^4=H, R^3 =Cl R^5=Me	136	56	0:1

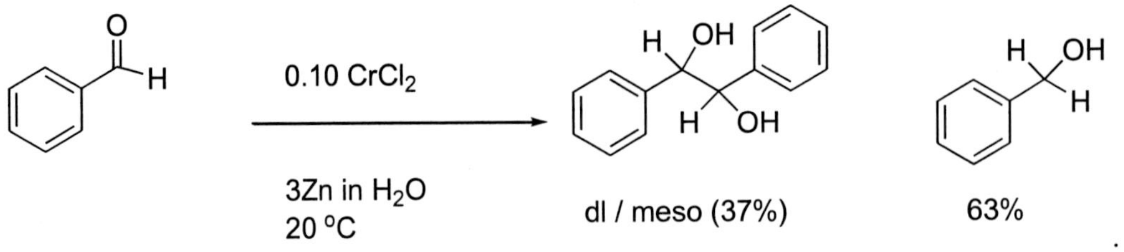

Figure 3.Pinacol coupling of benzaldehyde by $CrCl_2$ / Zn catalyst in water.

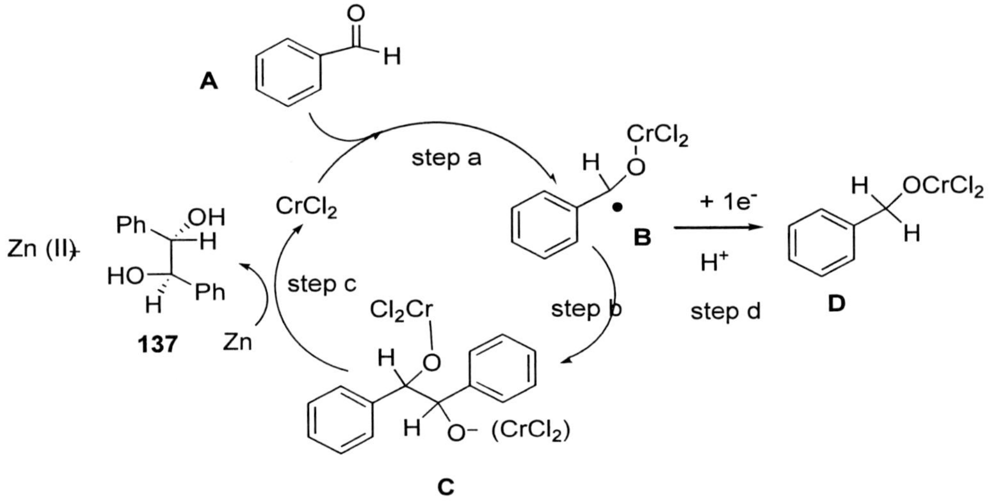

Figure 4. Catalytic cycle for Cr(II)-Zn-catalyzed pinacol reaction with competitive benzaldehyde reduction.

Similar results were obtained with other chalcones and aldehydes (Table 18). With benzylidene acetone and aldehydes, only one *trans*-isomer was formed whereas with other α.β-unsaturated ketones and aldehydes a mixture of *syn* and *anti trans* isomers was obtained.

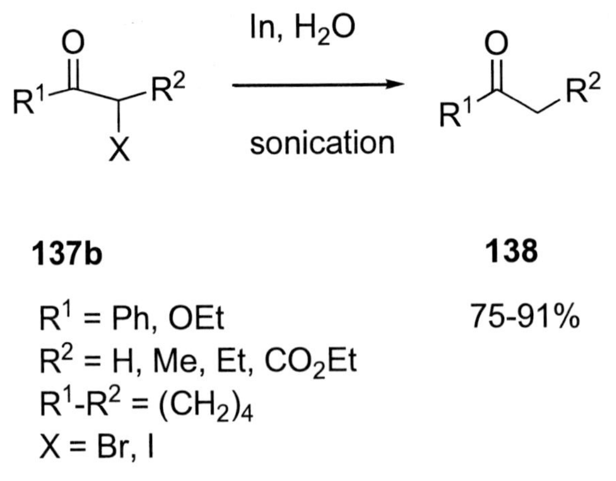

Scheme 59. Dehalogenation of α-halo carbonyl compounds.

Halterman and collaborators have achieved the pinacol coupling of benzaldehydes in water mediated by $CrCl_2$.[155]

The pinacol coupling of benzaldehyde (0.25 M or 1.25 M) in water was catalyzed by 5 – 25 mol % $CrCl_2$ in the presence of Zn-dust or Al-dust at 20ºC. In all cases at most 50% of the pinacol coupling product,1,2-diphenyl-1,2-ethanediol, was obtained with the major product, benzyl alcohol, being formed by a competitive 2e - reduction of the carbonyl. The *dl* -to *meso* -diastereoselectivity of the pinacol products ranged from 0.6:1 to nearly 1:1 (Figure 3).

According to the catalytic scheme depicted in Figure 4, Cr(II) can initially reduce the aldehyde (step a) to form radical intermediate B .The reactive carbon site in B can combine with a second aldehyde unit (either before or after its reduction) as in step b to form coupled product C . Hydrolysis to release the 1,2-diol and reduction of the chromium species back to a lower valent metal as in step c completes the desired catalytic cycle. However, intermediate B can be further reduced by a second electron from coordinated chromium or by an electron from an external metal as in step d. This competitive side reaction can lead to the formation of the undesired reduced benzyl alkoxide D that can hydrolyze to form benzyl alcohol. In terms of the chemical selectivity for the pinacol coupling versus reduction to form benzyl alcohol, the benzyl alcohol was always the major product under all conditions studied. Under most conditions, the ratio of pinacol to benzyl alcohol varied from 1:1 to 1:2.We noted that at higher temperature with a higher ratio of starting chromium catalyst, a lower selectivity for the pinacol coupling was obtained.

The catalytic reactions produced both diastereomeric pinacol products *dl* -137 and *meso* -137 with the *meso* -product favored in ratios from 0.6:1 to nearly 1:1. The stereoselectivity in the presence of aluminum as the stoichiometric reductant was similar to when zinc-dust was used.

3.12. Dehalogenation Reactions in Water

Indium is currently used as a reducing agent in water for organic halides.

A first systematic study on the dehalogenating power of indium was carried out by Ranu et al. This group provided an efficient and general methodology for the chemoselective reduction of α-halocarbonyl compounds and benzyl halides by indium metal in water under sonication. A wide range of structurally different α-iodo- and α-bromoketones and esters 137b underwent reduction, leading to the corresponding dehalogenated carbonyl compounds 138 (Scheme 59).

Brominated substrates were reduced slower than iodinated substrates. In fact, alkyl and aryl iodides remained inert although benzyl iodides and α-iodo-ketones were reduced. Selective deiodination was observed at the benzylic position *vs.* the aromatic carbon-iodine bond in the same substrate. The use of indium metal in aqueous medium was extended to the stereoselective reduction of aryl-substituted *gem*-dibromides to vinyl bromides. The reaction was performed in ethanol and saturated ammonium chloride solution under reflux, providing primarily the corresponding (E)-vinyl bromides (50:50 to 95:5 *E/Z* ratio) in high yields (70-95%). The compatibility with several sensitive functional groups (OMe, OBz, Cl, OTBDMS, and *o*-allyl) and the absence of over reduction processes are the main advantages of this methodology. However, thiophene- and furan-substituted *gem*-dibromides did not show any stereoselectivity, whereas low effectiveness was observed for alkyl-substituted *gem*-

dibromides. The use of micellar solutions as reaction media has shown an enhancement in the reactivity of certain processes.[156] Such is the case of the indium-mediated dehalogenation of α-halocarbonyl compounds 137c in water and in the presence of a catalytic amount of the surfactant sodium dodecyl sulfate (SDS).[157] These conditions were applied by Kim et al. to α-haloketones, esters, carboxylic acids, amides, and nitriles (Scheme 60).

For α-chlorocarbonyl compounds, the reaction was rather slow in comparison with that of the bromo derivatives, and a slightly higher temperature was required. In the absence of SDS, the reaction proceeded slowly and most of the starting materials were recovered unaltered after prolonged reaction times. The same group reported the efficient reductive conversion of 3-iodomethylcephalosporin into the corresponding 3-methylcephems by indium in an aqueous system.[158]

The capability of powdered zerovalent iron to dechlorinate DDT and related compounds at room temperature was investigated by Sayles et al.[159] Specifically, DDT (139), DDD [1,1-dichloro-2,2-bis-(*p*-chlorophenyl)ethane] (140), and DDE [1,1-dichloro-2,2- bis(*p*-chlorophenyl)ethylene] (141) were successfully dechlorinated by powdered zerovalent iron in buffered anaerobic aqueous solution at 20 °C, with without the presence of nonionic surfactant Triton X-114 (Scheme 61). The rates of dechlorination of DDT and DDE were independent of the amount of iron, with or without surfactant, though rates with surfactant were much higher than without. A mechanistic model was constructed that quantitatively fits the observed kinetic data, indicating that the rate of dechlorination of the solid-phase reactants was limited by the rate of dissolution into the aqueous phase.

Granular iron metal was found to cause the reductive dechlorination of two important chloracetanilide herbicides, alachlor and metolachlor,[160] used for broadleaf weeds and annual grasses in domestic soybean and corn crops. The reaction was performed with granular cast iron in aqueous solutions at room temperature. A two-site, rate-limited sorption and first-order degradation model was applied to both batch data sets, with excellent agreement for alachlor and fair agreement for metholachlor.

The products of the reaction were chloride ion (84% mass balance for alachlor and 68% for metholachlor) and the corresponding dechlorinated acetanilides. The N-dealkylated acetanilide was a minor byproduct (9%) in the case of alachlor.

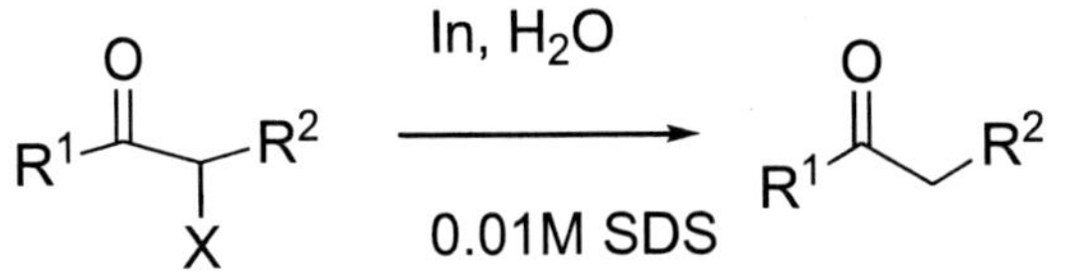

In, H_2O

0.01M SDS

137c **138b**

R^1 = alkyl, aryl, OH, OR, NHR, NR_2 83-99%

R^2 = H, alkyl, aryl

R^1-R^2 = $(CH_2)_4$

X = Br,Cl

Scheme 60. Dehalogenation of α-halo carbonyl compounds.

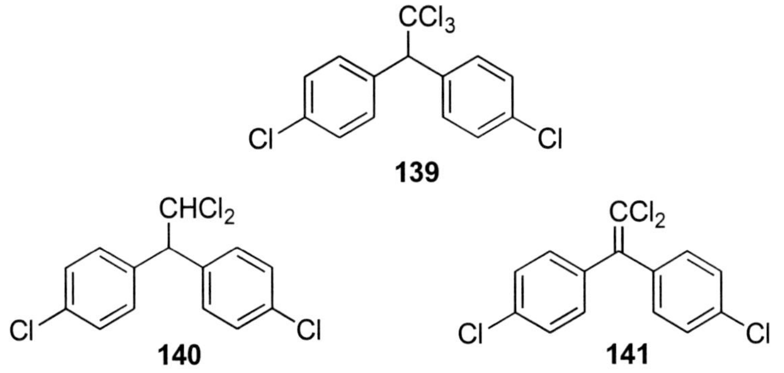

Scheme 61. Reductive dechlorination of DDT, DDD, and DDE by powdered Fe in aqueous solution.

Atrazine (2-chloro-4-ethylamino-6-isopropylamino-1,3,5-triazine) (142) is a herbicide used extensively in corn, sorghum, and sugarcane fields for the last 30 years,[161] with a long half-life in the environment (up to one year).[162] The possible water contamination, combined with the uncertainty of atrazine's carcinogenic and toxicological effects, has spurred interest in techniques that might more rapidly degrade atrazine and its metabolites. Batch aqueous experiments using fine-grained (100 mesh) zerovalent iron as an electron donor resulted in reductive dechlorination of atrazine to give 2-ethylamino-4-isopropylamino-1,3,5- triazine (143) (Scheme 62).[163] Identification of this compound initiated the development of analytical methods using HPLC, GC/MS, and HPLC/MS, to simultaneously quantify atrazine (142) and dechlorinated atrazine (143).

The dechlorination of atrazine (142) with metallic iron under low-oxygen conditions was studied at different reaction mixture pH values (2.0, 3.0, and 3.8).[164] The pH control was achieved by addition of sulfuric acid throughout the duration of the reaction. The lower the pH of the reaction, the faster the degradation of atrazine. The observed products of the degradation reaction were dechlorinated atrazine (143) and possibly hydroxyatrazine (2-ethylamino-4- isopropylamino-6-hydroxy-1,3,5-triazine). Triazine ring protonation was proposed to account, at least in part, for the observed effect of pH on atrazine by metallic iron.

Although the mechanisms of these reductions with zerovalent iron are not well elucidated, it appears that, generally, a two-electron transfer occurs either directly at the iron surface (by absorption of the organic halide) or through some intermediary [Scheme 63].[165] In a different mechanistic context, numerous studies have shown that dissociative adsorption of water takes place at clean iron metal surfaces, resulting in surface-bound hydroxyl, atomic oxygen, and atomic hydrogen ("nascent hydrogen").[166] The latter species can combine with itself, accounting for the formation of molecular hydrogen, or react with other compounds in the system, resulting in their hydrogenation [Scheme 63]. A third possibility would be the reduction by iron(II), resulting from corrosion of the metal [Scheme 63]. A debate over the relative importance of these mechanisms[167] has gone on for many years, but the electron-transfer model is generally preferred.

Sugimoto, Tanji, et al. investigated the indium-mediated reduction of halohetero-aromatics in water.[168] The authors found that the deiodination of iodoheteroaromatics using indium in water was very effective. The proposed mechanism for the deiodination of iodoquinolines is depicted in Scheme 64.

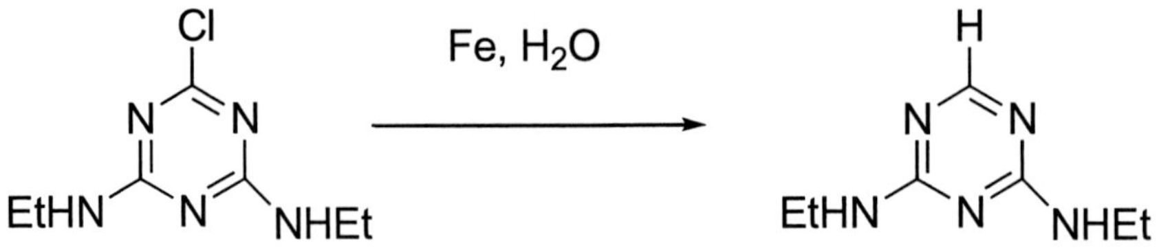

Fe, H_2O

142 **143**

Scheme 62. Dechlorination of atrazine by Fe in water.

$$RX + Fe + H^+ \longrightarrow RH + Fe^{+2} + X^- \quad (1)$$

$$Fe + 2H_2O \rightleftharpoons Fe^{+2} + H_2 + 2HO^- \quad (2)$$

$$RX + H_2 \longrightarrow RH + H^+ + X^- \quad (3)$$

$$RX + 2Fe^{2+} + H^+ \longrightarrow RH + 2Fe^{3+} + X^- \quad (4)$$

Scheme 63. Proposed reaction mechanism.

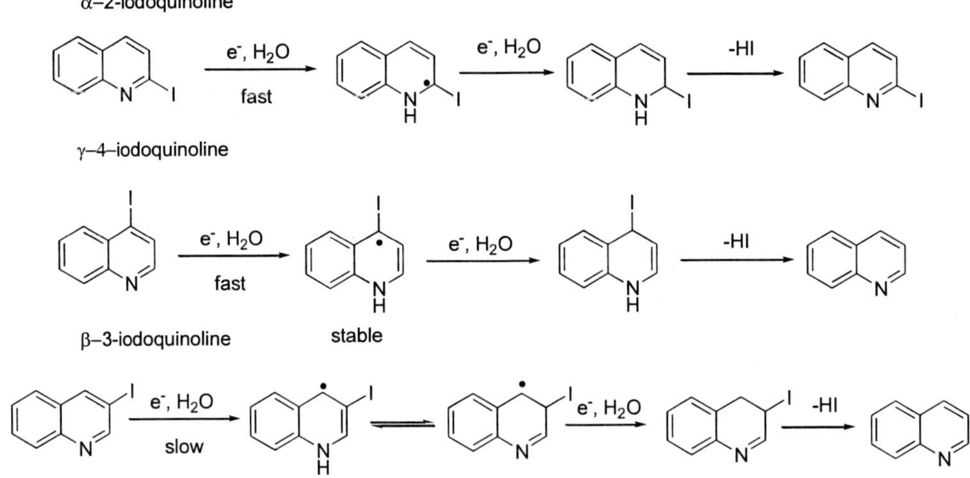

Scheme 64. Proposed deiodination mechanism of iodoquinolines in water mediated by indium metal.

When α or γ iodoquinoline is used as substrate, the dihydroquinoline radical generates smoothly since the radical is stabilized by iodine atom. On the other hand, β-iodoquinoline reacts with indium in water more slowly than α or γ iodo-quinoline because the dihydroquinoline radical is not stabilized by iodine atom. Several haloheteroaromatics were successfully dehalogenated by indium metal in water, such as iodopyridines, and iodoquinoline derivatives.[168]

3.13. Miscellaneous Reactions

3.13.1. Oxidation Reactions in Water

In an aqueous micelle system using *tert*-butyl hydroperoxide (TBHP) as oxidant in the presence of O_2 (eq 22) it is possible to obtain oxygenated cycloalkanes from their hydrocarbon precursors.[169] Use of a surfactant was necessary to create the micelles, and no reaction occurred in its absence. The reaction gave a mixture of cyclohexanol, cyclohexanone, and *tert*-butylperoxycyclohexane in the case of cyclohexane and 2-cyclohexen-1-ol and 2-cyclohexen-1-one and 3-(*tert*-butylperoxy) cyclohexene in the case of cyclohexene. The product ratio is dependent upon the amount of TBHP and starting material used. A radical mechanism, in which the favorable redox chemistry of the iron complexes in the aqueous micelle system provided *t*-BuO• and *t*-BuOO• radicals as initiators (Harber-Weiss process), was proposed.

[fe]; TBHP, O_2, H_2O; OH; O; OOBu; + ; (22)

A simple and effective method for the transformation, under mild conditions and in aqueous medium, of various cycloalkanes (cyclopentane, cyclohexane, methylcyclohexane, *cis* -and *trans* -1,2- dimethylcyclohexane, cycloheptane, cyclooctane and adamantane) into the corresponding cycloalkanecarboxylic acids bearing one more carbon atom, has been achieved by Pombeiro et al.[170].This method is characterized by a single- pot, low-temperature hydrocarboxylation reaction of the cycloalkane with carbon monoxide, water and potassium peroxodisulfate in water/acetonitrile medium, proceeding either in the absence or in the presence of a metal promoter (Scheme 65). The influence of various reaction parameters, such as type and amount of metal promoter, solvent composition, temperature, time, carbon monoxide pressure, oxidant and cycloalkane, is investigated, leading to an optimization of the cyclohexane and cyclopentane carboxylations. The highest efficiency is observed in the systems promoted by a tetracopper(II)triethanolaminate-derived complex, which also shows different bond and stereoselectivity parameters (compared to the metal- free systems) in the carboxylations of methylcyclohexane and stereoisomeric 1,2-dimethylcyclohexanes.

A free radical mechanism is proposed for the carboxylation of cyclohexane as a model substrate, involving the formation of an acyl radical, its oxidation and consequent hydroxylation by water. Relevant features of the present hydrocarboxylation method, besides the operation in aqueous medium, include the exceptional metal-free and acid-solvent-free reaction conditions, a rare hydroxylating role of water, substrate versatility, low temperatures (*ca.*50 °C) and a rather high efficiency (up to 72% carboxylic acid yields based on cycloalkane).

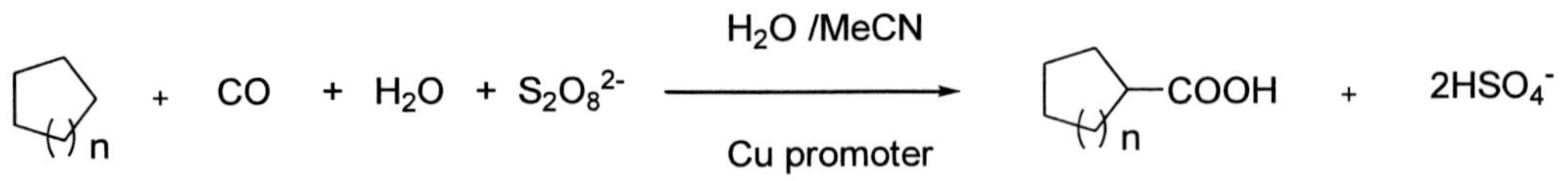

Scheme 65. Hydrocarboxylation of cycloalkanes to the corresponding cycloalkanecarboxylic acids in water / acetonitrile.

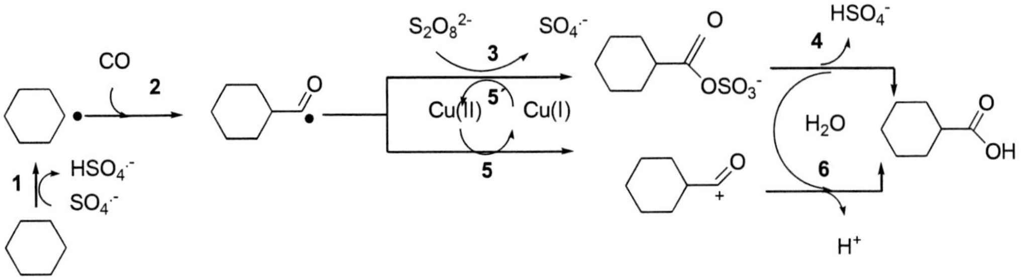

Scheme 66. Proposed simplified mechanism for the hydrocarboxylation of cyclohexane in water / acetonitrle.

For both metal-free and copper-promoted carboxylations of cyclohexane, it involves the formation of a free cyclohexyl radical, which is generated by H atom abstraction from $C_6 H_{12}$ (reaction 1, Scheme 66) by the sulfate radical SO_4•-The latter is derived from the thermolytic and copper-promoted decomposition of $K_2 S_2 O_8$

This involvement of cyclohexyl radical is confirmed by performing the carboxylations (both metal-free and copper-promoted) in the presence of the carbon-centred radical trap $CBrCl_3$,what results in the full suppression of cylcohexanecarboxylic acid formation and the appearance of cyclohexyl bromide as the main product. The radical pathway is also supported by the inhibiting effect of O_2 , acting as a cyclohexyl trap to give the $C_6 H_{11}$ COO• peroxyl radical.

Subsequent carbonylation of the cyclohexyl radical by carbon monoxide results in the acyl radical $C_6 H_{11}$ CO• (reaction 2, Scheme 66) that upon oxidation by $S_2 O_8^{2-}$ generates the acyl sulfate $C_6 H_{11} C(O)OSO_3^-$ (reaction 3, Scheme 66). This is hydrolyzed by water (reaction 4, Scheme 66) furnishing the cyclohexanecarboxylic acid. In the copper-promoted process, an alternative route (reaction 5) can occur, where the tetracopper(II)complex can behave as an oxidant of the acyl radical (reaction 5, Scheme 66).

This route involves the Cu(II)/Cu(I)redox couple and requires $K_2 S_2 O_8$ for regeneration (reaction 5) 'of the Cu(II) form. The highest activity of copper(II) in comparison with the other tested metal compounds can be accounted for by its particular effectiveness in the oxidation of carbon-centred radicals. Hydrolysis of the thus formed acylation $C_6 H_{11} CO^+$ ultimately leads to the $C_6 H_{11}$ COOH product (reaction 6, Scheme 66), via protonated cyclohexanecarboxylic acid $C_6 H_{11} C(OH)_2^+$ which is deprotonated by water, as supported by theoretical calculations on the corresponding species derived from the ethyl radical.

The hydroxylating role of water is played in both metal-free (3 →4) and copper-promoted (5 →6) pathways, as confirmed by experiments with $H_2{}^{18}O$ leading to $C_6 H_{11} CO\ {}^{18}OH$ as the main product. Less favorable routes include the formation of unlabeled $C_6 H_{11}$ COOH, proceeding through the mixed anhydride $C_6 H_{11} C(O)OSO_3$ H that is obtained by protonation of the acyl sulfate by HSO_4 ,or by coupling of C_6 H_{11} CO $^+$with HSO_4 .This anhydride would undergo intramolecular H-transfer with elimination of SO_3 , thus furnishing the C_6 H_{11} COOH product.

3.13.2. Oxidative Cleavage in Water

Oxidation of styrene and styryl derivatives can be accomplished by Fe-catalyzed oxidative cleavage in aqueous mixtures.[171] Oxidative cleavage of styrene yields benzaldehyde, as shown in Scheme 67.

Complete mechanistic interpretation of these results requires consideration of two main pathways in which the O–O bond of hydrogen peroxide can be cleaved upon reaction with the catalyst (Scheme 68). Hydrogen peroxide typically reacts with a metal complex to form an initial metal–allylperoxo intermediate (a). The O–O bond of the coordinated peroxide can then cleave heterolytically to form high-valent metal oxo complex and water (b) or homolytically to form OH radicals and a metal hydroxide complex (c).

In the proposed mechanism (Scheme 69), the active oxidizing species (formed upon reaction of hydrogen peroxide with the catalyst) is described as high-valent Fe(V)]O.[172] It can add onto the double bond leading to a carbon radical intermediate (proposed by Tuynman et al.).[173] This carbon radical intermediate is trapped by molecular oxygen followed by the abstraction of hydrogen or by the reaction between the carbon radical and activated hydrogen peroxide, which finally rearranges to give benzaldehyde as the sole product.

Biaryl coupling of 2-naphthols and substituted phenols was efficiently promoted by a supported Ru catalyst using O_2 as an oxidant in water (eq 23).[174] The supported catalyst can be reused seven times without losing catalytic activity. The big advantages of this method are that an environmentally friendly oxidant (O_2) and solvent (H_2O) can be used. The studies on the mechanism behind the reaction showed that the Ru-catalyzed biaryl coupling reaction proceeds through the radical coupling mechanism. When $FeCl_3$ was used as a stoichiometric oxidation reagent and catalyst, homocoupling of 2-naphthols and substituted phenols successfully occurred in water.[175]

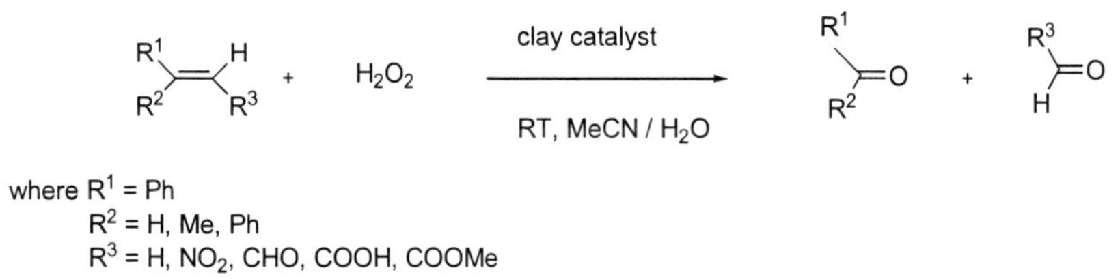

clay catalyst; RT, MeCN / H_2O

where R^1 = Ph
R^2 = H, Me, Ph
R^3 = H, NO_2, CHO, COOH, COOMe

Scheme 67. Oxidative cleavage of styrene in aqueous systems.

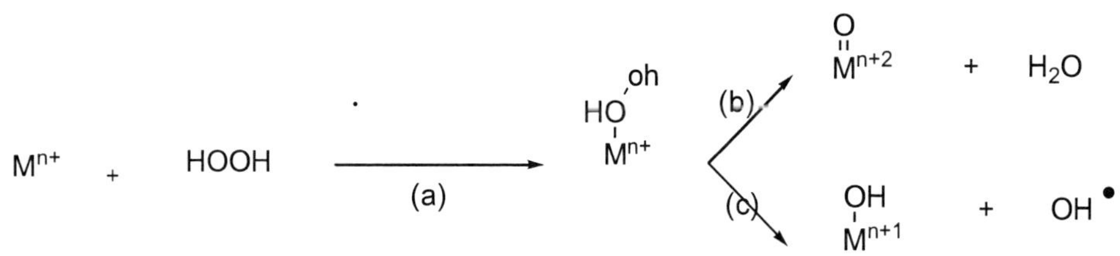

M^{n+} + HOOH →(a) HO–oh / M^{n+} →(b) $O{=}M^{n+2}$ + H_2O ; →(c) OH–M^{n+1} + $OH^{\bullet}$

Scheme 68. Proposed reaction mechanism for the oxidation process.

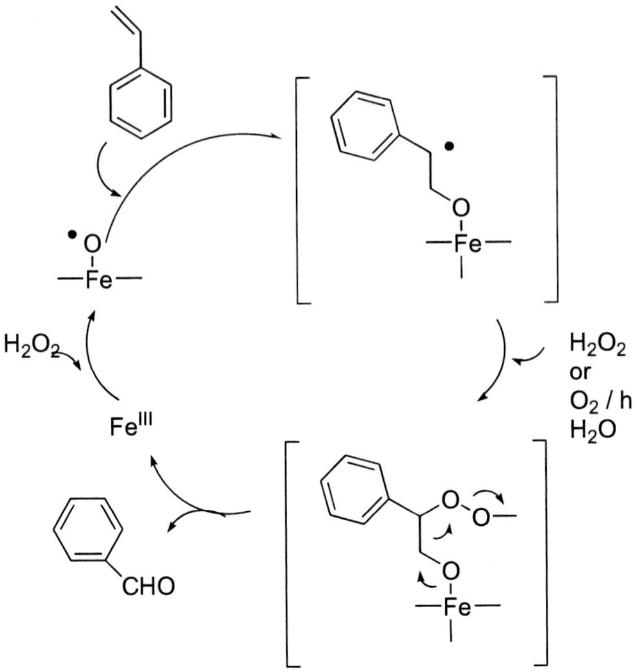
O
—Fe—
H2O2
FeIII
CHO
O
—Fe—
H2O2
or
O2 / h
H2O
O
O
O
—Fe—

Scheme 69. Proposed mechanism for the oxidation of styrene derivatives to benzaldehyde derivatives.

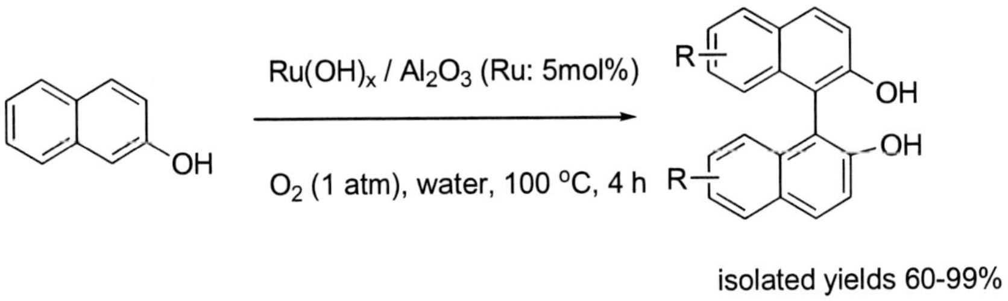
OH
Ru(OH)x / Al2O3 (Ru: 5mol%)
O2 (1 atm), water, 100 °C, 4 h
R
OH
OH
R
isolated yields 60-99%

(23)

Benzidine derivatives were obtained via oxidative coupling of *N,N*-dialkylarylamines using CuBr as a catalyst and H_2O_2 as an oxidant in water.[176] When CAN was used as oxidant, homocoupling of *N,N*-dialkylarylamines was also effectively promoted using water as solvent (eq 24).[177] A rationale for the mechanism of this coupling reaction is proposed via dimerization of diradical cations. Unlike homocoupling of 2-naphthols and substituted phenols which gave ortho products to the OH group, *para*-substituted products were selectively formed for *N,N*-dialkylarylamines substrates.

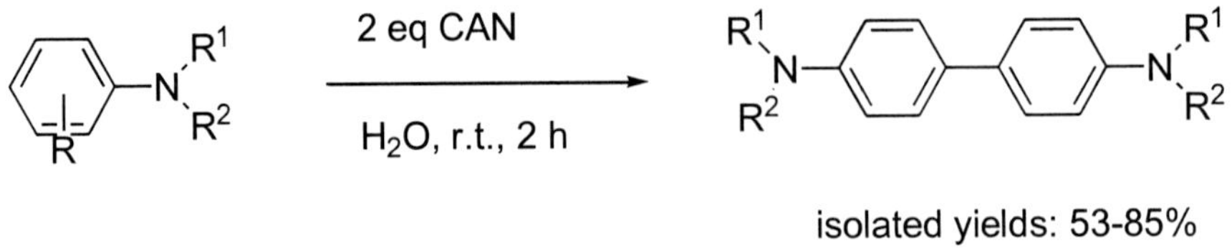
R1
N
R2
R
2 eq CAN
H2O, r.t., 2 h
R1
N
R2
R1
N
R2
isolated yields: 53-85%

(24)

3.13.3. Alkyl Radical Addition to Arene-Metal Complexes

Merlic and coworkers demonstrated that ketyl radical addition to $Cr(CO)_3$-complexed benzene is at least 100 000 times faster than attack on free benzene.[178] Alkyl radical addition to arene complexes have been accomplished using indium metal in water.

Treatment of $Mn(CO)_3$-PF_6-arene complex with an alkyl iodide and indium (3 equiv) in pure water at room temperature for 10 h afforded the addition product in very good yields. Compared to Zn , indium is obviously more effective in initiating the radical reaction.[179]

As was observed from these reactions, radical addition to cationic arene complexes proceed smoothly for all the primary, secondary, and tertiary carbon radicals, and the corresponding addition products are achieved in moderate to excellent yields.

There is an inherent advantage of the above radical methodology over the conventional ionic processes. Although the addition of functionalized nucleophiles to arene –$Mn(CO)_3$ complexes gave the corresponding products in reasonable to high yields, stabilized carbanions such as ketone enolates or α-ester carbanions were usually employed. On the other hand, the nucleophilic addition of functionalized zinc reagents such as $IZn(CH_2)_2CO_2Et$ or $IZn(CH_2)_2CN$ to complex $Mn(CO)_3$-PF_6-arene afforded the alkylated product rather than the expected β-alkylated product because of deprotonation of the acidic proton α to the ester or cyano group .

Table 19. Reduction of nitroarene derivatives to anilines in water in the presence of $InCl_3$

R_1, R_2, NO_2-substituted diarylalkyne with R_3 → (In, $InCl_3$; THF / H_2O) → R_1, R_2, NH_2-substituted diarylalkyne with R_3

entry	R^1	R^2	R^3	Time	Product yield (%)
1	H	H	H	5	87
2	Me	H	H	4	92
3	OMe	H	H	3	90
4	Cl	H	H	6	86
5	H	Cl	H	7	81
6	F	H	H	10	86
7	H	H	Me	5	83
8	Me	H	Me	4	82
9	OMe	H	Me	3	83
10	Cl	H	Me	14	80
11	H	Cl	Me	14	84
12	F	H	CF3	15	72
13	H	H	CF3	6	83
14	Me	H	CF3	3	86
15	OMe	H	CF3	14	76
16	Cl	H	CF3	14	88
17	H	Cl	CF3	14	86
18	F	H	CF3	16	89

A plausible mechanism could be drawn for In(0)-mediated radical addition. The interaction of an alkyl iodide with indium generates an alkyl radical , which adds to (□^{6}-benzene)Mn$(CO)_3$ cation to give the corresponding 17-valence-electron intermediate. The intermediate is then further reduced presumably by another molecule of indium metal via single-electron transfer process to give an 18-valence electron product. The reactivity pattern of alkyl iodides observed above strongly implies that generation of an alkyl radical is the rate determining step for the reaction.[179]

3.13.4. Reduction of C=Oand NO_2 Groups in Water

Kim and collaborators[180] undertook the reduction of nitroarene derivatives to anilines in the presence of 4 equiv of indium and 0.4 equiv of $InCl_3$ in THF/water (v/v = 5/1) at 50 ºC (Table 19).

Hilmersson and collaborators have accomplished the reduction of nitroalkanes and α,β-unsaturated nitroalkenes in water mediated by SmI_2 / water / amine.[181]

Initial experiments revealed that addition of a dilute solution (0.1 M) of the nitro compound to a premixed THF solution of SmI_2 (0.1 M), isopropylamine (0.3 M) and water (0.6 M) gave a clean and almost quantitative conversion of aliphatic nitro compounds to the respective amines, Table 20. All the reactions were instantaneous.

As a result of the successful reduction of the nitro group, the possibility to reduce α,β-unsaturated nitroalkenes directly to amines using the SmI_2 / H_2O / amine reagent was investigated. GC analysis indicated clean and instantaneous conversion to saturated amines. However, the isolated yields after workup were only 22 –75%, see Table 21. The dimethoxy derivative (entry 6) and the aliphatic α,β- unsaturated nitroalkene (entry 8) were isolated in fairly high chemical yields (75% and 70%, respectively). Again, the competing reduction of the aryl bromide was observed with the aryl bromide substrate (entry 4).

Table 20. Reduction of nitroalkanes in water by SmI_2

Entry	Starting material	Product	Yield, %
1	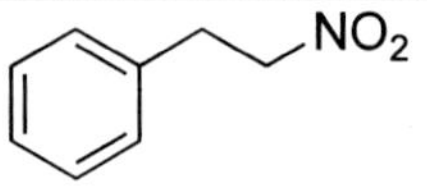 NO_2	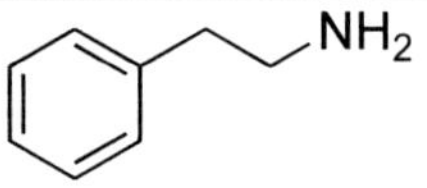 NH_2	99
2	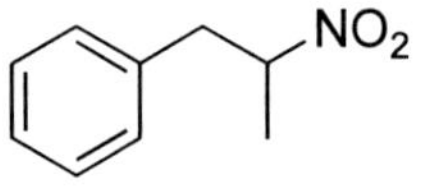 NO_2	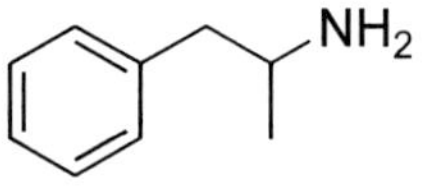 NH_2	92
3	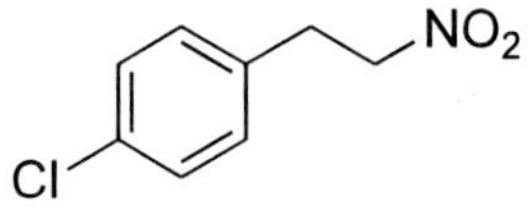 Cl, NO_2	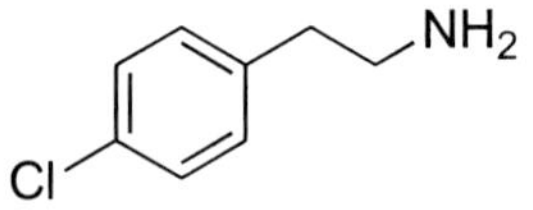 Cl, NH_2	95
4	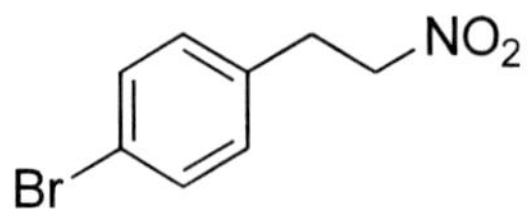 Br, NO_2	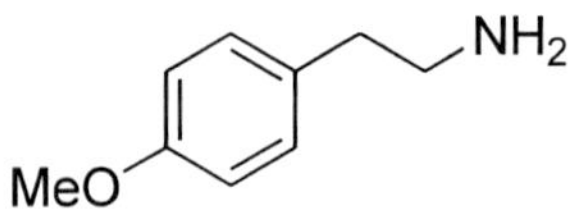 MeO, NH_2	60
5	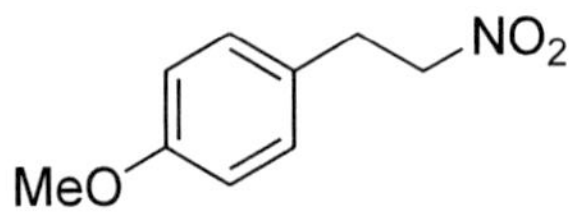 MeO, NO_2	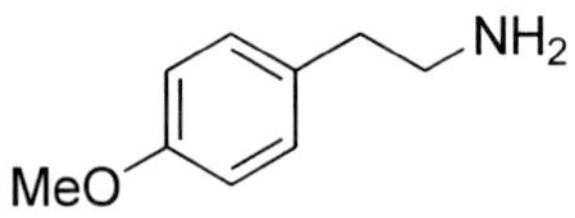 MeO, NH_2	96

Table 21. Reduction of nitroalkenes in water by SmI_2

entry	Starting material	Product	% yield
1	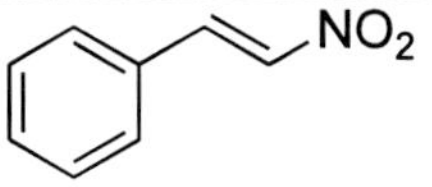 NO_2	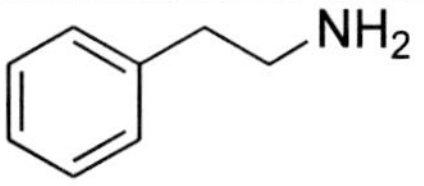 NH_2	60
2	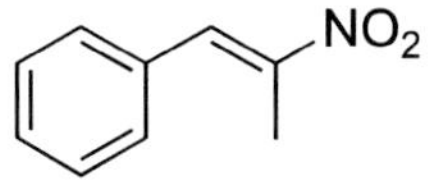 NO_2	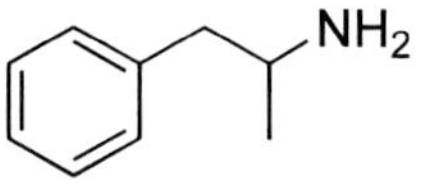 NH_2	52
3	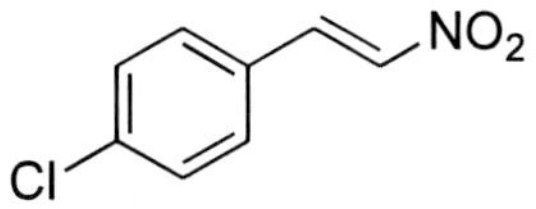 Cl, NO_2	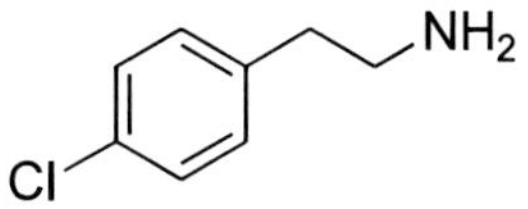 Cl, NH_2	47
4	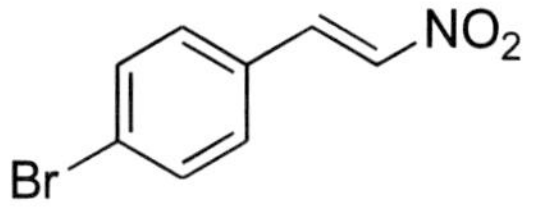 Br, NO_2	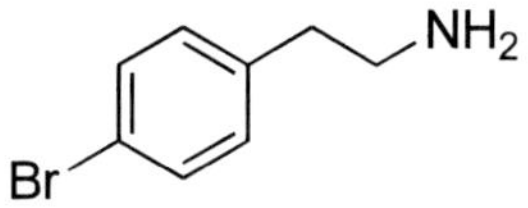 Br, NH_2	22
5	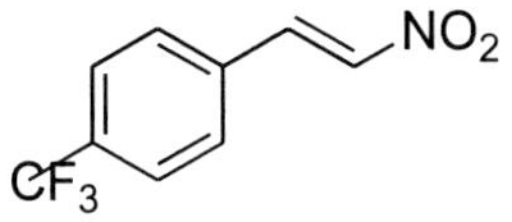 CF_3, NO_2	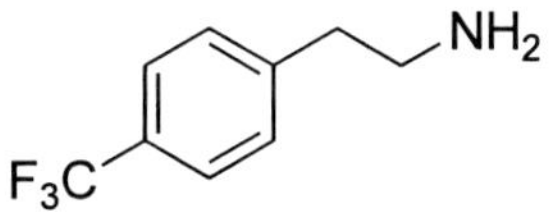 F_3C, NH_2	45
6	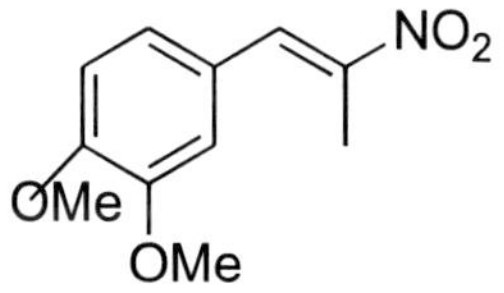 OMe, OMe, NO_2	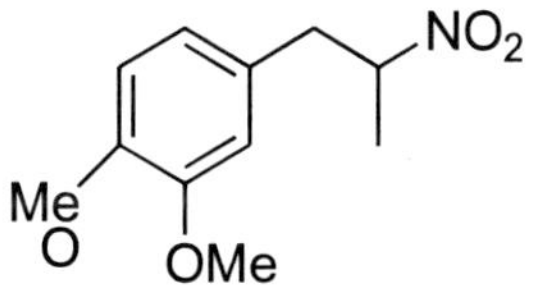 MeO, OMe, NO_2	75

Pan and collaborators[182] developed an $InCl_3$-catalyzed reduction of anthrones and anthraquinones by using aluminum powder in aqueous media (Scheme 70).

As R groups, alkyl and halide substituents can be present in the substrates.The yields of anthracene derivatives range from 72 to 92%.

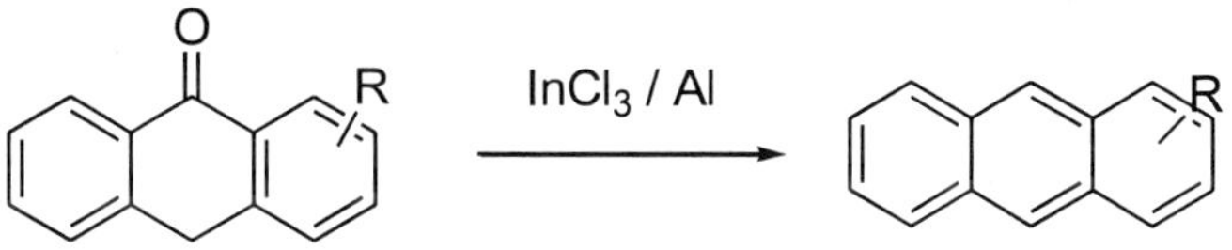

Scheme 70. Reduction of anthrone derivatives in water by $InCl_3$ / Al.

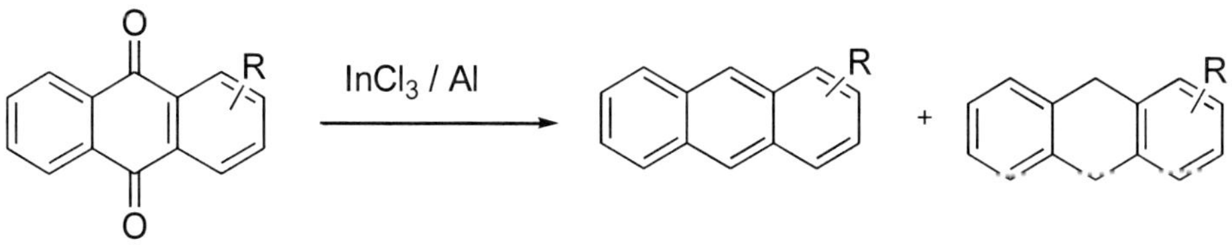

Scheme 71. Reduction of anthraquinones in water by $InCl_3$ / Al.

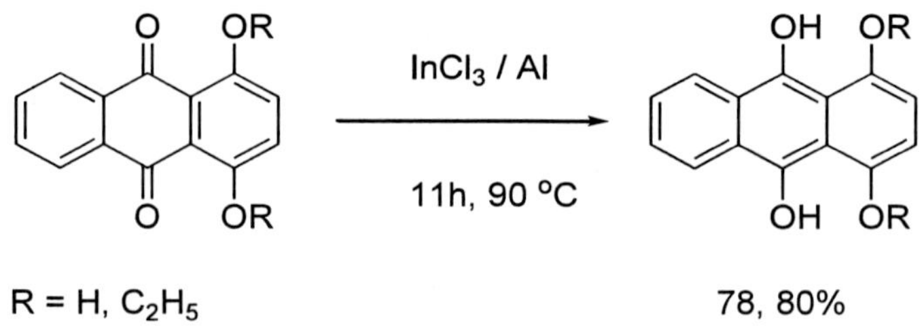

Scheme 72. Reduction of anthraquinone derivatives in water by $InCl_3$ / Al.

The reaction of 1,4-disubstituted anthraquinones (144), in which R is H or C_2H_5, with $InCl_3$ / Al gave different products. When the substituted anthraquinone (1 mmol), indium chloride (0.2 mmol), Al powder (4 mmol), and AcOH (1 mL) were mixed in 50% aqueous alcohol and stirred at reflux for 11 h, it gave compound 5 in good yields. The new carbonyl groups in compound 145 were not reduced(Scheme 71).

Based on the results from experiments, proposed mechanisms were provided. It was thought that the mechanism of reduction of anthrones is similar to the reduction at metal surfaces involving ketyl radical anions (Scheme 73). Protonated anthrone obtained an electron to form the intermediate 149. Intermediate 149 could react in two directions: anthracene 147 and 9,10-dihydroanthracene 148 was obtained, respectively. In these experiments, it was found that the intermediate 149 is easy to react along route 146, which just requires room temperature. Higher temperature is required in route 147. Therefore, when the temperature was reduced from 90 °C to rt, anthracene 147 was obtained as single product in $InCl_3$-catalyzed reduction of anthrone 146.

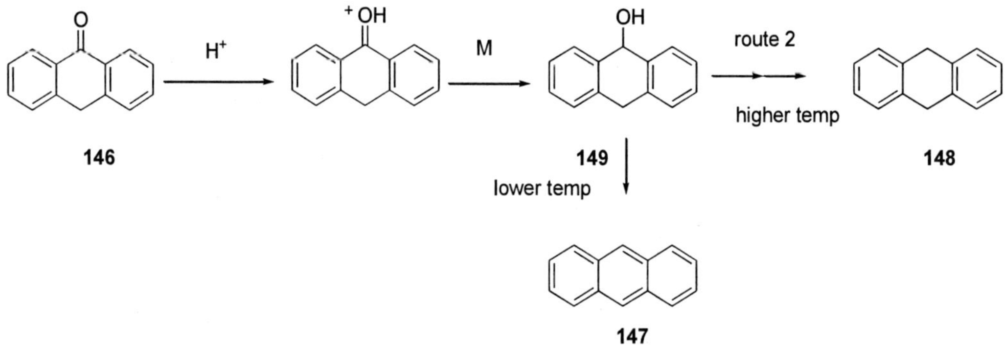

Scheme 73. Proposed reaction mechanism for the reduction of anthrone.

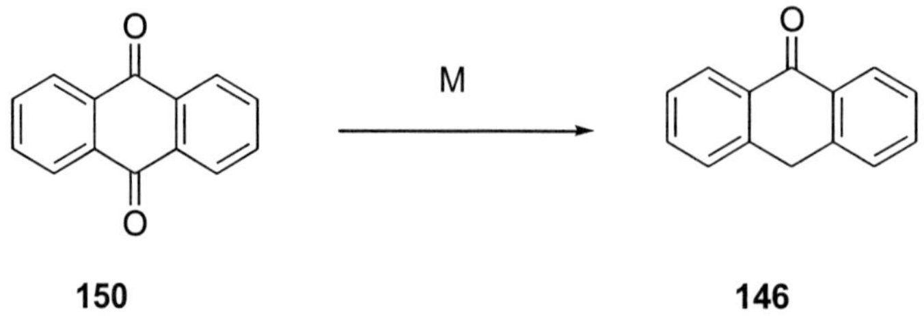

Scheme 74. Reduction of anthraquinone.

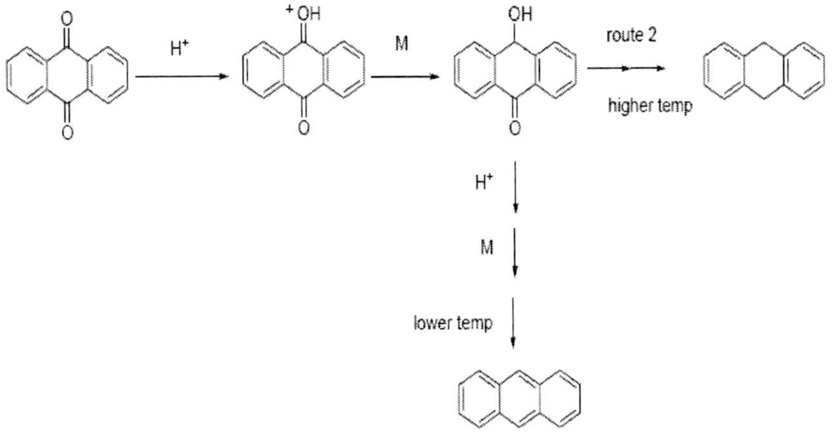

Scheme 75. Proposed reaction mechanism for the reduction of anthraquinone derivatives.

For the reduction of 9,10-anthraquinone 150, another possible mechanism was also proposed (Scheme 74). 9,10-Anthraquinone 150 was reduced to anthrone 146 firstly. Then anthrone formed from 9,10-anthraquinone reacted in the way described in Scheme 73. Higher temperature is required during reduction of compound 150 to compound 146, which is in agreement with experimental findings. At higher temperature, anthracene together with dihydroanthracene was obtained from anthrone. So anthracene could not be obtained as single product in $InCl_3$-catalyzed reduction of 9,10-anthraquinone 150. In order to further verify the proposed mechanism, $InCl_3$-catalyzed reduction process of 9,10-anthraquinone 150 was monitored by ESI-MS. Finally, the mechanism of $InCl_3$ -catalyzed reduction of 1,4-disubstituted anthraquinones (R ¼ H, C_2 H_5) (Scheme 72) was proposed (Scheme 75). Firstly, compound 150 was hydrolyzed to 1,4-dihydroxy-9,10-anthraquinone. Then compound 151 was reduced to intermediate 152 by two possible routes. Finally, the product 153 was formed through the intermediate 152.

Procter and collaborators have accomplished the reduction of lactones and cyclic 1,3-diesters in water mediated by SmI_2.[183]

The authors reported on the first reduction of lactones to diols using SmI_2 H_2O. The reagent system is selective for the reduction of lactones over esters, furthermore, it displays complete ring size-selectivity in that only 6-membered lactones are converted to the corresponding diols. Experimental and computational studies suggest the selectivity originates from the initial electron-transfer to the lactone carbonyl and that anomeric stabilization of the radical-anion formed is an important factor in determining reactivity. In addition to the selectivity of the reagent system, SmI_2 is commercially available, or convenient to prepare, easy to handle, operates at ambient temperature, and does not require toxic cosolvents or additives, making the transformation an attractive addition to the portfolio of reductions.

Lactones can also be used in reductive carbon-carbon bond formation through cyclization of the radicals formed by one electron reduction, generating cyclic ketones (or ketals) often with high diastereoselectivity. The cyclizations constitute the A series of competition experiments has been carried out to illustrate further the chemoselectivity observed with the SmI_2 H_2O reagent system.

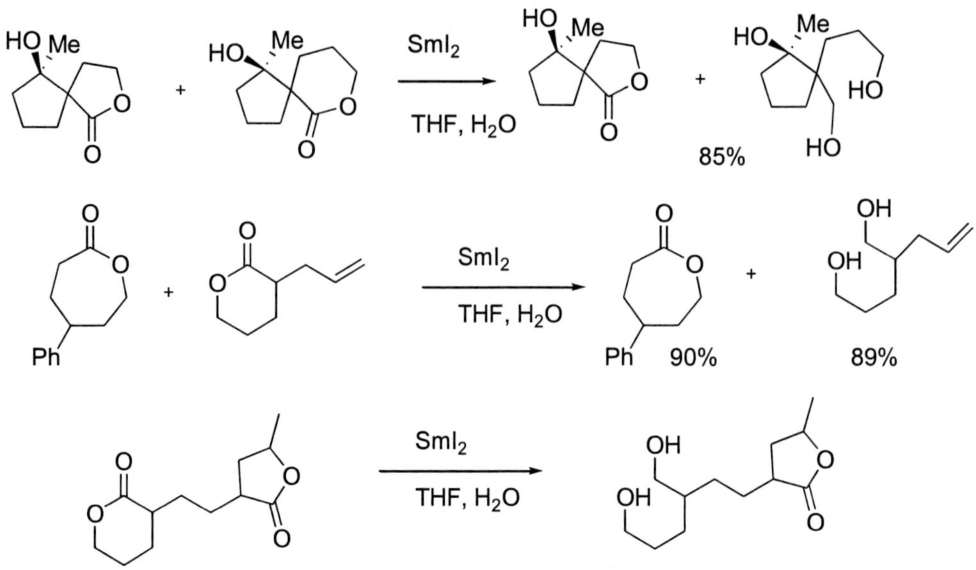

Scheme 76. Reduction of 6-membered lactones.

Mixtures of lactones were prepared and treated with SmI_2 H_2O. In all cases, no reduction products arising from 5, 7 and 8-membered lactones were observed while 6-membered lactones were reduced smoothly (Scheme 76).

Modified SmI_2 reagent systems employing additives (HMPA, DMPU, LiBr) were also ineffective for the reduction of other lactones.

A possible mechanism for the transformation is given in Scheme 77.

Activation of the lactone by coordination to Sm(II) and electron-transfer generates radical anion 156 that is then protonated. A second electron transfer generates carbanion 158 that is quenched by the H_2O cosolvent. Lactol 159 is in equilibrium with hydroxy aldehyde 160 and is reduced by a third electron-transfer from Sm(II) to give a ketyl radical anion 161. A final electron-transfer from Sm(II) gives an organosamarium that is protonated by H_2O. The amount of SmI_2 (approximately 7 equiv) required experimentally is consistent with the amount predicted by the proposed mechanism (4 equiv).

For 6-membered lactones, the authors believe that reduction generates a radical anion intermediate 156 (Scheme 77) that is stabilized by interaction with the lone-pairs on both the endocyclic and exocyclic oxygens. Such interactions are known to be more pronounced in 6-membered rings than in other, conformationally more labile, ring systems. It appears that the greater stability of the radical anion 156, compared to analogous radicals formed from the reduction of 5, 7 and 8-membered lactones, promotes the initial reduction step. This hypothesis is supported by the observation that 2-oxabicyclo[2,2,2]octan-3-one 162 (Scheme 78), from which an intermediate radical-anion would be unable to adopt the chair conformation necessary for optimal stabilization, is not reduced by SmI_2 H_2O (Scheme 78).

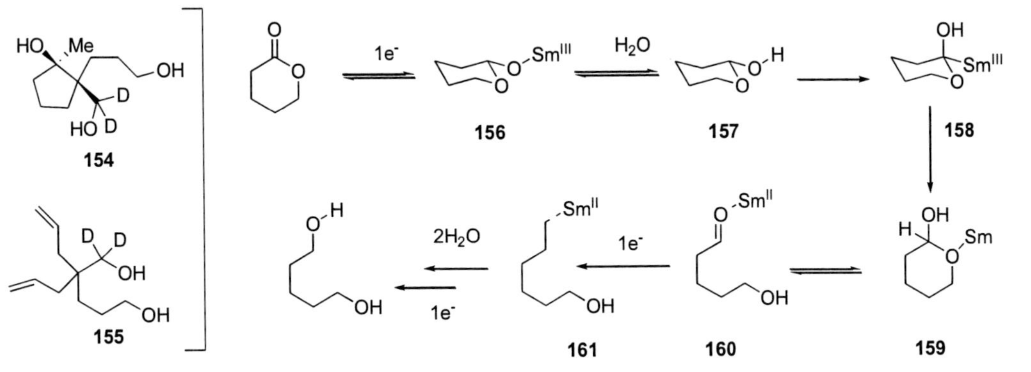

Scheme 77. Reduction mechanism of lactones.

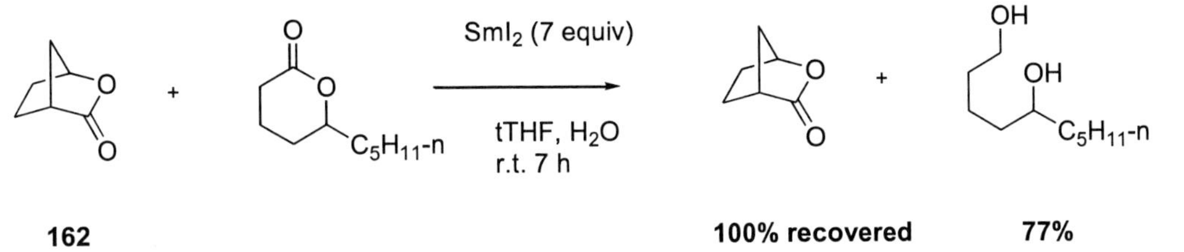

Scheme 78. Reduction of lactones by SmI_2 in water.

Calculations suggest that the first electron-transfer to the lactone carbonyl is endothermic (100 kJ mol $^{-1}$) in all cases. The relative reaction energy of this step for 6-membered lactones, however, is calculated to be 116 kJ mol $^{-1}$, about 25 26 kJ mol $^{-1}$ lower than those involving 5- and 7-membered rings. The relative reaction energy for the first electron-transfer to bicyclic lactone 162 is calculated to be 147.4 kJ mol^{-1}. The second electron transfer is lower in energy and similar for all systems, agreeing with kinetic studies showing that the first electron-transfer is the rate-determining step. The calculated lowest energy conformation of the radical anion derived from a 6-membered anion suggests that the radical does indeed adopt a pseudoaxial orientation apparently enjoying stabilization by an anomeric effect. Activation of the lactone by coordination to Sm(II) and electrostatic stabilization of the product radical-anion by coordination to Sm(III) is likely to render these reductions more favorable than the calculated, relative reaction energies suggest. (Scheme 79).

The same authors [183] also accomplished the reduction of cyclic 1,3-diesters employing SmI_2 / H_2O, as shown in Table 22 below.

The mechanism proposed also involves radical-ionintermediates as shown in Scheme 80.

The reduction of 172 with SmI_2 /D_2O gave 165-D,*D* (see Scheme 80) suggesting that anions are generated and protonated by H_2O during a series of electron transfer steps. A possible mechanism for the transformation is given in Scheme 80. Activation of the ester carbonyl by coordination to Sm(II) and electron transfer generate radical anion 166 that is then protonated. A second electron transfer generates carbanion 168 that is quenched by H_2O. Hemiacetal 169 is in equilibrium with aldehyde 170, which is reduced by a third electron transfer from Sm(II) to give a ketyl-radical anion 171. A final electron transfer from Sm(II) gives an organosamarium that is protonated. The amount of SmI_2 (approximately 7 equiv) required experimentally is consistent with the amount predicted by the proposed mechanism (4 equiv) (Scheme 80).

Hilmersson and collaborators [184] also investigated the mechanistic details of reduction of organic halides mediated by SmI_2/H_2O/amine system. The kinetics of the SmI_2 / H_2O / amine-mediated reduction of 1-chlorodecane was the subject of this study undertaken by Hilmersson, andstudied in detail. The rate of reaction was found to be first order in amine and 1-chlorodecane, second order in SmI_2, and zero order in H_2O. Initial rate studies of more than 20 different amines show a correlation between the base strength (p $K_{BH}+$) of the amine and the logarithm of the observed initial rate, in agreement with Brønsted catalysis rate law. To obtain the activation parameters, the rate constant for the reduction was determined at different temperatures. Additionally, the ^{13}C kinetic isotope effects (KIE) were determined for the reduction of 1-iododecane and 1-bromodecane. Primary ^{13}C KIEs (k_{12}/k_{13},20 °C) of 1.037 ± 0.007 and 1.062 ± 0.015, respectively, were determined for these reductions. This shows that cleavage of the carbon-halide bond occurs in the rate-determining step.

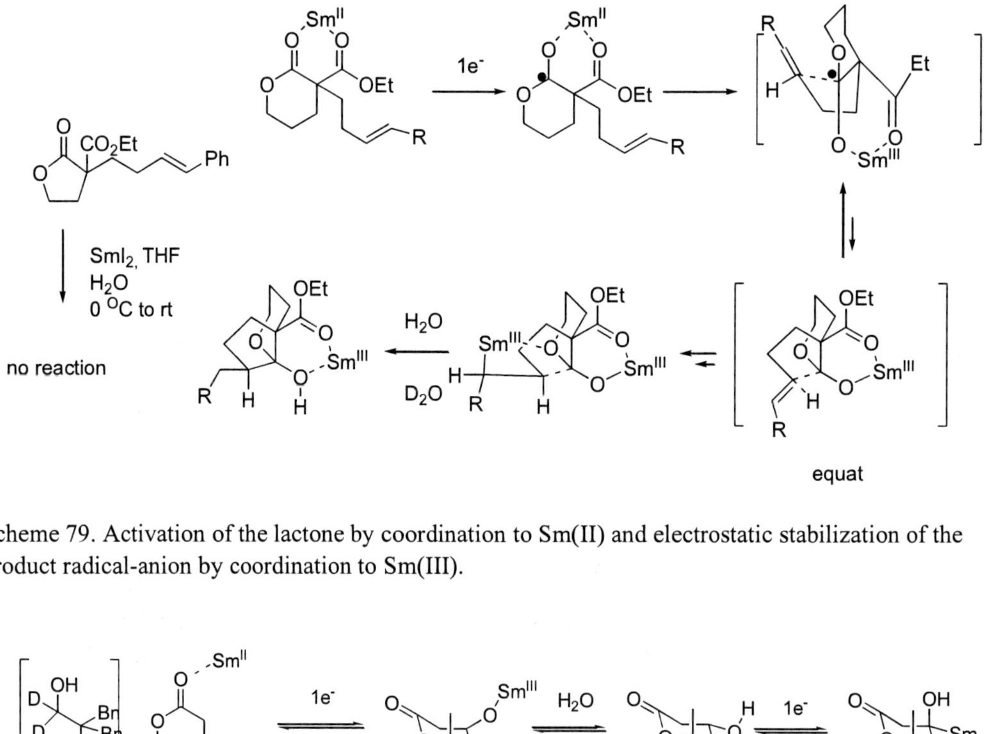

Scheme 79. Activation of the lactone by coordination to Sm(II) and electrostatic stabilization of the product radical-anion by coordination to Sm(III).

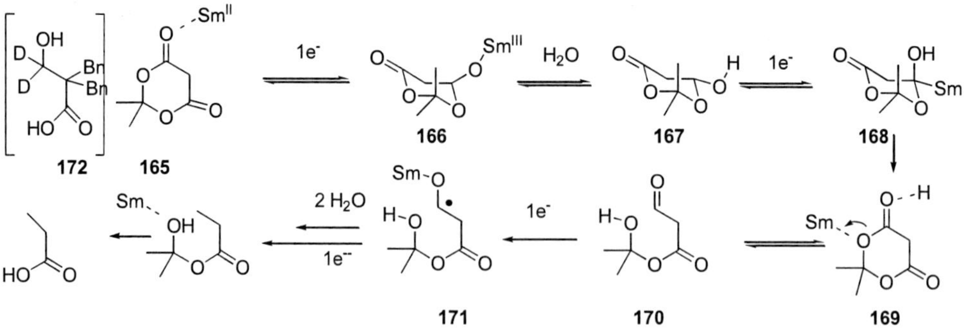

Scheme 80. Mechanism proposed for the reduction of lactones.

Table 22. Reduction of lactones

SmI_2 -H_2O, r.t.

163 → **164**

R^1	R^2	R^1	R^2	YIELD (%) **164**
Bn	Bn	Bn	Bn	88
	-$(CH_2)_4$-		-$(CH_2)_4$-	81
H	Bn	H	Bn	68
H	4-$MeOC_6H_4$	H	4-$MeOC_6H_4$	78
H	4-BrC_6H_4	H	4-BrC_6H_4	77
H	*i*-Bu	H	*i*-Bu	94
Me	Bn	Me	Bn	98
H	Ph	H	Ph	72
	=CH*i*Pr	H	*i*-Bu	87
	-(CH_2CH_2)-	H	Et	75

ACKNOWLEDGMENTS

Thanks are given to Conicet, Argentina (National Council of Scientific and Technical Research), and to Agencia Nacional Científica y Técnica for financial support.

REFERENCES

[1] Heiba, E. I.; Dessau, R. M. Oxidation by metal salts. VII. Syntheses based on the selective oxidation of organic free radicals. *J. Am. Chem. Soc.* 1971, *93*, 524-527.

[2] (a) Baciocchi, E.; Civatarese, G.; Ruzziconi, R. Synthesis of 1,4-dicarbonyl compounds by the ceric ammonium nitrate promoted reaction of ketones with vinyl and isopropenyl acetate. *TetrahedronLett.* 1987, *28*, 5357-5360. (b) Baciocchi, E.; Ruzziconi, R. Synthesis of 3-acyl and 3-carboalkoxyfurans by the ceric ammonium nitrate promoted addition of 1,3-dicarbonyl compounds to vinylic acetates. *Synth. Commun.* 1988, *18*, 1841-1846.

[3] Nair, V.; Balagopal, L.; Rajan, R.; Mathew, J. Recent Advances in Synthetic Transformations Mediated by Cerium(IV) Ammonium Nitrate. *Chem.Rev.* 2004, *37*, 21-30.

[4] Nair, V.; Mathew, J.; Facile synthesis of dihydrofurans by the cerium(IV) ammonium nitrate mediated oxidative addition of 1,3- dicarbonyl compounds to cyclic and acyclic alkenes. Relative superiority over the manganese (III) acetate mediated process. *J. Chem. Soc, Perkin Trans.* 1 1995, 187-188.

[5] Nicolau, K.C.; Gray, D. Total synthesis of hybocarpone. *Angew.Chem.,Int.Ed.* 2001, *40*, 761-763.

[6] Nair, V.; Panicker, S. B.; Nair, L. G.; George, T. G.; Augustine, A.Carbon-heteroatom bond-forming reactions mediated by cerium-(IV) ammonium nitrate: An Overview. *Synlett* 2003, 156-165.

[7] Trahanovsky, W. S.; Robbins, M. D. Oxidation of organic com-pounds with cerium(IV). XIV. Formation of α-azido-β-nitratoalkanes from olefins, sodium azide, and ceric ammonium nitrate. *J. Am. Chem. Soc.* 1971, *93*, 5256-5258.

[8] Lemieux, R. U.; Ratcliffe, R. M. The azidonitration of tri-o-acetyl-D-galactal *Can. J. Chem.* 1979, *93*, 1244-1251.

[9] Nair, V.; Panicker, S. B.; Augustine, A.; George, T. G.; Thomas, S.; Vairamani, M. An efficient bromination of alkenes using cerium(IV) ammonium nitrate (CAN) and potassium bromide. *Tetrahedron* 2001, *57*, 7417-7422.

[10] (a) Nair, V.; Panicker, S. B.; Mathai, S. Bromination of cyclopropanes using potassium bromide and cerium(IV) ammonium nitrate (CAN): synthesis of 1,3-dibromides. *Res. Chem. Intermed.* 2003, *29*, 227-231. (b) Su, C.; .Huang, X.; Liu, Q. CAN-Mediated Highly Regio- and Stereoselective Oxidation of Vinylidenecyclopropanes: A Novel Method for the Synthesis of Unsymmetrical Divinyl Ketone and Functional Enone Derivatives. *J.Org.Chem.* 2008, *73*, 6421-6424.

[11] Li, C-J. Aqueous Barbier-Grignard-type reaction: scope, mechanism and synthetic applications. *Tetrahedron* 1996, *52*, 5643-5668.

[12] Bieber, L.W.; Malvestiti, I.; Storch, E.C. Reformatsky reaction in water: evidence for a radical-chain process. *J.Org.Chem.* 1997, *62*, 9061-9064.

[13] Chan, T.H.; Li, C-J.; Wei, Z.Y. 1993-Lemieux award lecture. Organometallic-type reactions in aqueous media. A new challenge in organic synthesis. *Can.J.Chem.* 1994, *72*, 1181-1192.

[14] Keh, C.C.K.; Wei, C.; Li, C-J.The Barbier-Grignard-type carbonyl alkylation using unactivated alkyl halides in water. *J.Am.Chem.Soc.* 2003, *125*, 4062-4063.

[15] (a) Arimitsu, S.; Jacobsen, J.M.; Hammond, G.B. gem-Difluoropropargylation of aldehydes using catalytic In/Zn in aqueous media. *TetrahedronLett.* 2007, *48*, 1625-1627. (b) Arimitsu, S.; Hammod, G.B. Lanthanide Triflate-Catalyzed Preparation of β,β-Difluorohomopropargyl Alcohols in Aqueous Media. Application to the Synthesis of 4,4-Difluoroisochromans.*J.Org.Chem.* 2006, *71*, 8665-8668.

[16] (a) Cannella, R.; Clerici, A.; Pastori, N.; Regolini, E.; Ports, O. Mediated Alkyl Radical Addition to Imines Generated in Situ. *Org.Lett.* 2005, *7*, 645-648. (b) Bieber, L.W.; Storch, E.C.; Malvestiti, I.; da Silva, M.F. Silver catalyzed zinc Barbier reaction of benzylic halides in water. *TetrahedronLett.* 1998, *39*, 9393-9396.

[17] Shen, Z-L.; Loh, T-P. Indium-copper-mediated Barbier-Grignard-type alkylation reactions of imines in aqueous media. *Org.Lett.* 2007, *9*, 5413-5416.

[18] Friestad, G.K. Addition of carbon-centered radicals to imines and related compounds. *Tetrahedron* 2001, *57*, 5461-5496.

[19] (a) Miyabe, H.; Ueda, M.; Naito, T. Free radical reactions of imine derivatives in water. *J.Org.Chem.* 2000, *65*, 5043-5047. (b) Miyabe, M.; Ueda, M.; Nishimura, A.; Naito, T. Indium as a radical initiator in aqueous media: intermolecular alkyl radical addition to C=N and C=C bond. *Tetrahedron* 2004, *60*, 4227-4235.

[20] Miyabe, H.; Ueda, M.; Nishimura, A.; Naito, T. Indium-mediated intermolecular alkyl radical addition to electron-deficient C=N bond and C=C bond in water. *Org.Lett.* 2002, *4*, 131-134.

[21] Yang, Y-S.; Shen, Z-L.; Loh, T-P. Indium (zinc)-copper-mediated Barbier-type alkylation reactions of nitrones in water: synthesis of amines and hydroxylamines.*Org.Lett.* 2009, *11*, 1209-1212.

[22] Ueda, M. Development of radical reactions in water aimed at environmentally benign synthetic reactions. *Yakagaku Zasshi.* 2004, 124, 311-319.

[23] Srikanth, G.S.C.; Castle, S.L. Advances in radical conjugate additions. *Tetrahedron* 2005, 61, 10377-10441. Jang, D.O.; Cho, D. H.; Chung, C.-M. 1-Ethylpiperidium Hypophosphite: A Practical Mediator for Radical Carbon-Carbon Bond Formation*Synlett* 2001, 1923-1925.

[24] Jang, D. O.; Cho, D. H. N-Ethylpiperidine Hypophosphite Mediated Intermolecular Radical Carbon-Carbon Bond Forming Reaction in Water.*Synlett* 2002, 1523-1526.

[25] Cho, D. H.; Jang, D. O. Radical deoxygenation of alcohols and intermolecular carbon-carbon bond formation with surfactant-type radical chain carriers in water.*Tetrahedron Lett.* 2005, *46*, 1799-1801.

[26] Mascareñas, J. L.; Perez-Sestelo, J.; Castedo, L.; Mouriño, A. A short, flexible route to vitamin D metabolites and their side chain analogues*Tetrahedron Lett.* 1991, *32*, 2813-2815.

[27] Perez Sestelo, J.; Mascareñas, J. L.; Castedo, L.; Mouriño, A. Ultrasonically induced conjugate addition of iodides to electron-deficient olefins and its application to the synthesis of side-chain analogs of the hormone 1.alpha.,25-dihydroxyvitamin D3.*J. Org. Chem.* 1993, *58*, 118-122.

[28] Perez-Sestelo, J.; Mascareñas, J. L.; Castedo, L.; Mouriño, A short, flexible approach to vitamin D3 analogues with modified side chains. A. *Tetrahedron Lett.* 1994, *35*, 275-277.

[29] Suarez, R. M.; Perez Sestelo, J.; Sarandeses, L. A. Diastereoselective Ultrasonically Induced Zinc–copper Conjugate Addition to Chiral α-β-Unsaturated Carbonyl Systems in Aqueous Media. *Synlett* 2002, 1435-1438.

[30] Suarez, R. M.; Perez Sestelo, J.; Sarandeses, L. A. Diastereoselective Conjugate Addition to Chiral α,β-Unsaturated Carbonyl Systems in Aqueous Media: An Enantioselective Entry to α- andβ-Hydroxy Acids andβ-Amino Acids.*Chem. Eur.J.* 2003, *9*, 4179-4185.

[31] Suarez, R. M.; Perez Sestelo, J.; Sarandeses, L. A. Practical and efficient enantioselective synthesis of α-amino acids in aqueous media.*Org.Biomol. Chem.* 2004, *2*, 3584-3593.

[32] Perez Sestelo, J.; Cornella, I.; de Uña, O.; Mouriño, A.; Sarandeses, L. A. Stereoselective Convergent Synthesis of 24,25-Dihydroxyvitamin D3 Metabolites: A Practical Approach.*Chem. Eur. J.* 2002, *8*, 2747-2753.

[33] Perez Sestelo, J.; de Uña, O.; Mouriño, A.; Sarandeses, L. A. Diastereoselective Conjugate Addition to a Chiral Methyleneoxazolidinone in Aqueous Media.*Synlett* 2002, 719-722.

[34] Cornella, I.; Suarez, R. M.; Mouriño, A.; Perez Sestelo, J.; Sarandeses, L. A. J. *Steroid Biochem. Mol. Biol.* 2004, 89-90.

[35] Cornella, I.; Perez Sestelo, J.; Mouriño, A.; Sarandeses, L. A. Synthesis of New 18-Substituted Analogues of Calcitriol Using a Photochemical Remote Functionalization. *J. Org. Chem.* 2002, *67*, 4707-4714.

[36] (a) Miura, K.; Tomita, M.; Ichikawa, J.; Hosomi, A. Indium(III) Acetate-Catalyzed Intermolecular Radical Addition of Organic Iodides to Electron-Deficient Alkenes *Org.Lett.* 2008, *10*, 133-136. (b) Movassagh, B.; Navidi, M. One-pot synthesis of α-hydroxysulfides from styrenes and disulfides using the Zn/$AlCl_3$ system. *TetrahedronLett.* 2008, *49*, 6712-6714.

[37] (a) Petrier, C.; Luche, J-L. Allylzinc reagents additions in aqueous media. *J. Org. Chem.* 1985, *50*, 910–912. (b) Li, C. J.; Chan, T. H. Organometallic reactions in aqueous media. 2. Convenient synthesis of methylenetetrahydrofurans. *Organometallics* 1991, *10*, 2548–2549.

[38] Wada, M.; Ohki, H.; Akiba, K-Y. Bismuth(III) chloride/aluminium-promoted allylation of aldehydes to homoallylic alcohols in aqueous solvent. *J. Chem. Soc., Chem. Commun.* 1987, 708–709.

[39] Bonini, B. F.; Comes- Franchini, M.; Fochi, M.; Mazzanti, G.; Nanni, C.; Ricci, A. Allylation reactions of acylsilanes as synthetic equivalents of aldehydes. Application to a stereocontrolled synthesis of (1S,2S,5S)-(10S)-and-(10R)-allyl myrtanol. *Tetrahedron Lett.* 1998, *39,* 6737–6740.

[40] Fukuma, T.; Lock, S.; Miyoshi, N.; Wada, M. Water Accelerated Allylation of Aldehydes by Using Water Tolerant Grignard-Type Allylating Agents from Allylmagnesium Chloride and Various Metallic Salts. *Chem. Lett.* 2002, 376–377.

[41] Li, C. J.; Meng, Y.; Yi, X. H. Manganese-Mediated Carbon-Carbon Bond Formation in Aqueous Media: Chemoselective Allylation and Pinacol Coupling of Aryl Aldehydes. *J. Org. Chem.* 1998, *63*, 7498–7504.

[42] Li, L-H.; Chan, T.H. Organometallic reactions in aqueous media. Antimony-mediated allylation of carbonyl compounds with fluoride salts. *TetrahedronLett.* 2000, *41*, 5009-5012.

[43] Zhou, J. Y.; Jia, Y.; Sun, G. F.; Wu, S. H. Barbier-Type Allylation of Aldehydes and Ketones with Metallic Lead in Aqueous Media. *Synth. Commun.* 1997, *27*, 1899–1906.

[44] Chan, T. H.; Yang, Y. Organometallic reactions in aqueous media. Allylation of aldehydes with diallylmercury or allylmercury bromide. *Tetrahedron Lett.* 1999, *40*, 3863–3866

[45] Cintas, P. Synthetic Organoindium Chemistry: What Makes Indium so Appealing?. *Synlett* 1995, 1087–1096.

[46] Zhang, W-C.; Li, C-J. Magnesium-mediated carbon-carbon bond formation in aqueous media: Barbier-Grignard allylation and pinacol coupling of aldehydes.*J.Org.Chem.* 1999, *64*, 3230-3236.

[47] (a) Lawrence, L.M.; Whitesides, G.M. Trapping of free radical intermediates in the reaction of alkyl bromides with magnesium. *J.Am.Chem.Soc.* 1980, *102*, 2493-2494. (b) Walborsky, H.M.; Zimmermann, C. The surface nature of Grignard formation reagent.Cyclopropylmagnesium bromide. *J.Am.Chem.Soc.* 1992, *114*, 4996-5000.

[48] Alexander, E.R. Principles of Ionic Organic Reactions, *John Wiley & Sons*, 1950, 188.

[49] Hygum Dam, J.; Fristrup, P.; Madsen, R. Combined experimental and theoretical mechanistic investigation of Barbier allylation in aqueous media. *J.Org.Chem.* 2008, *73*, 3228-3235.

[50] Li, C-J.; Meng, Y.; Yi; X-H. Manganese-mediated reactions in aqueous media: chemoselective allylation and pinacol coupling of aryl aldehydes. *J.Org.Chem.* 1997, *62*, 8632-8633.
[51] Chan, T.H.; Isaac, B.M. Organometallic-type reactions in aqueous media mediated by indium: application to the synthesis of carbohydrates. *PureAppl.Chem.* 1996, *68*, 919-924.
[52] Gajewski, J.J.; Bocian, W.; Brichford, N.L.; Henderson, J.L. Secondary deuterium kinetic isotope effects in irreversible additions of allyl reagents to benzaldehyde. *J.Org.Chem.* 2002, *67*, 4236-4240.
[53] (a) Nokami, J.; Otera, J.; Sudo, T.; Okawara, R. Allylation of aldehydes and ketones in the presence of water by allylic bromides, metallic tin, and aluminum.*Organometallics* 1983, *2*, 191-195.
[54] Nokami, J.; Wakabayashi, S.; Okawara, R. Intramolecular allylation of carbonyl compounds. A new method for five and six membered ring formation.*Chem. Lett.* 1984, 869-872.
[55] Wu, S. H.; Huang, B. Z.; Zhu, T. M.; Yiao, D. Z.; Chu, Y. L. *Acta Chim. Sinica* 1990, *48*, 372-374.
[56] Einhorn, C.; Luche, J. L. Selective allylation of carbonyl compounds in aqueous media. *J. Organomet. Chem*. 1987, *322*, 177-180.
[57] See: (a) Pereyre, M.; Quintard, J.-P.; Rahm, A. *Tin in Organic Synthesis*; Butterworth: London, 1987. (b) Smith, P. J. *Toxicological Data on Organotin Compounds*; International Tin Research Inst.: London, 1978.
[58] Chan, T. H.; Li, C. J.; Wei, Z. Y. Crossed aldol type condensation reactions in aqueous media.*J. Chem. Soc., Chem. Commun.* 1990, 505-509.
[59] (a) Kim, E.; Gordon, D. M.; Schmid, W.; Whitesides, G. M. Tin- and indium-mediated allylation in aqueous media: application to unprotected carbohydrates.*J. Org. Chem.* 1993, *58*, 5500. (b) Kundu, A.; Prabhakar, S.; Vairamani, M.; Roy, S. A Novel Copper(II)/Tin(II) Reagent for Aqueous Carbonyl Allylation:In Situ Diagnostics of Reactive Organometallics in Water. *Organometallics* 1997, *16*, 4796-4799.
[60] Diallyltin dibromide has been found an efficient allylation reagent of carbonyl compounds in organic solvents. See: Kobayashi, Nishio, K. *Tetrahedron Lett.* 1995, *36*, 6729-6732, and reference cited therein.
[61] In ref [53], Nokami et al. showed that diallyltin dibromide (4) reacted with benzaldehyde in ether-water mixed solvent, but the intermediacy of 4 in the aqueous Barbier reaction was assumed but not demonstrated.
[62] Chan, T.H.; Yang, Y.; Li, C.J. Organometallic reactions in aqueous media. The nature of the organotin intermediate in the tin-mediated allylation of carbonyl compounds. *J.Org.Chem.* 1999, *64*, 4452-4455.
[63] Bian, Y-J.; Xue, W-L.; Yu, X-G. The allylation reactions of aromatic aldehydes and ketones with tin dichloride in water. *Ultrasonic Sonochemistry* 2010, *17*, 58-60.
[64] Chan, T.H.; Yang, Y. Indium-Mediated Organometallic Reactions in Aqueous Media:The Nature of the Allylindium Intermediate. *J.Am.Chem.Soc.* 1999, *121*, 3228-3229.
[65] Chung, W.J.; Higashiya, S.; Oba, Y.; Welch, J.T. Indium- and zinc-mediated allylation of difluoroacetyltrialkylsilanes in aqueous media. *Tetrahedron* 2003, *59*, 10031-10036.

[66] (a).-Podlech, J.; Maier, T. C. Indium in Organic Synthesis. *Synthesis* 2003, 633–655. (b) Paquette, L. A. Fundamental Mechanistic Characteristics and Synthetic Applications of Allylations Promoted by Indium Metal in Aqueous Media. *Synthesis* 2003, 765–774.

[67] Shen, Z-L-; Cheong, H-L.; Loh, T-P. Indium/copper-mediated conjugate addition of unactivated alkyl iodides to α β unsaturated carbonyl compounds in water. *TetrahedronLett.* 2009, *50*, 1051-1054. Paquette, L.; Mitzel, T,M. Addition of Allylindium Reagents to Aldehydes Substituted at α or β with Heteroatomic Functional Groups. Analysis of the Modulation in Diastereoselectivity Attainable in Aqueous, Organic, and Mixed Solvent Systems. *J.Am.Chem.Soc.* 1996, *118*, 1931-1937.

[68] Araki, S.; Jin, S.-J.; Idou, Y.; Butsugan, Y. Allylation of Carbonyl Compounds with Catalytic Amount of Indium.*Bull. Chem. Soc. Jpn.* 1992, *65*, 1736-1738.

[69] Reetz, M. T.; Jung, A. 1,3-Asymmetric induction in addition reactions of chiral .beta.-alkoxy aldehydes: efficient chelation control via Lewis acidic titanium reagents.*J. Am. Chem. Soc.* 1983, *105*, 4833-4837.

[70] Narasaka, K.; Pai, H. C. Stereoselective synthesis of meso (or erythro) 1,3-diols from .BETA.-hydroxyketones.*Chem. Lett.* 1980, 1415-1418. (b) Narasaka, K.; Pai, H. C. Stereoselective reduction of β- hydroxyketones to 1,3-diols highly selective 1,3-asymmetric induction via boron chelates.*Tetrahedron* 1984, *12*, 2233-2236. (c) Evans, D. A.; Chapman, K. T.; Carreira, E. M. Directed reduction of beta.-hydroxy ketones employing tetramethylammonium triacetoxyborohydride.*J. Am. Chem. Soc.* 1988, *110*, 3560-3565.

[71] Chen, X.; Hortelano, E. R.; Eliel, E. L.; Frye, S. V. Are chelates truly intermediates in Cram's chelate rule?*J. Am. Chem. Soc.* 1990, *112*, 6130-6131; *idem* 1992, *114*, 1778-1784. (b) Frye, S. V.; Eliel, E. L.; Cloux, R. Rapid-injection nuclear magnetic resonance investigation of the reactivity of α- and β-alkoxy ketones with dimethylmagnesium: kinetic evidence for chelation.*J. Am. Chem. Soc.* 1987, *109*, 1862-1865.

[72] Corcoran, R. C.; Ma, J. Geometrical aspects of the activation of ketones by Lewis acids.*J. Am. Chem. Soc.* 1992, *114*, 4536-4539.

[73] Mori, S.; Nakamura, M.; Nakamura, E.; Koga, N.; Morukuma, K. Theoretical Studies on Chelation-Controlled Carbonyl Addition. Me_2Mg Addition to .alpha.- and .beta.-Alkoxy Ketones and Aldehydes.*J. Am. Chem. Soc.* 1995, *117*, 5055-5061.

[74] (a) Poll, T.; Metter, J. O.; Helmchen, G. Concerning the Mechanism of the Asymmetric Diels-Alder Reaction: First Crystal Structure Analysis of a Lewis Acid Complex of a Chiral Dienophile.*Angew. Chem., Int. Ed. Engl.* 1985, 24, 112-114. (b) Buhro, W. E.; Georgiou, S.; Fernández, J. M.; Patton, A. T.; Strouse, C. E.; Gladysz, J. A. Synthesis, structure, and reactivity of the formaldehyde complex $[(.\eta.^5\text{-}C_5H_5)Re(NO)(PPh_3)(\eta.^2\text{-}H_2C:O)]^+$ $PF6^-$.*Organometallics* 1986, *5*, 956-965. (c) Arai, M.; Kawasuji, T.; Nakamura, E. 1,2-Asymmetric induction in the S_N2'-allylation of organocopper and organozinc reagents.*J. Org. Chem.* 1993, *58*, 5121-5129.

[75] Petrier, C.; Luche, J. L. Allylzinc reagents additions in aqueous media.*J. Org. Chem.* 1985, *50*, 910. (b) Einhorn, C.; Luche, J. L. Selective allylation of carbonyl compounds in aqueous media.*J. Organomet. Chem.* 1987, *322*, 177-183.

[76] Wilson, S. R.; Guazzaroni, M. E. Synthesis of homoallylic alcohols in aqueous media.*J. Org. Chem.* 1989, *54*, 3087-3091.

[77] Chan, T.-H.; Li, C. J.; Lee, M. C.; Wei, Z. Y. 1993 R.U. Lemieux Award Lecture Organometallic-type reactions in aqueous media—a new challenge in organic synthesis.*Can. J. Chem.* 1994, *72*, 1181-1192.

[78] (a) Petrier, C.; Luche, J. L. Allylzinc reagents additions in aqueous media.*J.Org. Chem.* 1985, *50*, 910-912.

[79] Smadja, W. Acyclic Diastereofacial Selection in Intermolecular Radical Reactions Steric vs Electronic Controls.*Synlett* 1994,1.

[80] Araki, S.; Ito, H.; Butsugan, Y. Synthesis of β-Hydroxyesters by Reformatsky Reaction Using Indium Metal.*Synth. Commun.* 1988, *18*, 453-458.

[81] Marshall, J. A.; Hinkle, K. W. Synthesis of anti-Homoallylic Alcohols and Monoprotected 1,2-Diols through $InCl_3$-Promoted Addition of Allylic Stannanes to Aldehydes.*J. Org. Chem.* 1995, *60*, 1920-1921.

[82] Lu, W.; Ma, J.; Yang, Y.; Chang, T.H. Organometallic Reactions in Aqueous Media. Indium-Mediated 1,3-Butadien-2-ylation of Carbonyl Compounds. *Org.Lett*, 2000, *2*, 3469-3471.

[83] Yi, X-H.; Meng, Y.; Li, C-J. Indium mediated reactions in water: Synthesis of β-hydroxyl esters. *TetrahedronLett.* 1997, *38*, 4731-4734.

[84] (a) Usugi, S..i; Yorimitsu, H.; Oshima K. Triethylborane-induced radical allylation of α-halo carbonyl compounds with allylgallium reagent in aqueous media. *TetrahedronLett.* 2001, *42*, 4535-4538. (b) Chan, T-H.; Li, C-J. Organometallic reactions in aqueous medium. Conversion of carbonyl compounds to 1,3-butadienes or vinyloxiranes. *Organomet*allics 1990, *9*, 2649-2650.

[85] Araki, S.; Kamei, T.; Hirashita, T.; Yamamura, H.; Kawai, . A New Preparative Method for Allylic Indium(III) Reagents by Reductive Transmetalation of α-Allylpalladium(II) with Indium(I) Salts.*Org. Lett.* 2000, *2*, 847-849.

[86] Miyabe, H.; Yamaoka, Y.; Naito, T.; Takemoto, Y. Regioselectivity in palladium-indium iodide-mediated allylation reaction of glyoxylic oxime ether and N-sulfonylimine. *J.Org.Chem.* 2003, *68*, 6745-6751.

[87] Miyabe, H.; Yamoaka, Y.; Naito, T.; Takemoto, Y. Palladium -Indium Iodide-Mediated Allylation and Propargylation of Glyoxylic Oxime Ether and Hydrazone:The Role of Water in Directing the Diastereoselective Allylation. *J.Org.Chem.* 2004, *69,* 1415-1418.

[88] Petrier, C.; Dupuy, C.; Luche, J. L. Conjugate additions to α,β-unsaturated carbonyl compounds in aqueous media *Tetrahedron Lett.* 1986, *27*, 3149-3152.

[89] Luche, J. L.; Allavena, C. Ultrasound in organic synthesis 16. Optimization of the conjugate additions to α,β-unsaturated carbonyl compounds in aqueous media.*Tetrahedron Lett.* 1988, *29*, 5369-5372.

[90] Dupuy, C.; Petrier, C.; Sarandeses, L. A.; Luche, J. L. Ultrasound in Organic Syntheses. XIX. Further Studies on the Conjugate Additions to Electron Deficient Olefins in Aqueous Media.*Synth.Commun.* 1991, *21*, 643-645.

[91] Luche, J. L.; Allavena, C.; Petrier, C.; Dupuy, C. Ultrasound in organic synthesis 17 mechanistic aspects of the conjugate additions to α-enones in aqueous media.*Tetrahedron Lett.* 1988, *29*, 5373-5374.

[92] Blanchard, P.; El Kortbi, M. S.; Fourrey, J.-L.; Robert-Gero, M. A conjugate addition of primary alkyl iodide derived species to electron deficient olefins.*Tetrahedron Lett.* 1992, *33*, 3319-3322.

[93] Da Silva, A. D.; Maria, E. J.; Blanchard, P.; Fourrey, J.-L.; Robert-Gero, M. A Synthetic Approach to Carbocyclic Sinefungin. *Nucleosides Nucleotides* 1998, *17*, 2175-2187.

[94] Maria, E. J.; Da Silva, A. D.; Fourrey, J.-L. A Radical-Based Strategy for the Synthesis of Higher Homologues of Sinefungin.*Eur. J. Org. Chem.* 2000, 627-631.

[95] Blanchard, P.; Da Silva, A. D.; El Kortbi, M. S.; Fourrey, J.-L.; Robert-Gero, M. Zinc-iron couple induced conjugate addition of alkyl halide derived radicals to activated olefins. *J. Org. Chem.* 1993, *58*, 6517-6519.

[96] Crich, D.; Davies, J. W.; Negrón, G.; Quintero, L. *J. Chem.Res., Synop.* 1988, *140*.

[97] Crich, D.; Davies, J. W. Free-radical addition to di- and tripeptides containing dehydroalanine residues.*Tetrahedron* 1989, *45*, 5641-5654.

[98] Yim, A.-M.; Vidal, Y.; Viallefont, P.; Martinez, J. Solid-phase synthesis of α-amino acids by radical addition to adehydroalanine derivative.*Tetrahedron Lett.* 1999, *40*, 4535-4538.

[99] Shen, Z-L.; Yeo, Y-L.; Loh, T-P. Indium-copper and indium-silver mediated Barbier-Grignard-type reaction of aldehydes using unactivated alkyl halides in water. *J.Org.Chem.* 2008, *73*, 3922-3924.

[100] Li, C.J.; Chan, T.H.Organic synthesis using indium-mediated and catalyzed reactions in aqueous media.*Tetrahedron* 1999, *55*, 11149-11176.

[101] a.-Lu, W.; Chan T.H. Indium-mediated organometallic reactions inaqueous media. Stereoselectivity in the crotylation of sulfonimines bearing a proximal chelating group. *J.Org.Chem.*, 2001, *66*, 3467-3473. b.-Fleming, F.F.; Gudipati, S. Alkenenitriles:Zn/Cu Promoted Conjugate Additions of Alkyl Iodides in Water. *Org.Lett.* 2006, *8*, 1557-1559.

[102] Kang, S.; Jang, T.S.; Keum G.; Kang, S.B.; So-Yeop Han, and Youseung Kim* In/$InCl_3$-mediated cross-coupling reactions of methyl vinyl ketone with benzaldehyde in aqueous media. *Org.Lett.* 2000, *2*, 3615-3617.

[103] Mahesh, M.; Murphy, J.A.; Wessel, H.P. Novel deoxygenation reactions of epoxides by indium. *J.Org.Chem.* 2005, *70*, 4118-4123.

[104] Bieber, L.W.; da Silva, M.F. Mild and efficient synthesis of propargylamines by copper-catalyzed Mannich reaction. *TetrahedronLett.* 2004, *45*, 8281-8283.

[105] Estevam, I.H.S.; Bieber, L.W. Barbier-type allylation of iminium ions generated in situ in aqueous medium. *TetrahedronLett.* 2003, *44*, 667-670.

[106] Clerici, A.; Porta, O. *Gazz.Chim.Ital.* 1992, *122*, 165-166.

[107] Dermican, A; Parsons, P.J. The synthesis of fused-ring systems utilizing the intramolecular addition of alkenyl radicals to furans. *Eur.J.Org.Chem.* 2003, 1729-1732.

[108] (a) Miyabe, H.; Ueda, M.; Fujii, K.; Goto, T.; Naito, T. Tandem carbon-carbon bond-forming radical addition-cyclization of oxime ether and hydrazone. *J.Org.Chem.* 2003, *68*, 5618-5626. (b) Miyabe, H.; Fujii, K.; Goto, T.; Naito, T. A new alternative to the Mannich reaction: tandem radical addition-cyclization for asymmetric synthesis of γ-butyrolactones and β-aminoacids. *Org.Lett.* 2000, *2*, 4071-4074.(c) Miyabe, H.; Fujii, K.; Tanaka, H.; Naito, T. Solid-phase tandem radical addition-cyclization reaction of oxime ethers. *Chem.Comm.* 2001, 831-832.

[109] Ueda, M.; Miyabe, H.; Nishimura, A.; Miyata, O.; Takemoto, Y.; Naito, T. Indium-mediated tandem radical addition-cyclization-trap reactions in aqueous media. *Org.Lett.* 2003, *5*, 3835-3838.
[110] Bailey, W.F.; Carson, M.W. Lithium-iodine exchange mediated atom transfer cyclization: catalytic cycloisomerization of 6-iodo-1-hexenes. *J.Org.Chem.* 1998, *63,* 361-365.
[111] Sakuma, D.; Togo, H. Facile and Environmentally Benign Zn-Mediated Cyclopropanation of Electron-Deficient 2-Iodoethyl-Substituted Olefins via Radical 3-*exo-trig* Manner.*Synlett* 2004, 2501-2504.
[112] Raussou, S.; Urbain, N.; Mangeney, P.; Alexakis, A.; Platzer, N. Preparation of chiral hexahydroquinolizinones and tetrahydroindolizinones by regio- and diastereoselective sonochemical cyclization of chiral dihydropyridines. *Tetrahedron Lett.* 1996, *37*, 1599-1602.
[113] Mangeney, P.; Hamon, L.; Raussou, S.; Urbain, N.; Alexakis, A. Radical cyclizations of 1,4-dihydropyridines. Synthesis of chiral fused nitrogen heterocycles. Synthesis of lupinine and epilupinine. *Tetrahedron* 1998, *54*, 10349-10362.
[114] Marshall, J. A.; Robinson, E. D. A mild method for the synthesis of furans. Application to 2,5-bridged furano macrocyclic compounds.*J. Org. Chem.* 1990, *55*, 3450-3451.
[115] Marshall, J. A.; Wang, X. J. Synthesis of furans by silver(I)-promoted cyclization of allenyl ketones and aldehydes.*J. Org. Chem.* 1991, *56*, 960-969.
[116] Marshall, J. A.; Sehon, C. A. Synthesis of Furans and 2,5-Dihydrofurans by Ag(I)-Catalyzed Isomerization of Allenones, Alkynyl Allylic Alcohols, and Allenylcarbinols.*J. Org. Chem.* 1995, *60*, 5966-5968.
[117] Marshall, J. A.; Bartley, G. S. Observations Regarding the Ag(I)-Catalyzed Conversion of Allenones to Furans.*J. Org. Chem.* 1994, *59*, 7169-7171.
[118] Hashmi, A. S. K.; Schwarz, L.; Bats, J. W. Mercury(II)-Catalyzed Synthesis of Spiro[4.5]decatrienediones from Allenyl Ketones and Comparison with Silver(I)-, Palladium(II)- and Brønsted Acid-Catalyzed Reactions.*J. Prakt. Chem.* 2000,*342*, 40-43.
[119] Panchaud, P.; Renaud, P. A convenient procedure for radical carboazidation and azidation.*J.Org.Chem.* 2004, *69*, 3205-3207.
[120] Sugi, M.; Sakuma, D.; Togo, H. Zinc- and indium-mediatedring-expansion reactions of α-halomethyl cyclic β-ketoesters in aqueous ethanol.*J.Org.Chem.* 2003, *68*, 7629-7633.
[121] Cao, L.; Li, C. *p*-methoxybenzenediazonium tetrafluoroborate / $TiCl_3$: a novel initiator for halogen atom transfer radical reactions in aqueous media. *TetrahedronLett.* 2008, *49*, 7380-7382.
[122] Hawthorne, M. F.; Reintjes, M. The preparation of alkylhydroquinones by the reductive alkylation of quinones with trialkylboranes. *J.Am.Chem.Soc.* 1965, *87*, 4585-4587.
[123] Kabalka, G. W.The reaction of trialkylboranes with 1,4-naphthoquinone: a new, convenient synthesis of 2-alkyl-1,4-naphthalenediols. Evidence for a free radical chain mechanism.*J.Organomet.Chem.* 1971, *33*, C25-C28.
[124] Tsuchiya, Y.; Izumisawa, Y.; Toho, H. 3-exo-tet Cyclization of 2,2-disubstituted 1,3-dihalopropanes with indium in aqueous and ionic liquid solvent system. *Tetrahedron* 2009, *65*, 7533-7537.

[125] Khan, F.A.; Satapathy, F.; Dash. J.; Savitha, G. *trans*-Hydrindane Ring System. *J.Org.Chem.* 2004, *69*, 5295-5301.

[126] Kim, B.H.; Jim, Y.; Jum, Y.M.; Han, R.; Baik, W.; Lee, B.M. Indium mediated reductive heterocyclization of 2-nitroacylbenzenes or 2-nitroiminobenzenes toward 2,1-benzisoxazoles in aqueous media. *TetrahedronLett.* 2000, *41*, 2137-2140.

[127] (a) Russell, G. A.; Jawdosiuk, M.; Makosza, M. Reactions of resonance stabilized anions. 35. Electron transfer processes. 18. Reaction of 2-halo-2-nitropropanes and 2,2-dinitropropane with resonance-stabilized carbanions and 1-alkynyllithiums. *J. Am. Chem. Soc.* 1979, *101*, 2355-2359. (b) Russell, G. A.; Metcalfe, A. R. Electron transfer processes. 19. Reaction of aryl radicals with anions in aqueous solution. *J. Am. Chem. Soc.* 1979, *101*, 2359-2362. (c) Russell, G. A.; Mydryk, B. Electron transfer processes. 31. Coupling of enolates of phenones with 2-chloro-2-nitropropane. *J. Org. Chem.* 1982, *47*, 1879-1884. (d) Russell, G. A.; Baik, W. *J. Chem. Soc., Chem. Commun.* 1988, 196-198.

[128] (a) Alcaide, B.; Almendros, P.; Aragoncillo, C.; Redondo, M.C.; Torres, M.R. Synthesis of Strained Tricyclic β-Lactams by Intramolecular [2+2] Cycloaddition Reactions of 2-Azetidinone-Tethered Enallenols: Control of Regioselectivity by Selective Alkene Substitution. *Chem.Eur.J.* 2006, *12*, 1539-1546. (b) Alcaide, B.; Almendros, P.; Aragoncillo, C.; Redondo, M.C. Carbonyl Allenylation/Free Radical Cyclization Sequence as a New Regio- and Stereocontrolled Access to Bi- and Tricyclic β-Lactams. *J.Org.Chem.* 2007, *72*, 1604-1608.

[129] Estévez, R.E.; Oller-López, J.L.; Robles, R.; Melgarejo, C.R.; Gansauer, A.; Cuerva, J.M.; Oltra, J.E. Stereocontrolled Coupling between Aldehydes and Conjugated Alkenals Mediated by Ti^{III}/H_2O.*Org.Lett.* 2006, *8*, 5433-5436.

[130] Fittig, R. *Liebigs Ann.* 1859, *110*, 23.

[131] (a) Bhar, S.; Guha, S. The remarkable role of water during the chemoselective reduction of ketones mediated by metallic aluminium. *Tetrahedron Lett.* 2004, *45,* 3775-3777; (b) Li, L. H.; Chan, T. H. Effect of Fluoride Salts on Metal-Mediated Reactions. Aluminum/Fluoride Salt-Mediated Reduction and Pinacol Coupling of Carbonyl Compounds in Aqueous Media. *Org. Lett.* 2000, *2*, 1129-1132; (c) Sahade, D. A.; Mataka, S.; Sawada, T.; Tsukinoki, T.; Tashiro, M. [2.2]Metacyclophanes having hydroxy groups on the bridge. *Tetrahedron Lett.* 1997, *38*, 3745-3746; (d) Khurana, J. M.; Sehgal, A. Rapid pinacolization of carbonyl compounds with aluminium-KOH. *J. Chem. Soc., Chem. Commun.* 1994, 571.

[132] (a) Ueda, T.; Kanomata, N.; Machida, H. Synthesis of Planar-Chiral Paracyclophanes via Samarium(II)-Catalyzed Intramolecular Pinacol Coupling. *Org. Lett.* 2005, *7*, 2365-2368; (b) Aspinall, H. C.; Greeves, N.; Valla, C. Samarium Diiodide-Catalyzed Diastereoselective Pinacol Couplings. *Org. Lett.* 2005, *7*, 1919-1922; (c) Matsukawa, S.; Hinakubo, Y. Samarium(II)-Mediated Pinacol Coupling in Water:Occurrence of Unexpected Disproportionation and Action of Low-Valent Samarium as an Active Species. *Org. Le*tt. 2003, *5*, 1221-1223; (d) Wang, L.; Zhang, Y. M. Novel reductive coupling cyclization of 1,1-dicyanoalkenes promoted by metallic samarium in aqueous media. *Tetrahedron Lett.* 1998, *39*, 5257-5260; (e) Molander, G. A.; Kenny, C. Intramolecular reductive coupling reactions promoted by samarium diiodide. *J. Am. Chem. Soc.* 1989, *111*, 8236-8246; (f) Namy, J. L.; Souppe, J.; Kagan, H. B. Efficient

formation of pinacols from aldehydes or ketones mediated by samarium diiodide. *Tetrahedron Lett.* 1983, *24*, 765-766.
[133] (a) Xu, X. L.; Hirao, T. *J. Org. Chem.* 2005, *70*, 8594-8596; (b) Hirao, T.; Takeuchi, H.; Ogawa, A.; Sakurai, H. *Synlett* 2000, 1658-1660; (c) Hirao, T.; Hatano, B.; Imamoto, Y.; Ogawa, A. A Catalytic System Consisting of Vanadium, Chlorosilane, and Aluminum Metal in the Stereoselective Pinacol Coupling Reaction of Benzaldehyde Derivatives. *J. Org. Chem.* 1999, *64*, 7665-7667; (d) Kammermeier, B.; Beck, G.; Jendralla, H.; Jacobi, D. Vanadium(II)- and Niobium(III)-Induced, Diastereoselective Pinacol Coupling of Peptide Aldehydes toC2-Symmetrical HIV-Protease Inhibitors. *Angew. Chem., Int. Ed. Engl.* 1994, *33,* 685-689; (e) Annunziata, R.; Benaglia, M.; Cinquini, M.; Cozzi, F.; Giaroni, P. Synthesis of optically active 3-(1-hydroxyalkyl)phthalides by stereoselective pinacol cross-coupling. *J. Org. Chem.* 1992, *57*, 782-784; (f) Kempf, D. J.; Sowin, T. J.; Doherty, E. M.; Hannick, S. M.; Codavoci, L.; Henry, R. F.; Green, B. E.; Spanton, S. G.; Norbeck, D. W. Stereocontrolled synthesis of C2-symmetric and pseudo-C2-symmetric diamino alcohols and diols for use in HIV protease inhibitors. *J. Org. Chem.* 1992, *57*, 5692-5700; (g) Freudenberger, J. H.; Konradi, A. W.; Pederson, S. F. Intermolecular pinacol cross coupling of electronically similar aldehydes. An efficient and stereoselective synthesis of 1,2-diols employing a practical vanadium(II) reagent. *J. Am. Chem. Soc.* 1989, *111*, 8014-8016.
[134] Handy, S. T.; Omune, D. A Chelation Effect on the Pathway between Intramolecular Hydrodimerization and Pinacol Coupling. *Org. Lett.* 2005, *7*, 1553-1555.
[135] (a) Mukaiyama, T.; Yoshimura, N.; Igarashi, K.; Kagayama, A. Efficient diastereoselective pinacol coupling reaction of aliphatic and aromatic aldehydes by using newly utilized low valent titanium bromide and iodide species. *Tetrahedron* 2001, *57*, 2499-2506; (b) Li, T.-Y.; Cui, W.; Liu, J. G.; Zhao, J. Z. A highly *dl*-stereoselective pinacolization of aromatic aldehydes mediated by $TiCl_4$–Zn.*Chem. Commun.* 2000, 139-140; (c) Tsukinoki, T.;awaji, T.; Hashimoto, I.; Mataka, S.; Tashiro, M. Pinacolic Coupling of Aromatic Carbonyl Compounds Using Zn Powder in Aqueous Basic Media without Organic Solvents. *Chem. Lett.* 1997, 235-240; (d) Tanaka, K.; Kishigami, S.; Toda, F. A new method for coupling aromatic aldehydes and ketones to produce α.-glycols using zinc-zinc dichloride in aqueous solution and in the solid state.*J. Org. Chem.* 1990, *55*, 2981-2983.
[136] Takai, K.; Morita, R.; Matsushita, H.; Toratsu, C. Cross-pinacol-type coupling reactions between α,β-unsaturated ketones and aldehydes with low-valent metals. *Chirality* 2003, *15*, 17-19.
[137] David, S. H.; Gregory, C. F. Metal Hydride Mediated Intramolecular Pinacol Coupling of Dialdehydes and Ketoaldehydes. *J. Am. Chem. Soc.* 1995, *117*, 7283-7284.
[138] Yamamoto, Y.; Hattori, R.; Itoh, K. Highly *trans*-selective intramolecular pinacol coupling of dials catalyzed by bulky Cp_2TiPh. *Chem. Commun.* 1999, 825-826.
[139] (a) Groth, U.; Jeske, M. Diastereoselective Ce(III)-Catalyzed Pinacol Couplings of Aldehydes.*Synlett* 2001, 129; (b) Groth, U.; Jeske, M. Diastereoselective Ce(OiPr)3-Catalyzed Pinacol Couplings of Aldehydes. *Angew. Chem., Int. Ed.* 2000, *39*, 574-576.
[140] Khan, R. H.; Mathur, R. K.; Ghosh, A. C. Tellurium-KOH Mediated Rapid Pinacolization of Carbonyl Compounds. *Synth. Commun.* 1997, *27*, 2193-2196.

[141] Ephritikhine, M.; Maury, O.; Villiers, C.; Lance, M.; Nierlich, M.A mechanistic study of the reductive coupling of acetone with uranium compounds. *J. Chem. Soc., Dalton Trans.* 1998, 3021-3028.

[142] Svatos, A.; Boland, W. *Synlett* 1998, 126-128.

[143] Wang, Z. Y.; Yuan, S. Z.; Zha, Z. G.; Zhang, Z. D. Coupling reaction of carbonyl compounds mediated by gallium metal in aqueous media. *Chin. J. Chem.* 2003, *21*, 1231-1235.

[144] (a) Ohe, T.; Ohse, T.; Mori, K.; Ohtaka, S.; Uemura, S. Indium-Catalyzed Cross-Coupling Reactions between α,β-Unsaturated Carbonyl Compounds and Aromatic Aldehydes. *Bull. Chem. Soc. Jpn.* 2003, *76*, 1823-1827; (b) Nair, V.; Ros, S.; Jayan, C. N.; Rath, N. P. Indium/indium trichloride mediated pinacol cross-coupling reaction of aldehydes and chalcones in aqueous media: a facile stereoselective synthesis of substituted but-3-ene-1,2-diols? *Tetrahedron Lett.* 2002, *43*, 8967-8969; (c) Mori, K.; Ohtaka, S.; Uemura, S. $InCl_3$-Catalyzed Reductive Coupling of Aromatic Carbonyl Compounds in the Presence of Magnesium and Chlorotrimethylsilane. *Bull. Chem. Soc. Jpn.* 2001, *74*, 1497-1498; (d) Lim, H. J.; Keum, G.; Kang, S. B.; Chung, B. Y.; Kim, Y. Indium mediated pinacol coupling reaction of aromatic aldehydes in aqueous media. *Tetrahedron Lett.* 1998, *39*, 4367-4368.

[145] Wang, C.; Pan, Y.; Wu, A. $InCl_3$/Al mediated pinacol coupling reactions of aldehydes and ketones in aqueous media. *Tetrahedron* 2007, *63*, 429-434.

[146] Keller, C.L.; Dalessandro, J.D.; Hotz, R.P.; Pinhas, A.R. Reactions in water: Alkyl nitrile coupling reactions using Fenton´s reagent. *J.Org.Chem.* 2008, *73*, 3616-3618.

[147] Walling, C. Intermediates in the Reactions of Fenton Type Reagents.*Acc. Chem. Res.* 1998, *31*, 155-157.

[148] (a) See, for example: MacFaul, P. A.; Wayner, D. D. M.; Ingold, K. U. *Acc. Chem. Res.* 1998, *31*, 159-162. (b) Goldstein, S.; Meyerstein, D. Comments on the Mechanism of the Fenton-Like Reaction.*Acc. Chem. Res.* 1999, *32*, 547-550.

[149] See, for example: Sawyer, D. T.; Sobkowiak, A.; Matsushita, T. *Acc. Chem. Res.* 1996, *29*, 409-416.

[150] (a) Coffman, D. D.; Jenner, E. L.; Lipscomb, R. D. Syntheses by Free-radical Reactions. I. Oxidative Coupling Effected by Hydroxyl Radicals.*J. Am. Chem. Soc.* 1958, *80*, 2864-2872. (b) Also see: Walling, C.; El-Taliawi, G.; M, *J. Am. Chem. Soc.* 1973, *95*, 844-847.

[151] Attempted radical-coupling reactions of phenylacetonitrile were unsuccessfuldue to its insolubility in water.

[152] Yang, Y-S.; Shen, Z-L.; Loh, T-P. Zn/$InCl_3$-mediated pinacol cross-coupling reactions of aldehydes with α.β-unsaturated ketones in aqueous media. *Org.Lett.* 2009, *11*, 2213-2215.

[153] Kalyaman, N.; Rao, V. Novel reductive coupling of aldimines by indium under aqueous conditions. *TetrahedronLett.* 1993, *34*, 1647-1648.

[154] Nair, V.; Ros, S.; Jayan, C.N.; Rath, NP. Indium/indium trichloride mediated pinacol cross-coupling reaction of aldehydes and chalcones in aqueous media: a facile stereoselective synthesis of substituted but-3-ene-1,2-diols. *TetrahedronLett.* 2002, *43*, 8967-8969.

[155] Halternan, R.L.; Porterfield, J.P.; Mekala, S. Chromium-catalyzed pinacol coupling of benzaldehyde in water. *TetrahedronLett.* 2009, *50*, 7172-7174.

[156] Alonnso, F.; Beletskaya, I.P.; Yus, M. Metal-mediated reductive hydrodehalogenation of organic halides. *Chem.Rev.* 2002, *102*, 4009-4091. For a review, see: Taciolu, S. *Tetrahedron* 1996, *52*, 11113-11152.
[157] Park, L.; Keum, G.; Kang, S. B.; Kim, K. S.; Kim, Y. Indium-mediated reductive dehalogenation of α-halocarbonyl compounds in water.*J. Chem. Soc., Perkin Trans. 1* 2000, 4462-4464.
[158] Chae, H.; Sangwon, C.; Keum, G.; Kang, S. B.; Pae, A. N.; Kim, Y. A facile synthesis of 7-amino-3-desacetoxycephalosporanic acid derivatives by indium-mediated reduction of 3-iodomethylcephems in aqueous media.*Tetrahedron Lett.* 2000, *41*, 3899-3901.
[159] Sayles, G. D.; You, G.; Wang, M.; Kupferle, M. J. DDT, DDD, and DDE Dechlorination by Zero-Valent Iron.*Environ. Sci. Technol.* 1997, *31*, 3448-3454.
[160] Eykholt, G. R.; Davenport, D. T. Dechlorination of the Chloroacetanilide Herbicides Alachlor and Metolachlor by Iron Metal.*Environ. Sci. Technol.* 1998, *32*, 1482-1487.
[161] Bridges, D. C. In *Triazine Herbicides: Risk Assessment*; Bal-lantine, L. G., McFarland, J. E., Hackett, D., Eds.; ACS Symposium Series 683; American Chemical Society: Washington, DC, 1998; p 24.
[162] Wackett, L. Atrazine Pathway Map. The University of Minnesota Biocatalysis/Biodegradation Database. Available online at http:// www.labmed.umn.edu/umbbd.
[163] Monson, S. J.; Ma, L.; Cassada, D. A.; Spalding, R. F. Confirmation and method development for dechlorinated atrazine from reductive dehalogenation of atrazine with Fe(0).*Anal. Chim. Acta* 1998, *373*, 153-160.
[164] Dombek, T.; Dolan, E.; Schultz, J.; Klarup, D. Rapid reductive dechlorination of atrazine by zero-valent iron under acidic conditions.*Environ.Pollut.* 2001, *111*, 21-27.
[165] Weber, E. J. Iron-Mediated Reductive Transformations:Investigation of Reaction Mechanism.*Environ. Sci. Technol.* 1996, *30*, 716-719.
[166] Hung, W.-H.; Schwartz, J.; Bernasek, S. L. Sequential oxidation of Fe(100) by water adsorption: formation of an ordered hydroxylated surface. *Surf. Sci.* 1991, *248*, 332-342.
[167] (a) Matheson, L. J.; Tratnyek, P. G. Reductive Dehalogenation of Chlorinated Methanes by Iron Metal.*Environ. Sci. Technol.* 1994, *28*, 2045-2053. (b) Johnson, T. L.; Scherer, M. W.; Tratnyek, P. G. Kinetics of Halogenated Organic Compound Degradation by Iron Metal.*Environ. Sci. Technol.* 1996, *30*, 2634-2640. (c) Tratnyek, P. G.; Johnson, T. L.; Scherer, M. M.; Eykholt, G. R. *Ground Water Monit. Rem.* 1997, *17*, 109. (d) Tratnyek, P. G.; Scherer, M. M. *Proc. Natl. Conf. Environ. Eng.* 1998, 110. (e) Scherer, M. M.; Balko, B. A.; Gallagher, D. A.; Tratnyek, P. G. Correlation Analysis of Rate Constants for Dechlorination by Zero-Valent Iron.*Environ. Sci. Technol.* 1998, *32*, 3026-3033. (f) Johnson, T. L.; William, F.; Gorby, Y. A.; Tratnyek, P. G. Degradation of carbon tetrachloride by iron metal: Complexation effects on the oxide surface.*J. Contam. Hydrol.* 1998, *29*, 379-398.
[168] Hirasawa, N.; Takahashi, Y.; Fukuda, E.; Sugimoto, O.; Tanji, K-i. Indium-mediated dehalogenation of haloheteroaromatics in water. *TetrahedronLett.* 2008, *49*, 1492-1494.
[169] Rabion, A.; Buchanan, R. M.; Seris, J.-L.; Fish, R. H. Biomimetic oxidation studies. 10. Cyclohexane oxidation reactions with active site methane monooxygenase enzyme

models and t-butyl hydroperoxide in aqueous micelles: Mechanistic insights and the role of *t*-butoxy radicals in the Cbz.sbnd;H functionalization reaction.*J. Mol. Catal.A* 1997, *116*, 43-47.

[170] Kirillova, M.V.; Kirillov, A.M.; Pombeiro, A.J.L. Metal-Free and Copper-Promoted Single-Pot Hydrocarboxylation of Cycloalkanes to Carboxylic Acids in Aqueous Medium. *Adv.Synth.Catal.* 2009, *351*, 2936-2948.

[171] Dhakshinamoorthy, A.; Pitchumani, K. Clay-anchored non-heme iron/salen complex catalyzed cleavage of C-C bond in aqueous medium.*Tetrahedron* 2006, *62*, 9911-9918.

[172] (a) Nam, W.; Han, H. J.; Oh, S.-Y.; Lee, Y. J.; Choi, M.-H.; Han, S.-Y.; Kim, C.; Woo, S. K.; Shin, W. tert-Alkyl Hydroperoxides by Iron(III) Porphyrin Complexes. *J.Am.Chem.Soc.* 2000, *122*, 8677-8684 and references cited therein; (b) Mansuy, D.;Leclaire, J.; Fontecave, M.; Dansette, P. Regioselectivity of olefin oxidation by iodosobenzene catalyzed by metalloporphyrins : control by the catalyst.*Tetrahedron* 1984, *40*, 2847-2857 and references cited therein.

[173] Tuynman, A.; Spelberg, J. L.; Kooter, I. M.; Schoemaker, H. E.; Wever, R. Enantioselective Epoxidation and Carbon-Carbon Bond Cleavage Catalyzed by Coprinus cinereus Peroxidase and Myeloperoxidase.*J.Biol. Chem.* 2000, *275*, 3025-3030.

[174] Matsushita, M.; Kamata, K.; Yamaguchi, K.; Mizuno, N. Heterogeneously Catalyzed Aerobic Oxidative Biaryl Coupling of 2-Naphthols and Substituted Phenols in Water.*J. Am.Chem. Soc.* 2005, *127*, 6632-6640.

[175] Wallis, P. J.; Booth, K. J.; Patti, A. F.; Scott, J. L. Oxidative coupling revisited: solvent-free, heterogeneous and in water.*Green Chem.* 2006, *8*, 333-337.

[176] Jiang, Y.; Xi, C.; Yang, X. CuBr/H_2O_2-Mediated Oxidative Coupling of *N,N*-Dialkylarylamines in Water: A Practical Synthesis of Benzidine Derivatives.*Synlett* 2005, 1381-1384.

[177] Xi, C.; Jiang, Y.; Yang, X. Remarkably efficient oxidative coupling of N,N-dialkylarylamines in water mediated by cerium(IV) ammonium nitrate.*Tetrahedron Lett.* 2005, *46*, 3909-3911.

[178] Merlic, C.A.; Miller, M.M.; Hietbrink, B.N.; Houk, K.N. 6-Arene)tricarbonylchromium Complexes toward Additions of Anions, Cations, and Radicals. *J.Am.Chem.Soc.* 2001, *123*, 4904-4918.

[179] Cao, L.; Shen, M.; Li, C. 6-Arene)tricarbonylmanganese Complexes in Aqueous Media. *Organometallics* 2005, *24*, 5983-5988.

[180] Kim, J.S.; Jan, J.H.; Lee, J.J.; Jun, Y.M.; Lee, B.M.; Kim, B.H. Indium-HI-mediated one-pot reaction of 1-(2-arylethynyl)-2-nitroarenes to 2-arylindoles. *TetrahedronLett.* 2008, *49*, 3733-3738.

[181] Ankner, T.; Hilmersson, G. Instantaneous SmI2/H2O/amine mediated reduction of nitroalkanes and α,β-unsaturated nitroalkenes. *TetrahedronLett.* 2007, *48*, 5707-5710.

[182] Wang, C.; Wan, J.; Zheng, Z.; Pan, Y. A new InCl3-catalyzed reduction of anthrones and anthraquinones by using aluminum powder in aqueous media. *Tetrahedron* 2007, *63*, 5071-5075.

[183] (a) Parmar, D.; Duffy, L.A.; Sadasivam, D.V.; Matsubara, H.; Bradly, P.A.; Flowers II, R.A.; Procter, D.J. Studies on the Mechanism, Selectivity, and Synthetic Utility of Lactone Reduction Using SmI_2 and H_2O.*J.Am.Chem.Soc.* 2009, *131*, 15467-15473. (b) Guazelli, G.; De Grazia, S.; Collins, K.D.; Matsubara, H.; Spain, M.; Procter,

D.J.Selective Reductions of Cyclic 1,3-Diesters Using SmI_2 and H_2O. *J.Am.Chem.Soc.* 2008, *131*, 7214-7215.

[184] Dahlén, A.; Hilmersson, G. Mechanistic Study of the SmI_2/H_2O/Amine-Mediated Reduction of Alkyl Halides: Amine Base Strength (pK_{BH+}) Dependent Rate. *J.Am.Chem.Soc.* 2005, *127*, 8340-8347.

In: Organic Radical Reactions in Water and Alternative Media ISBN 978-1-61209-648-3
Editor: J. Alberto Postigo, pp. 171-205

Chapter 4

CLASSICAL SYNTHETIC FREE RADICAL TRANSFORMATIONS IN ALTERNATIVE MEDIA: SUPERCRITICAL CO_2, IONIC LIQUIDS AND FLUOROUS MEDIA

***Ioannis N. Lykakis** and V. Tamara Perchyonok*#**

Department of Chemistry, University of Crete, Voutes 71003 Iraklion, Greece
VTPCHEM PTY Ltd, Melbourne 3163, Australia

ABSTRACT

Solvent usage is often an integral part of manufacturing process, whether it is chemical or another industrial sector. Thus, this unavoidable choice of a specific solvent for a desired manufacturing process can have profound economic, environmental, and societal implications. Some of the impacts are long lasting especially from an environmental perspective, which has been well documented in the scientific literature. The pressing need to develop alternative solvents for manufacturing processes originates, in part, from these implications and constitutes an essential strategy under an emerging field of green chemistry. Whereas there have been excellent advances in developing several alternative clean solvents, it is unlikely that the one solvent will be a panacea for various chemical protocols. This chapter highlights the substantial progress, which has been made in the last decade to the reactive free radical species in alternative media, such as ionic liquids (supported and non-supported), fluorous solvents and supercritical CO_2. It shows that, armed with an elementary knowledge of kinetics and some common sense, it is possible to harness radicals into tremendously powerful tools for solving synthetic problems and broad range of applications. The issue is a valuable and informative entry for growth and development of green free radical chemistry in aqueous and alternative media.

---

* Tel +61414596304, fax +61395725223 email: tamaraperchyonok@gmail.com

Tel +302810545041, fax +302810545001, email lykakis@chemistry.uoc.gr.

1. INTRODUCTION

Chemists are now moving away from volatile, environmentally harmful, and biologically incompatible organic solvents. With their low cost, ready availability, and capacity to remove environmentally unfriendly by-products, water and alternative biocompatible reaction media such as ionic liquids, fluorous solvents, or on solid support have become obvious replacements. Recent advances on free radical chemistry in water have expanded the versatility and flexibility of bond formations in aqueous media and alternative media [1-3]. This chapter highlights the substantial progress, which has been made in the last decade on the reactive free radical species in alternative media, such as ionic liquids (supported and non-supported), fluorous solvents and supercritical CO_2 [4]. It shows that, armed with an elementary knowledge of kinetics and some common sense, it is possible to harness radicals into tremendously powerful tools for solving synthetic problems and acquire a broad range of applications. The issue is a valuable and informative entry for growth and development of green free radical chemistry in aqueous and alternative media.

2. RADICAL CHEMISTRY: BRIEF INTRODUCTION

Radicals are species with at least one unpaired electron, which in contrast to organic anions and cations, react easily with themselves in bond forming reactions. In the liquid phase most of these reactions occur with diffusion controlled rates. Radical-radical reactions can be slowed down only if radicals are stabilized by electronic effects (stable radicals) or shielded by steric effects (persistent radicals). But these effects are not strong enough to prevent diffusion controlled recombination of, for example benzyl radicals or *tert*-butyl radicals [5]. Only in extreme cases, the radical or di-*tert*-butylmethyl radical, recombination rates are low [6]. While the recombination rates of the triphenylmethyl radical is reduced due to both steric and radical stabilizing effects, the steric effect alone slows down the recombination of the di-*tert*-butyl methyl radical. Since neither of the radicals have C-H bonds β to the radical centre, disproportionation reaction, in which the hydrogen atom is transferred, cannot occur [1e, 7].

2.1. Reactions between Radicals

The fact that reactions between radicals are in most cases very fast could lead to a conclusion that direct radical combination is the most synthetically useful reaction mode. This, however, is not the case because direct radical-radical reactions have several disadvantages:

- In the recombination reactions, the radical character is destroyed so that one has to work with at least equivalent amounts of radical initiators.
- The diffusion controlled rates in radical-radical reactions give rise to low selectivity which cannot be influenced by reaction conditions
- The concentrations of radicals are so low that reactions with non-radicals, like solvents, which are present in high concentrations, are very hard to prevent.

2.2. Reaction between a Radical and Non-radical Species

Nevertheless, there are several classes of synthetically useful reactions involving free radicals with non-radicals [5-15]. It posses the following advantages:

- The radical character is not destroyed during the reaction; therefore, one can work with catalytic amounts of radical initiators.
- Most of the reactions are not diffusion controlled, and the selectivities can be influenced by variation of the substituents.
- The concentration of the non-radicals can be easily controlled.

In most cases, in order to apply reactions between radicals and non-radical species for synthesis, chain reactions have to be encouraged and established.For the successful use of radical chains two fundamental conditions have to be met:

- The selectivity of the radicals involved in the chain have to differ from each other
- The reaction between radicals and non-radicals must be faster than radical combination reaction.

In practice these rules can be best illustrated by chain reactions that have gained increasing synthetic application over the years and become one of the fundamental pillars of free radical chemistry [5,8,9]. In a chain reaction, alkylhalides and alkenes react in the presence of tributyltin hydride to give products.

For a successful application of the tin-method, alkyl radicals must attack alkenes to form radical adducts. Trapping of the newly formed radical yields the formation of the products and tributyltin radicals, which react with alkyl halides to give back radical adducts. The tin method can be synthetically useful only if these reactions are faster than all other possible reactions of formed radicals. Therefore, the radicals in the chain must meet certain selectivity and reactivity prerequisites.

2.3. Reactivity and Selectivity

As it was previously mentioned radicals can undergo a number of different and competitive reactions. These processes have different rates of reaction and if one reaction proceeds at a much faster rate than all of the rest we have selective and high-yielding processes. Alternatively, if a variety of reactions proceed at similar rates, the radical will react unselectively to produce a number of different products. The rates of these reactions can vary enormously and, for example, the rate constants of abstraction reaction can vary by a factor of at least 10000. The key factors that influence radical reactivity include enthalpy, entropy, steric effects, stereo-electronic effects, polarity and redox potential (Figure 1) [6,16].

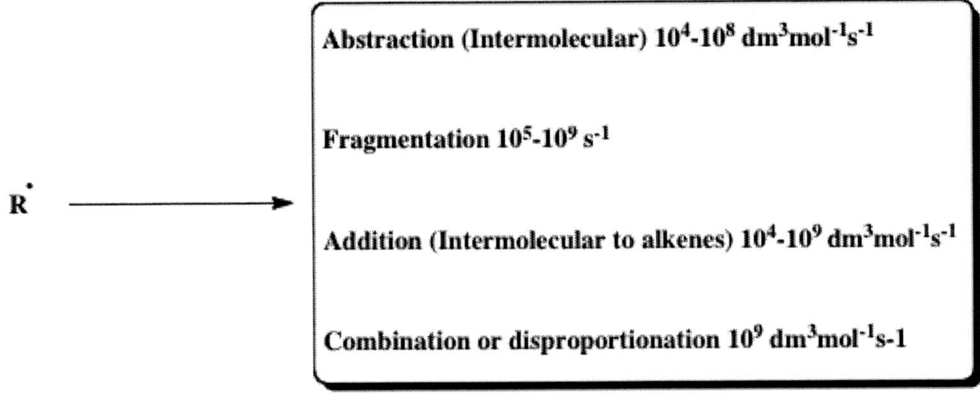

Figure 1. Rates offundamental radical reactions.

2.4. Chain *versus*Non-chain Free Radical Processes in Brief

The tremendous advantages and achievements in the field of free radical chain reactions were highlighted, with the desired end products being predetermined in the propagation steps. The faster the propagation step, the greater the observed efficiency: less initiator is needed, fewer unwanted side reactions can compete; and radical-radical interactions constituting the termination step become negligible. In summary, the whole approach was aimed at reducing radical-radical interaction by keeping the steady-state concentration of the intermediate radical species as low as possible. Since the interactions are usually diffusion controlled and therefore unselective, it might be seen (from the first glance) as an impossible task to gain control on inter-radical reaction for synthetic purpose. However it is not the case at all, due to the elegant and ingenious solution to this problem based on the persistent radical effect, also known as the Fischer-Ingold effect. This phenomenon, which has only recently been understood, underlies several reactions occurring in nature as well as novel synthetic applications recently discovered and elegantly applied in the special field of living radical polymerization.

3. Radical Reactionsin SupercriticalFluids

3.1. Radical Reactions and Supercritical CO_2: Is there a Hidden Advantage?

The use of supercritical fluids as a reaction medium offers the chemical and pharmaceutical industries the opportunity to replace conventional hazardous organic solvents and simultaneously optimize and control more precisely the effect of the solvent on reactions. Supercritical fluids, unlike conventional liquid solvents, can be "pressure- tuned" to exhibit gas-like to liquid-like properties.Supercritical fluids have liquid-like local densities and solvent strength, which can be tuned by adjusting the pressure in the reactor, allowing thus the control of the solubility of the reactants along with density-dependant properties such as dielectric constant, viscosity and diffusivity.Additionally, solubility control through pressure can allow for easy separation of products and catalysts from the supercritical solvent.

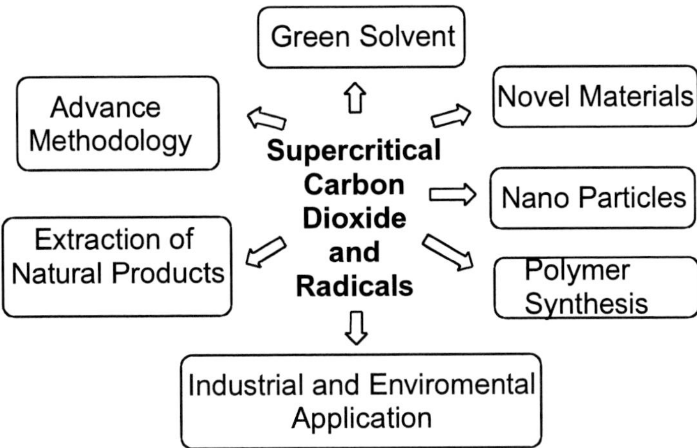

Figure 2. Summary of application of supercritical CO_2 in radical chemistry.

Supercritical fluids are an attractive medium for chemical reactions because of their unique properties. Most of the important physical and transport properties that influence the kinetics of chemical reactions are intermediate between those of a liquid and a gas in supercritical fluid medium. The reactants and the supercritical fluids frequently form a single supercritical fluid phase. Supercritical fluids share many of the advantages of gas phase reactions including miscibility with other gases, low viscosity, high diffusivities, thereby providing enhanced heat transfer and the potential for fast reactions. Supercritical fluids are especially attractive as reaction medium for diffusion-controlled reagents (Figure 2).

Supercritical fluids (Scoffs, SCF), and supercritical carbon dioxide ($scCO_2$) in particular, are emerging as some of the most promising sustainable technologies to have been developed in recent years. They are not new however and can be traced back to original reports dated as long back as the existence of the critical point of alcohol using equipment originally designed by Denys Papen in 1960.

A SCF is defined as a substance above its critical temperature (T_C) and critical pressure (P_C) (Figure 3). This definition should arguably include the clause "but below the pressure required for condensation into a solid", however this is commonly omitted, as a pressure required for condensing a SCF into a solid is generally impracticably high. The critical point represents the highest temperature and pressure at which the substance can exist as a vapor and a liquid in equilibrium. The phenomenon can be easily explained with reference to the phase diagram for pure carbon dioxide. This shows the areas were carbon dioxide exists as a gas, liquid, solid or as a SCF. The curves represent the temperature and pressures where two phases coexist in equilibrium (at the triple point, the three phases co-exist). The gas-liquid coexistence curve is known as the boiling curve. If we move upwards along the boiling curve, increasing temperatures and pressure, then the liquid becomes less dense as the pressure rises. Eventually, the densities of the two phases converge and become identical, the distinction between gas and liquid disappears, and the boiling curve comes to an end at the critical point.

The critical point of carbon dioxide occurs at a pressure of 73.8 bar and a temperature of 31.1°C (Figure 3).

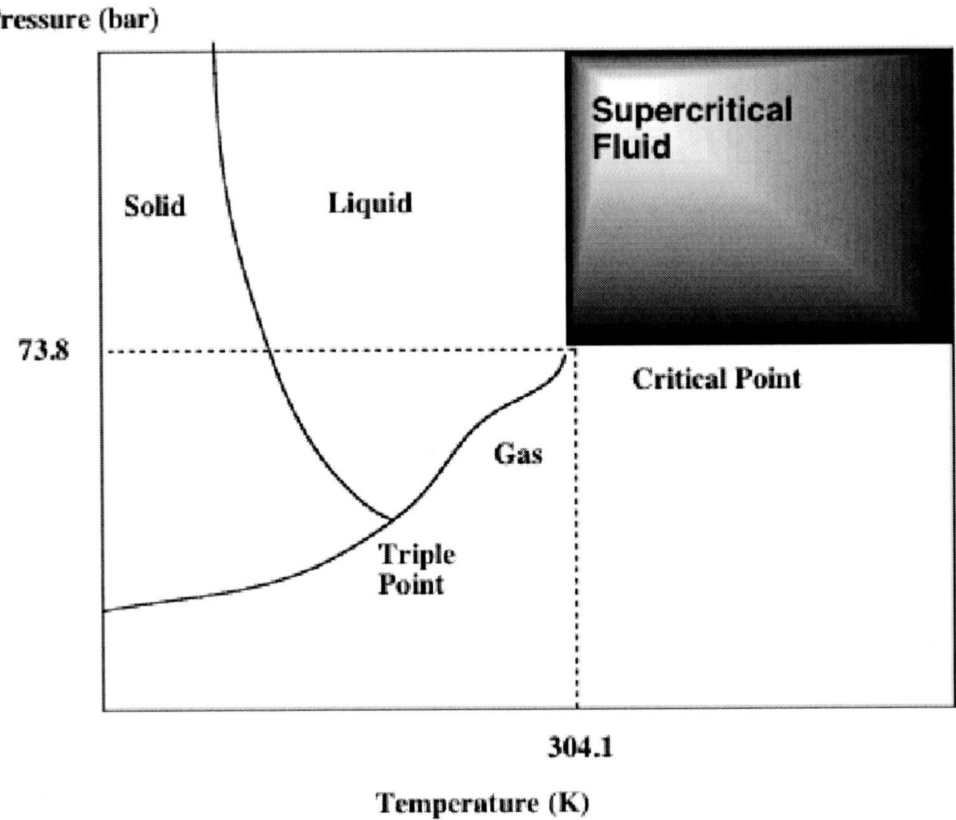

Figure 3. The phase diagram for pure carbon dioxide.

3.2. Radical Reactions in Supercritical Carbon Dioxide in Detail

As SCFs are low-density fluids with low viscosity resulting in high rates of diffusion, they offer potential benefits for radical processes. A significant proportion of the radical reactions studied have been in polymerization applications. Although radical reactions were some of the first synthetic processes to be considered in SCF, relatively few literature examples exist.

Many free radical halogenation reactions have historically been carried out in CCl_4, but $scCO_2$ is an attractive alternative. The reaction of molecular bromine with toluene and ethyl benzene, forms the corresponding benzylic bromides in good yields (Scheme 1) and Ziegler bromination using N-bromosuccinimide (NBS) was also successful [17].

$scCO_2$ is also a practical medium for free radical carbonylation of organic halides to ketones and aldehydes. Using a silane-mediated carbonylation of an alkyl halide, alkene and CO using AIBN initiator gave yields comparable to those obtained in benzene. Related intramolecular reactions also proceeded efficiently and showed interesting pressure dependent selectivity.

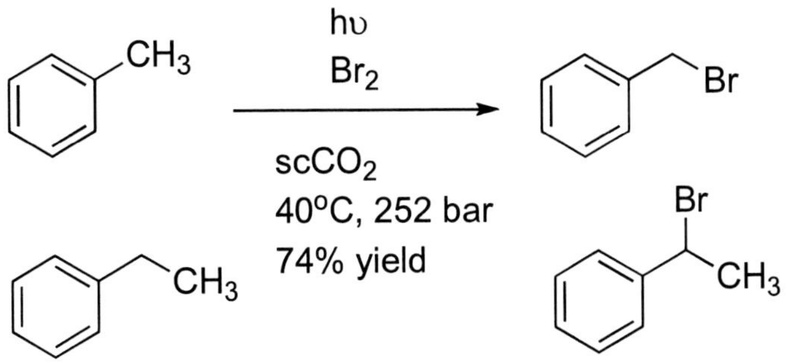

hυ
Br_2
CH_3
Br
$scCO_2$
40°C, 252 bar
74% yield
CH_3
Br
CH_3

Scheme 1.Radical bromination in $scCO_2$.

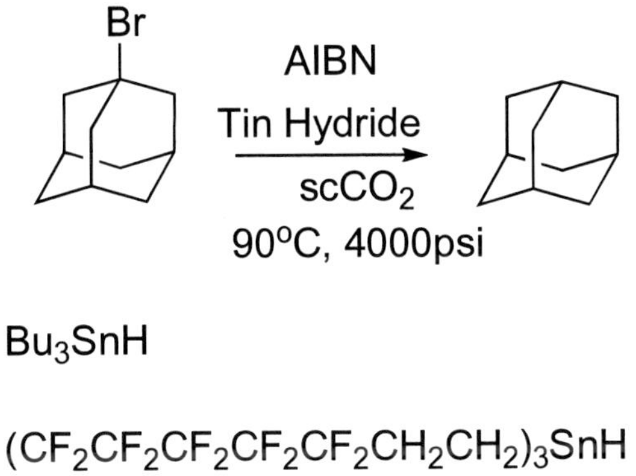

Br
AIBN
Tin Hydride
$scCO_2$
90°C, 4000psi

Bu_3SnH

$(CF_2CF_2CF_2CF_2CF_2CH_2CH_2)_3SnH$

Scheme 2. Tin hydride (Bu_3SnH) reduction in $scCO_2$.

The use of tin hydride reagents in $scCO_2$ has also been reported. Both tributyl tin hydride and tris(perfluorohexylethyl)tin hydride were investigated, where tris(perfluorohexylethyl)tin hydride being miscible under the reaction conditionswhereas tributyl tin hydride is not (Scheme 2). These reagents are also being applied in fluorous phase chemistry. Bromoadamantane was reduced by tris(perfluorohexylethyl)tin hydride (initiated by AIBN) under $scCO_2$ conditions to give the corresponding reduced product adamantane in 90% yield after 3 hours. The work up for this reaction is particularly clean by partitioning between benzene and perfluorohexane. Surprisingly despite the insolubility of tributyl tin hydride the reduction of 1-bromoadamantane to adamantane was also facilitated, and the reduced product was isolated in 80% yield. Reactions of steroidal bromides, iodides and selenides with each hydrogen donors respectively yielded the corresponding reduced products in high yields (85-98%) [18].

Several radical cyclization reactions were also studied (Scheme 3). Reduction of 1,1-diphenyl-6-bromo-1-hexene with tris(perfluorohexylethyl)tin hydridegave the 5-exo product in 87% yield. Similarly, reduction of aryl iodide with tris(perfluorohexylethyl)tin hydride gave quantitative conversion to cyclized product.Interestingly no reaction was observed with tributyltin hydride in both cases.

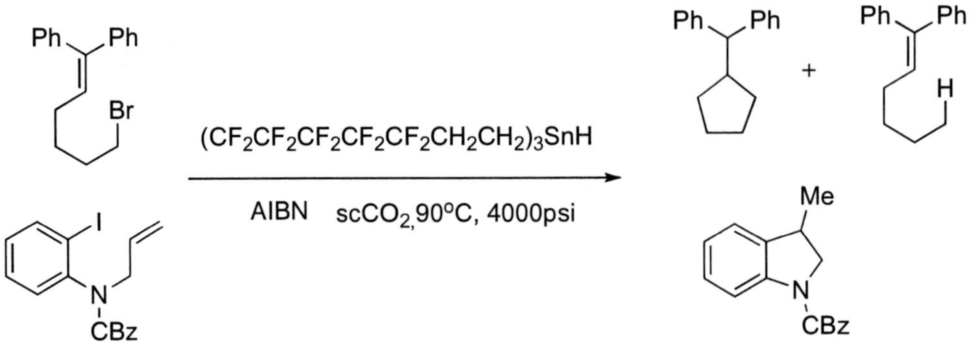

Ph Ph Br

$(CF_2CF_2CF_2CF_2CF_2CH_2CH_2)_3SnH$

AIBN $scCO_2$, 90°C, 4000psi

I N CBz

Ph Ph + Ph Ph H

Me N CBz

Scheme 3. Free radical cyclization reactions in $scCO_2$.

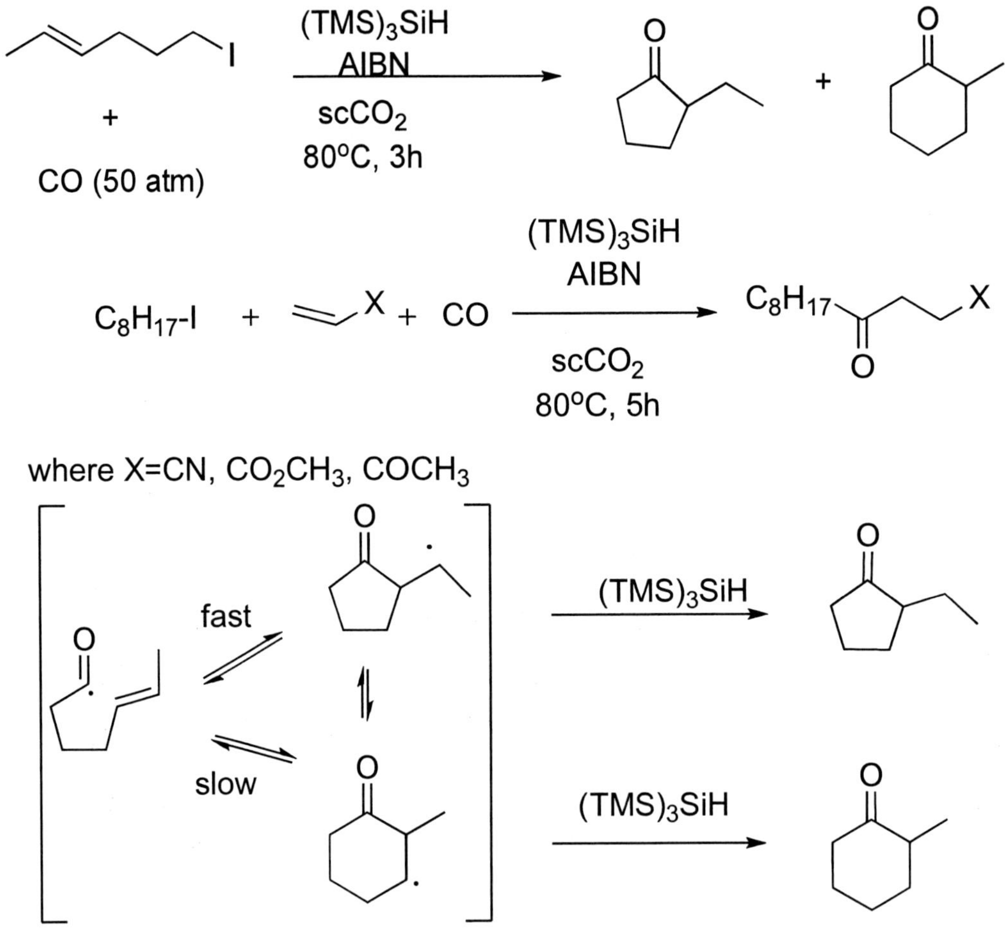

I + CO (50 atm)

$(TMS)_3SiH$ AIBN $scCO_2$ 80°C, 3h

O + O

C_8H_{17}-I + X + CO

$(TMS)_3SiH$ AIBN $scCO_2$ 80°C, 5h

C_8H_{17} X O

where X=CN, CO_2CH_3, $COCH_3$

O fast slow O O

$(TMS)_3SiH$ O

$(TMS)_3SiH$ O

Mechanism:

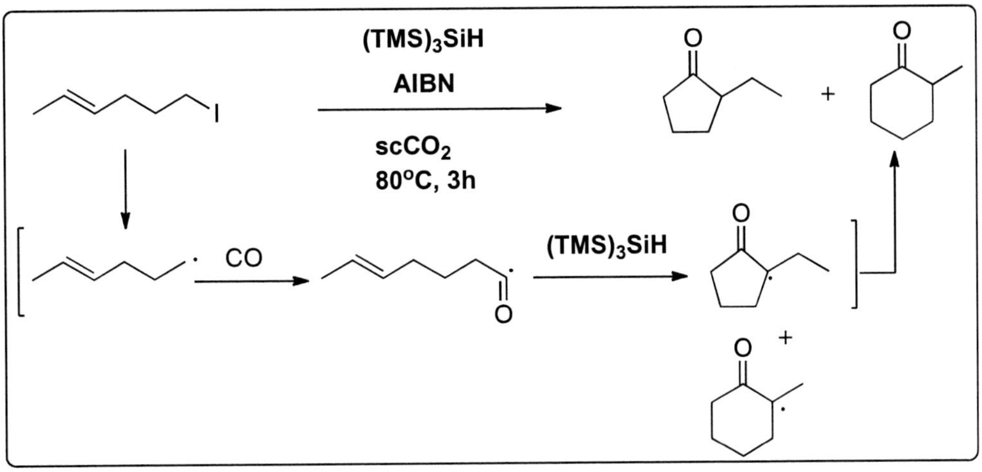

(TMS)$_3$SiH
AIBN
scCO$_2$
80°C, 3h
CO
(TMS)$_3$SiH

Scheme 4. Silane mediated efficient free radical carbonylation of the organic halides to ketones in scCO$_2$.

Free radical carbonylation reactions have extensively been investigated as a promising tool for the introduction of the carbonyl group because of its generality and scope of application in organic synthesis. Recently, Ikariya has applied the supercritical CO$_2$as a reaction medium for the silane-mediated efficient free radical carbonylation oforganic halides to ketones [19]. The reductive carbonylation of 1-iodooctane and acrylonitrile with CO$_2$-soluble (TMS)$_3$SiH in scCO$_2$ (CO$_2$ 20 atm, 350 atm total pressure) containing AIBN as radical initiator at 80°C proceeded smoothly in a 80% yield (Scheme 4). The radical carbonylative ring-closing reaction was expanded to intramolecular radical / carbonylation of6-iodohexyl acrylatewith 2 equivalents of(TMS)$_3$SiHin sc(CO$_2$) including CO (CO 50 atm, total pressure 295 atm) at 80 °C for 2 h affordedthe eleven-membered macrolide in 68% isolated yield. The yield of the macrolide is comparable to that achieved in benzene.

3.3. Future Directions

Using CO$_2$ as a solvent or a starting material for chemical synthesis and reactions has been a subject of extensive research since the early 90's. Some of the developments in this area have already led to new processes in chemical separation and manufacturing. scCO$_2$ can replace environmentally hazardous benzene in a broad range of free radical transformations. Carbon dioxide-based dry cleaning techniques and synthesis of fluoropolymers in SF-CO$_2$ are examples of industrial applications of these new supercritical fluid-based techniques.Demonstrations for the remediation of toxic metals in solid waste and reprocessing of spent nuclear fuel in supercritical CO$_2$ have also been initiated recently. It is likely that research and development in CO$_2$-based technology for chemical separation and material processing will continue to expand in this decade towards industrial scale flow reactors and development of novel and functional supercritical anti-solvents.

4. Radical Reactionsin IonicLiquids

4.1. Ionic Liquids and Alternative Media:General Introduction

The ionic liquids are more closely related to conventionally used solvents and therefore chemistry and applications could be developed rapidly. A history of ionic liquids including properties, availability, costs, application, preparation and green potential can be found in several reviews. An increasing range of ionic liquids is now commercially available [20,21].

The term "ionic liquid" is used to refer to a salt which exists in the liquid state at or around ambient temperature, hence the term "room temperature ionic liquids".The ionic liquid usually consists of an organic cation (often contains a nitrogen heterocycle) and an inorganic anion.These room temperature ionic liquids exhibit properties that make them potentially useful reaction media for synthesis and catalysis. They are good solvents for a diverse range of chemical transformations due to the thermal stability, negligible vapour pressure as well as ability to be used as *in-situ* Lewis acids and basis and therefore catalyse the desired transformations as well as recyclability and potential to reuse the ionic liquids after the transformation has been completed.It should be appreciatedhowever, that the challenge here is not really whether a reaction can be carried out in an ionic liquid as they are very good solventsfor many kinds of organic reactions, and its quite trivial to change from a conventional solvent to an ionic liquid. The important point of why one would want to change to an ionic liquid is to do something that would otherwise be very difficult with a conventional solvent. Such considerations are particularly relevant for synthetic chemistry. It is beyond the scope of this review to describe in details; however the practical examples will be presented and described.

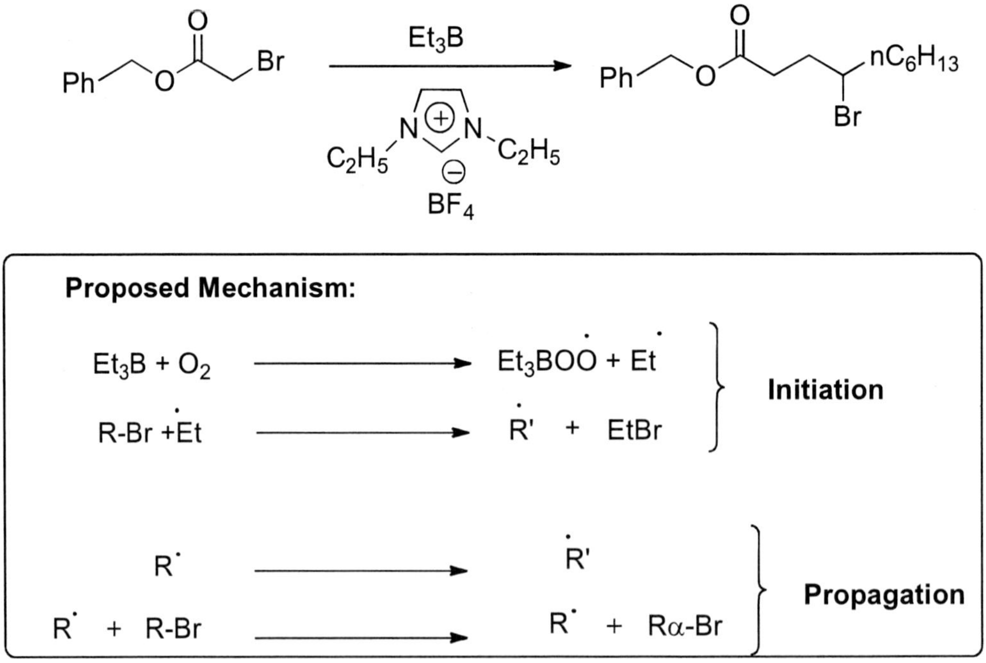

Proposed Mechanism:

$$Et_3B + O_2 \longrightarrow Et_3BO\dot{O} + Et^{\bullet}$$
$$R\text{-}Br + \dot{E}t \longrightarrow \dot{R}' + EtBr$$
(Initiation)

$$R^{\bullet} \longrightarrow \dot{R}'$$
$$R^{\bullet} + R\text{-}Br \longrightarrow R^{\bullet} + R\alpha\text{-}Br$$
(Propagation)

Scheme 5. Bromine atom transfer reaction in ionic liquid.

4.2. Radical Chain Reactions in Ionic Liquids:Triethylborane–Induced Radical Reactions

Some triethylborane induced radical reactions were found to proceed in ionic liquids by Oshimaet. al. [21b].The reactions include atom transfer radical cyclization reactions, hydrostannylation reactions of alkynes and atom transfer reactions (Scheme 5).

Benzyl bromoacetate participates in the bromine atom transfer reaction of 1-octene in 1-ethyl-3-methylimidazolium tetrafluoroborate to afford the corresponding adducts in excellent yields. The facile bromine atom transfer addition in the ionic liquid indicates that the ionic liquids may have a highly polar nature. The ionic liquids used in the investigations were attempted to be recycled with various degree of success [21b].

4.3. Radical Additions of Thiolsto Alkenes and Alkynes in Ionic Liquids

Nanniet. al. has investigated radical addition of thiols to double and triple carbon-carbon bonds examined in typical ionic liquids ([bmim][PF_6], [bmim][Tf_2N], and [bmim][BF_4]) under different temperature/ initiator conditions (i.e. 80-100 °C / AIBN-VAZO®, r.t./ triethylborane, r.t. / AIBN / UV radiation, r.t. / photoinitiator / UV radiation) [22]. All the addition products were usually obtained with high efficiencies and very good recyclability of the ionic liquid. In some cases, small but significant differences were noticed by changing the reaction medium from benzene to an ionic liquid (Scheme 6) [22e].

This outcome suggests that hydrothiolation of alkynes is somewhat faster in ionic liquid with respect to aromatic solvents, in line with what was suggested above for the hydrothiolation of alkenes. Of course, formation of the kinetic product is favored at lower temperatures, although kinetic control is never attained in any ionic liquid. By comparing the various results obtained at r.t., it seems that [bmim][PF_6] would be the best solvent for promoting formation of the *Z*-isomer, and hence the medium in which hydrothiolation of alkynes is faster.

Finally, Nanni briefly examined the possible use of ionic liquid solvents in click-chemistry reactions leading to biologically interesting molecules [22c].The preliminary results include addition of L-*N*-Fmoc-cysteine*tert*-butyl ester to 1-hexadecene and to an *O*-allylglucoside, and hydrothiolation of phenylacetylene with L-cysteine ethyl ester hydrochloride (Scheme 7).

The first adduct (*1*), which has been recently synthesized in 42% yield by Brunsveld and Waldmann through a radical thiol-ene, racemization-free procedure using thermal conditions, was chosen as a target compound for the recent interest in accessing hydrolysis-resistant non-natural *S*-alkylated cysteine derivatives. The reaction was carried out in [bmim][PF_6] under DMPA-promoted photolysis conditions with 3 equiv of 1-hexadecene for 5 h and, after usual workup followed by chromatography, similarly adduct *1* was yielded but in somewhat lower yield (30%).

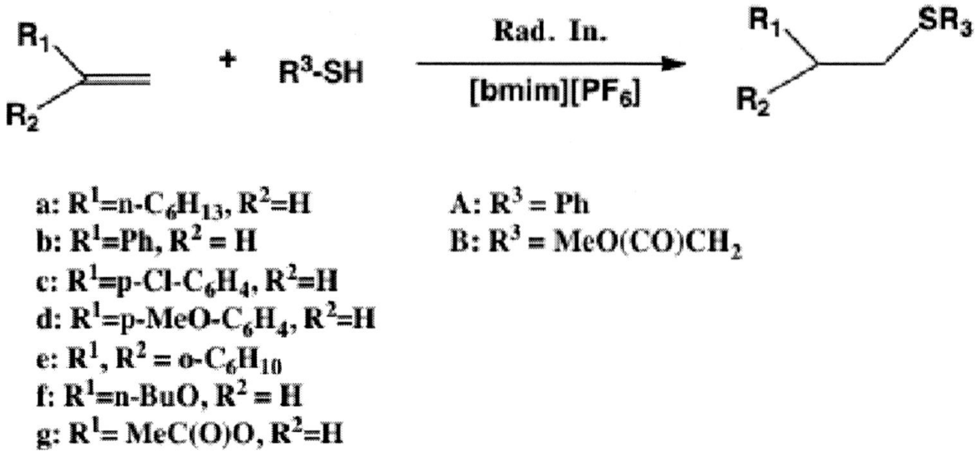

a: R^1=n-C_6H_{13}, R^2=H
b: R^1=Ph, R^2 = H
c: R^1=p-Cl-C_6H_4, R^2=H
d: R^1=p-MeO-C_6H_4, R^2=H
e: R^1, R^2 = o-C_6H_{10}
f: R^1=n-BuO, R^2 = H
g: R^1= MeC(O)O, R^2=H

A: R^3 = Ph
B: R^3 = MeO(CO)CH_2

Scheme 6. Radical addition of thiol to alkene in [bmim][PF_6].

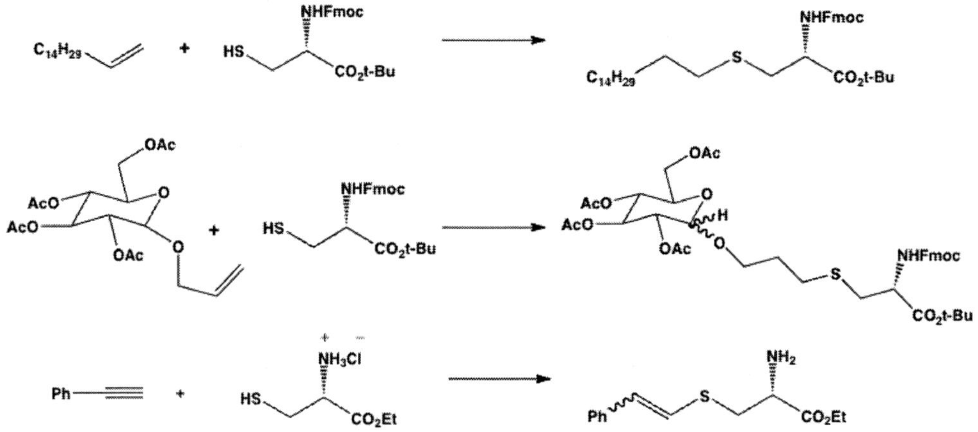

Scheme 7. Hydrothiolation as a tool for the synthesis of biologically interesting compounds.

The second sulfide (*2*) was selected in view of the ever growing importance of *O*-linked glycosides and their use for preparation of glyco- and/or peptidomimetics. Also in this case the reaction was performed by DMPA-promoted photolysis of the peracetylated*O*-allylglucoside (ca. 1:1 α/β mixture) in [bmim][PF_6] in the presence of orthogonally-protected L-*N*-Fmoc-cysteine*tert*-butyl ester (1 equiv). After workup and chromatography, target compound *2* was obtained in 50% yield.

The third reaction was simply a trial experiment to see whether mercapto-substituted aminoacids can add efficiently to an alkyne to give *3*-like vinyl sulfide adducts. The reaction between phenylacetylene and L-cysteine ethyl ester hydrochloride was carried out in [bmim][PF_6] by DMPA-promoted photolysis and yielded vinyl sulfide *3* in 78% yield (55:45 *E/Z* ratio). It is worth noting that this reaction gave *3* only to a slightly higher extent (86%) (55:45 *E/Z* ratio) when was repeated in a traditional solvent such as DMF.

In conclusion, Nanniet. al reported that radical hydrothiolation of alkenes and alkynes can occur in ionic liquids with at least the same efficiency as in traditional solvents. The ionic liquids are compatible with the use of different radical initiation conditions, i.e. thermal decomposition of azo-initiators (80-100 °C), reaction of triethylborane with dioxygen (r.t.), and UV-photolysis at r.t. in the presence of either an azo-initiator (AIBN) or a photosensitizer (DMPA). The reaction products can be efficiently isolated from the ionic liquid by centrifuge-mediated extraction with diethyl ether, and the ionic liquid can be usually recycled up to 3 times without any significant change in yields and byproducts under any reaction conditions. Some results show that the ionic liquids, if compared with traditional solvents, seem to favor the hydrothiolation reaction, probably through stabilization of the transition state for hydrogen transfer from the starting thiol to the intermediate alkyl or vinyl radical. Although the results are merely preliminary and yields are not optimized, it seems that this protocol could be successfully applied to the synthesis of biologically interesting molecules through click-chemistry procedures carried out in ionic liquids.

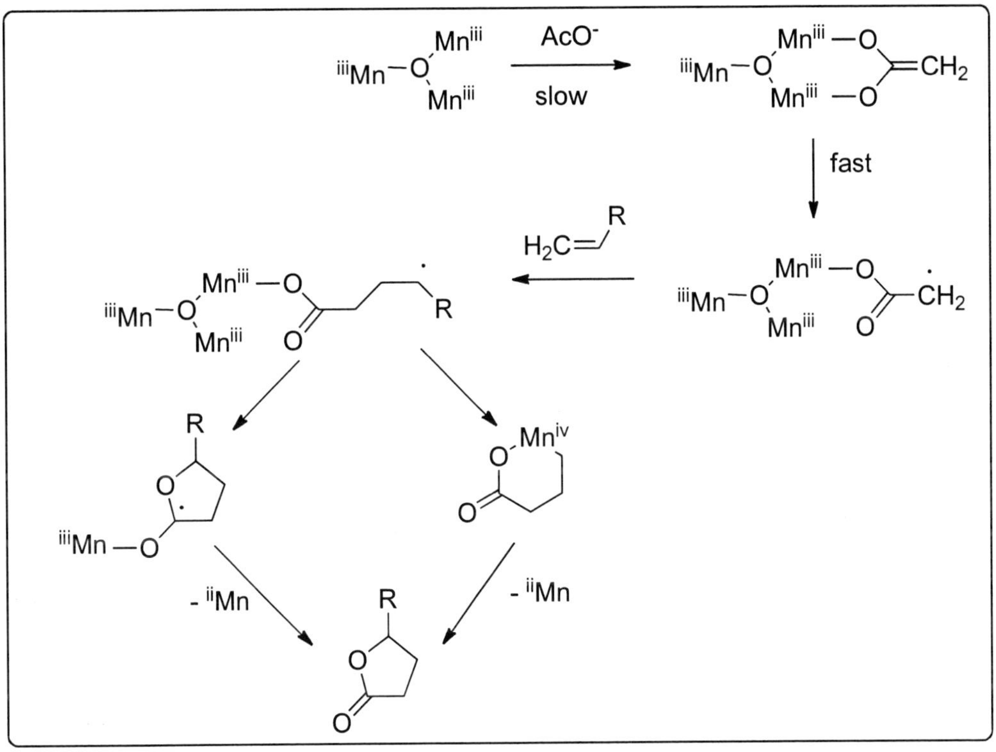

Scheme 8. Mechanism of $Mn(OAc)_3$ mediated radical oxidation reaction.

4.4. Radical Non-chain Reactions in Ionic Liquids

4.4.1. Formation of Radicalsby OxidationwithTransition MetalSalts: General Perspective

Transition metals in high oxidation states are often capable of extracting one electron from electron rich organic substances. Ketones, esters, nitriles and various other "carbon acids" that can form enols, enolates and related structures are by far the most commonly used substrates. Their oxidation can lead to a free radical, which then follows one or more of the pathways available. It is important to take into account that the rate of radical production will depend on the exact structure of the substrate, its propensity to exist as the corresponding enol or enolate in the medium, the pH, the solvent, the temperature, and of course the redox potential of the metallic salt (which can be strongly affected by the nature of the ligand around the metal) and the exact mechanism by which electron transfer actually occurs (i.e. inner or outer sphere).

4.4.2.OxidationsInvolvingMn(III)in IonicLiquids

Radical generation through oxidation of enolizable substrates using Mn(III) salts is by far the most common and the field is rapidly expanding. $Mn(OAc)_3xH_2O$, is the usual oxidant: which is actually a trimer made up of an oxo-centre triangle of Mn(III) ions bridged by acetates units, however for simplicity and convenience we shall use the simplified formula of $Mn(OAc)_3$ throughout the rest of the examples [23]. These reactions are considered to proceed via a free radical process where Mn(III) initiates the oxidation of the carbonyl compound and the newly formed radical undergoes an intermolecular addition to the olefins to produce a new radical. The addition of a radical to an aromatic ring gives rise to a stabilized radical, which is then oxidized with Mn(III) to restore the aromaticity and yield the corresponding furan (Scheme 8) [23].

Many of the methods that were previously employed for $Mn(OAc)_3$-mediated radical reactions involved the use of acetic acid as a solvent.Because of the poor solubility of $Mn(OAc)_3$ in organic solvents the need for high temperatures for many reactions, and the use of acetic acid limited the range of substrates that could be employed.In order to improve this drawback, Parson investigated the elegant way of using ionic liquids to establish milder reaction conditions in $Mn(OAc)_3$- mediated reactions [24]. They showed that ionic liquids, such as 1-butyl-3-methylimidazolium tetrafluoroborate ([bmim][BF_4]), which is miscible with polar solvents (e.g. methanol, dichloromethane) could be used in $Mn(OAc)_3$-mediated radical reactions (Scheme 9) [24].

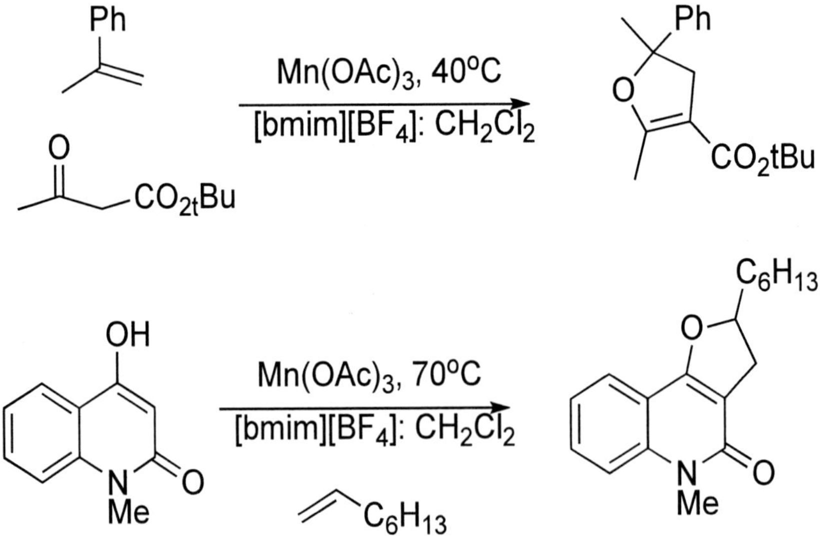

Scheme 9.Mn(III) mediated reactions in ionic liquids.

Cerium(IV) ammonium nitrate-mediated oxidative radical reactions are carried out in the presence of ionic liquids, including 1-butyl-3-methylimidazolium tetrafluoroborate, for the first time. The presence of the ionic liquid not only increases the rate and yield of reactions in dichloromethane but also extends the range of 1,3-dicarbonyl precursors, which can be utilized in these carbon-carbon bond-forming reactions (Scheme 10) [25].

Cerium(IV) ammonium nitrate (or CAN) has been widely used to oxidize numerous organic compounds, including 1,3-dicarbonyls, to form radicals.

The resulting α-dicarbonyl radicals can add intra- or inter-molecularly to a range of electron-rich alkenes to form radical adducts, which can be oxidized using a second equivalent of Ce(IV).

The carbocations can then undergo nucleophilic attack or deprotonation to form, for example, nitrates or alkenes. Indeed, the use of CAN in intermolecular carbon-carbon bond forming reactions has been shown to have advantages over the more commonly used manganese(III) acetate. These oxidative radical reactions are synthetically appealing as they permit the formation of functionalized products in one-pot reactions using inexpensive CAN.

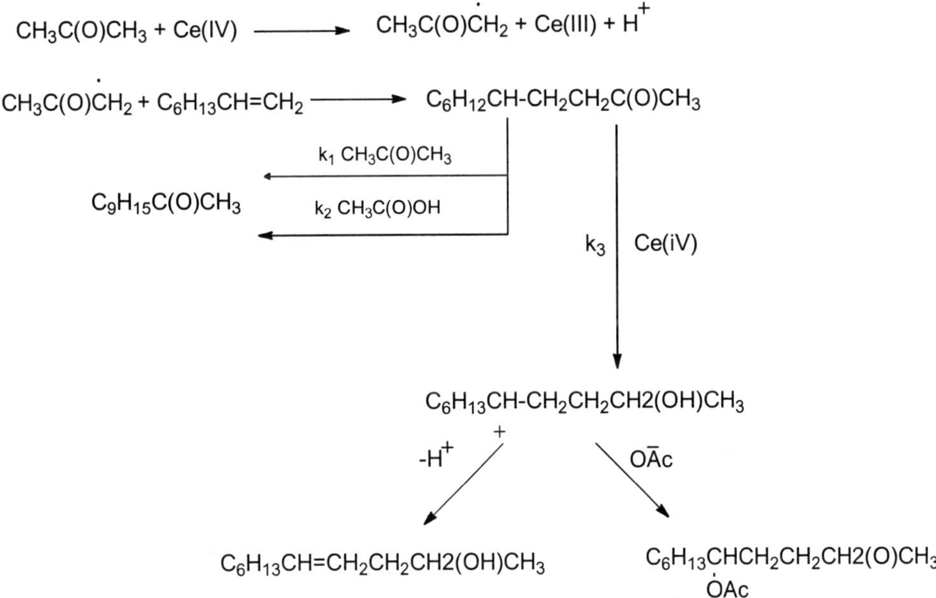

Scheme 10. General oxidation mechanism for CAN-mediated radical reactions.

The most common solvents employed in CAN oxidation reactions are acetonitrile or methanol. However, these polar solvents are not always ideal for non-polar substrates and alcohols can react with intermediate carbocations to produce unwanted ethers. With the aim of extending the range of solvents, which can be employed in CAN oxidative cyclizations, Parson has investigated the novel use of Cerium(IV) ammonium nitrate-mediated oxidative radical reactions in the presence of ionic liquids, including 1-butyl-3- methylimidazo-liumtetrafluoroborate. The presence of the ionic liquid not only increases the rate and yield of reactions in dichloromethane but also extends the range of 1,3-dicarbonyl precursors, which can be utilized in these carbon–carbon bond-forming reactions (Scheme 11) [26].

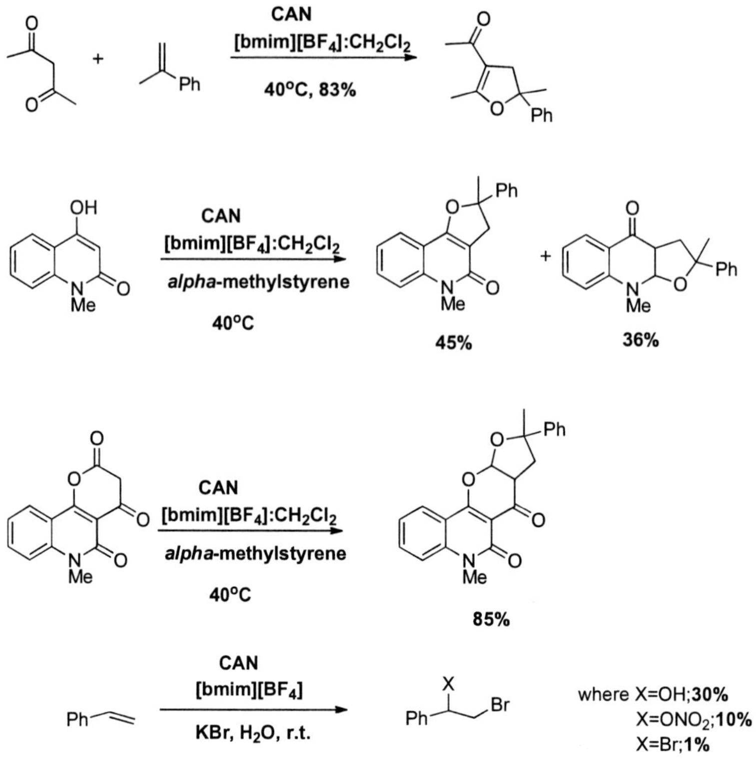

Scheme 11. CAN-mediated oxidative radical reactions in ionic liquids.

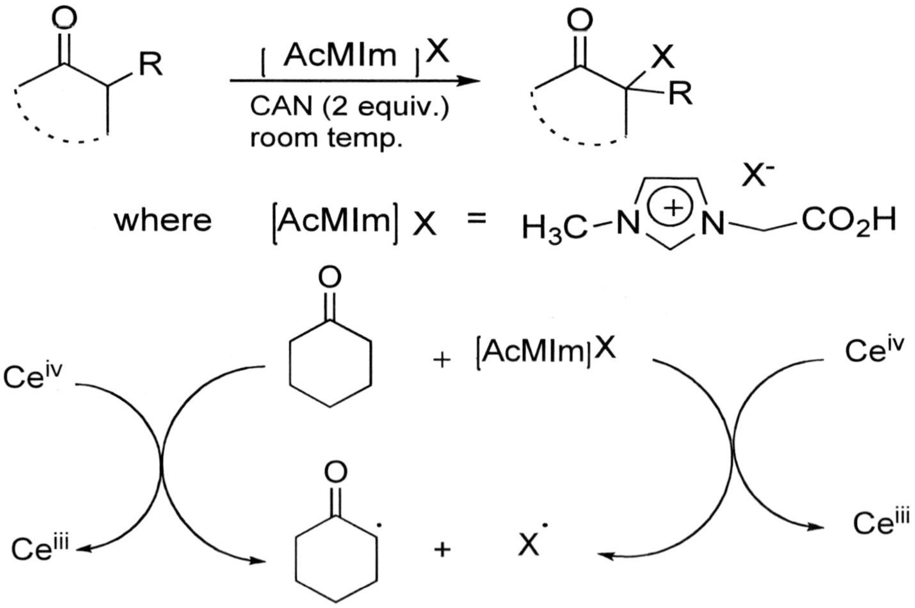

Scheme 12. Use of [AcMIm]X and CAN in the α-halogenation of carbonyl compounds and plausible mechanism for the α-halogenation.

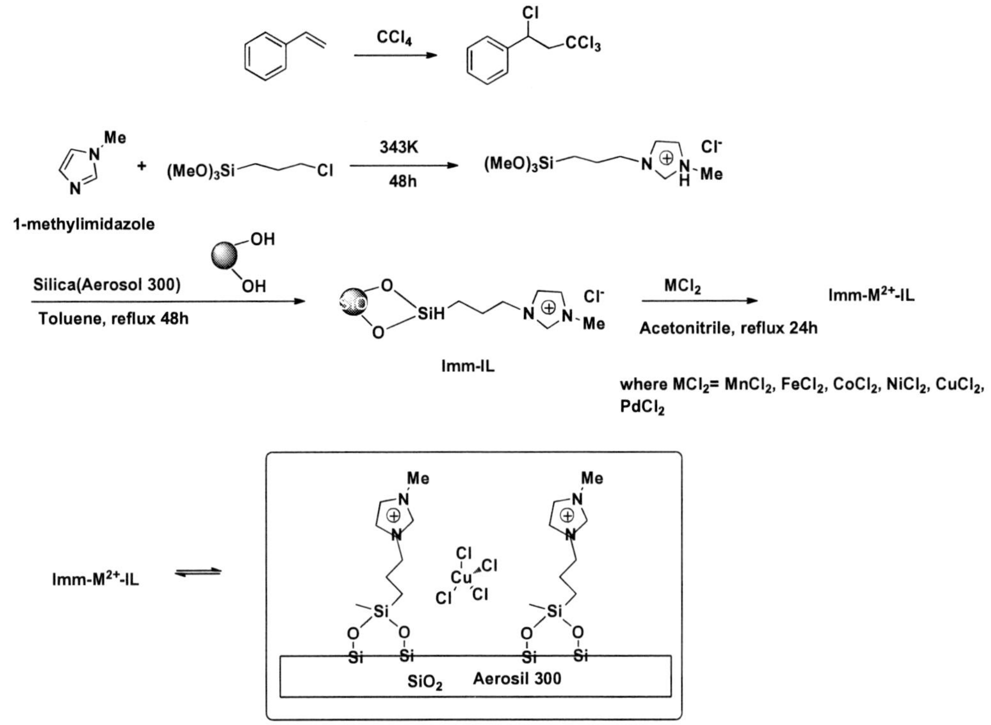

Scheme 13. Karasch reaction catalyzed by Imm-M^{2+}-IL.

Another important reaction, which was explored in ionic liquids is the α-halogenation of carbonyl compounds. Ionic liquids, acetylmethylimidazolium halide, in combination with ceric ammonium nitrate promoted halogenation of a wide variety of ketones and 1,3-ketoesters at the α-position. The ionic liquid acts here as a reagent as well as a reaction medium, and thus the reaction does not require any organic solvent or conventional halogenating agent. The reaction is completely stopped when the radical quencher TEMPO is used (Scheme 12) [27]. The mechanism proposed for the transformation, suggests that CAN plays a vital role as a one electron oxidant. The preferential halogenation in the non-substituted α-position of α-substituted ketone is also in accordance with various radical stabilities of newly formed radicals.

4.5. Supported Ionic Liquids: Versatile Reaction and Separation Media. The Latest Development

As previously mentioned, ionic liquids have attracted growing interest as alternative reaction media to replace volatile organic solvents in catalysis. Their ionic nature, non-volatility and thermal stability make them highly suitable for biphasic ionic liquids-organic transition metal catalysis. The almost unlimited combination of cation-anion pairs further allows the synthesis of Taylor-made ionic liquids that can stabilize catalytic species. However, traditional biphasic ionic liquids-organic systems require larger amounts of ionic liquids which make them unattractive based on economics considerations since ionic liquids still are expensive solvents, even though being commercially available by now. In addition, the high viscosity of ionic liquids can induce mass transfer limitations if the chemical reaction is fast, in which case the reaction takes place only within the narrow diffusion layer and not in the bulk of the ionic liquid catalyst solution. Hereby, only a minor part of the ionic liquid and the dissolved precious transition metal catalyst are utilized. It is therefore the interest in supported ionic liquids that brought attention as the versatile reaction media for variety of synthetically useful transformations including Kharasch reaction, i.e. solvent free addition reaction of styrene and carbon tetrachloride to produce 1-(1,3,3,3-tetrachloropropyl)benzene (Scheme 13) [28].

Normally the reaction is carried out using homogeneous catalysts, thus making this approach one of the first examples using a heterogeneous metal complex system. Of the system examined, only Fe^{2+} and Cu^{2+} containing catalyst were found to be active, and when optimizing the copper (II) system a conversion of 98% and an excellent product selectivity of 95% could be obtained. However, during recycling experiments the yield decreased from 93% to 80%. The existence of the immobilized imidazolium group was found, by studies of analogous silica-supported $CuCl_2$ catalyst (I.e. without ionic liquid), to be essentially important as copper ions for the catalytic activity. The immobilization of the complex on the silica was assumed to restrict the conformation of reactants and thereby facilitate the high selectivity by favoring the formation of the chlorinated addition products rather than oligomerization of styrene.

4.4. Conclusions and Future Directions

Since its birth over a decade ago, the fieldof green chemistry has seen rapid expansion,with numerous innovative scientificbreakthroughs associated with the productionand utilization of chemicalproducts. The concept and ideal ofgreen chemistry now goes beyond chemistryand touches subjects ranging fromenergy to societal sustainability. The keynotion of green chemistry is ''efficiency'',including material efficiency, energy efficiency,man-power efficiency, and propertyefficiency (e.g., desired function vs.toxicity). Any ''wastes'' aside from theseefficiencies are to be addressed throughinnovative green chemistry means.''Atom-economy'' and minimization ofauxiliary chemicals, such as protectinggroups and solvents, form the pillar ofmaterial efficiency in chemical productions.By far, the largest amount of''auxiliary wastes'' in most chemicalproductions are associated with solventusage. In a classical chemical process, solvents are used extensively for dissolving reactants, extracting and washingproducts, separating mixtures, cleaning reaction apparatus, and dispersing products for practical applications. While the invention of various exotic organicsolvents has resulted in some remarkable advances in chemistry, the legacy of such solvents has led to various environmental and heath concerns. Consequently, as part of green chemistry efforts, a variety of cleaner solvents has been evaluated as replacements. However, an ideal and universal green solvent for all situations does not exist. Among the most widely explored greener solvents are ionic liquids, supercritical CO_2, and water. These solvents complement each other nicely both in properties and applications. Importantly, the study of green solvents goes far beyond just solvent replacement. The use of green solvents has led science to uncharted territories. For example, the study of ionic liquids made large-scale supported synthesis possible for the first time and the utilization of supercritical CO_2 has led to breakthroughs in microelectronics and nanotechnologies.

5. FLUOROUSCHEMISTRYAS AN ALTERNATIVE REACTION MEDIUM FOR FREE RADICAL TRANSFORMATIONS

5.1.FluorousSeparation Techniques: from "Liquid-Liquid" to "Solid-Liquid" and "Light Fluorous"

"What is fluorous chemistry?" That is a question still too common amongst the chemical community, which becomes more involved with supercritical media, ionic liquids as well as aqueous media as quite common media for a broad range of free radical synthetically useful transformations [29-40]. "Fluorous" was the term coined for highly fluorinated (or perfluorinated) solvents, in an analogous way to "aqueous" for water based systems.Fluorous solvents are immiscible with both organic and aqueous solvents, thus hexane (C_6H_{14}) and perfluorohexane (C_6F_{14}, commonly known as FC-72^{TM}) are immiscible.

Fluorous chemistry has been developed as a broad-based technology platform that addresses may challenges in the modern drug discovery environment [29]. Fluorous technologies can be divided into two broad groups: Fluorous Tagging Strategy and Fluorous separation [30]. The advantage of fluorous tagging methods lies in the opportunity to combine solution-phase reaction with phase tag-based separation. Perfluorinated (fluorous) chains such as C_6F_{13} and C_8F_{17} are employed as phase tags to facilitate the separation process. The fluorous chain is usually attached to the parent molecule through a $(CH_2)_m$segment to insulate the reactive site from electron withdrawing fluorines. A fluorous chain $C_nF_{2n+1}(CH_2)_m$ can be abbreviated to Rfnhm when it is presented in the reaction equation. In principle, any synthetic development in conventional solution-phase or in polymer-bound chemistry can be adapted to fluorous chemistry. Many fluorous groups are used to tag reagents (including scavengers and catalysts) or to tag substrates. Fluorous reagents are commonly used for single or short-step parallel synthesis, while fluorous-tagged substrates are more suitable for multistep parallel synthesis (Scheme 14).

Fluorous separation relies on the strong and selective affinity interaction between fluorous molecules and fluorous separation media [32]. The separation medium can be fluorous solvents which are immiscible with common organic solvents at room temperature and thus can be used for liquid-liquid extractions. A "heavy fluorous" tag (60% or more fluorine by molecular weight) is required to drive the fluorous molecule to a fluorous phase from non-fluorous phase. Perfluorinated alkanes such as FC-72 (perfluorohexanes) and highly fluorinated ethers such as HFC-7100 ($C_4F_{17}OCH_3$) are good solvents for fluorous extractions.

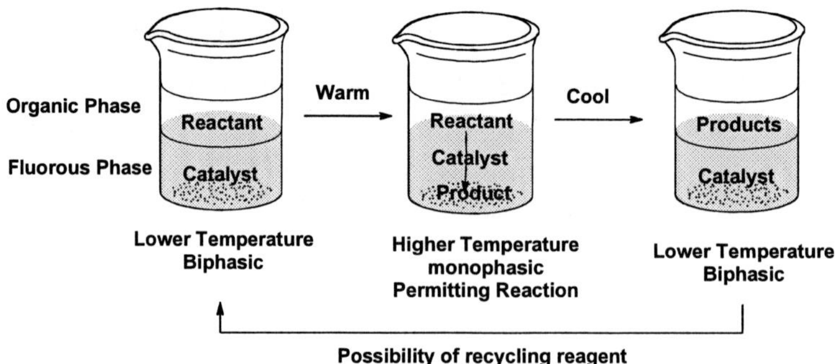

a. Tagging reagents/scavengers/catalysts

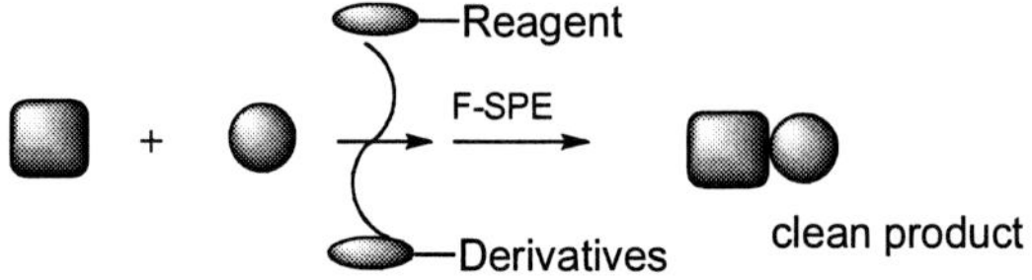

b. Tagging substrates for parallel synthesis

Scheme 14.Fluorous tagging strategies: a) Tagging reagent/scavengers/ catalyst; b) Tagging substrates for parallel synthesis.

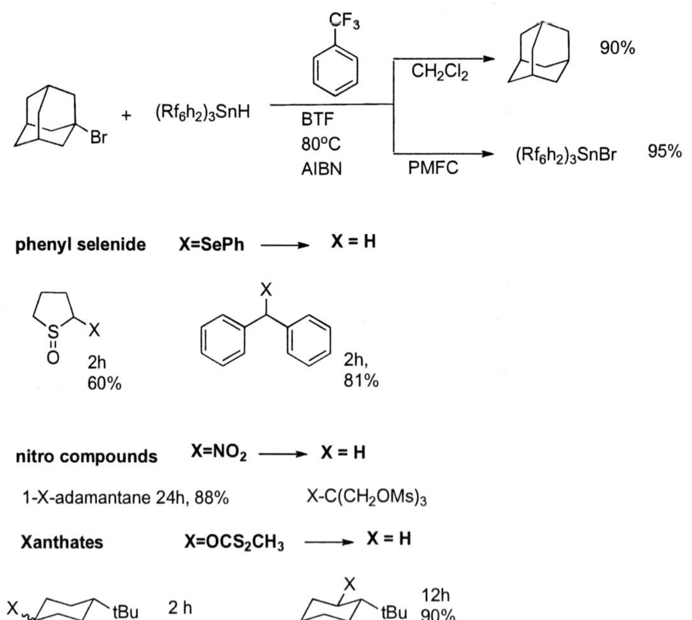

Scheme 15.Scope of free radical transformations in the presence of fluorous tin hydride and tributyl tin hydride.

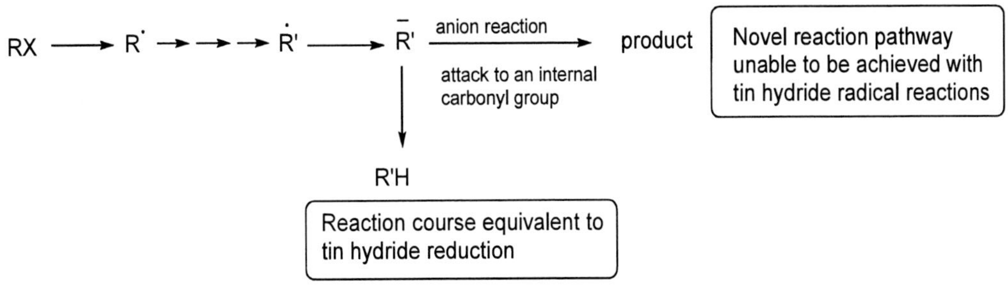

RX → R˙ → R˙' → $\bar{R}'$ —anion reaction→ product

Novel reaction pathway unable to be achieved with tin hydride radical reactions

attack to an internal carbonyl group

R'H

Reaction course equivalent to tin hydride reduction

Mechanism:

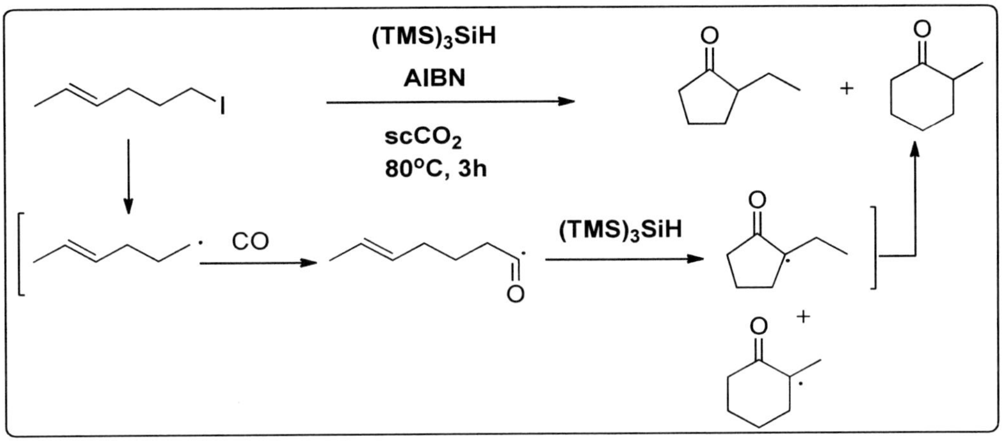

$(TMS)_3SiH$
AIBN
$scCO_2$
80°C, 3h

CO

$(TMS)_3SiH$

Scheme 16. Schematic diagram of Tandem Carbonylation Reaction with hydride species in one electron reductive system.

The important features which makes fluorous solvents/synthesis so easily integrated into synthetic applications is the synergy achieved through the solution-phase reactions and solid phase separations as well as good combinatorial capabilities and offers the following advantages:

- Possibility to follow reactions by TLC, HPLC, IR and NMR
- Applicability to utilize the fluorous and non-fluorous separations to maximize the recover of reactants and products.
- Good solubility in the range of conventional organic solvents
- More than one fluorous reagent can be used in a single reaction
- Easily adaptable methods from the non-fluorous procedures
- Recovery of fluorous materials after separation.

The most important characteristic that makes fluorous synthesis superior to solid-supported synthesis is the favourable reaction kinetics associated with the solution-phase reaction.

All chemical reactions are limited both by the efficiency of the transformation and also the ease of purification of the reaction mixture. Chemists have traditionally concentrated on the former of these two problems (the conversion of starting materials to products), to detriment of the latter purification issue.In recent years, several new techniques have appeared in order to address this issue of fluorous phase chemistry. The aim of this chapter is to highlight the development of fluorous chemistry from what was often considered to be an expensive laboratory "curiosity" to a technique that is now an independent and valid scientific tool ready for adoption by a chemical community in general, especially in free radical mediated synthesis: chain and non-chain reactions.

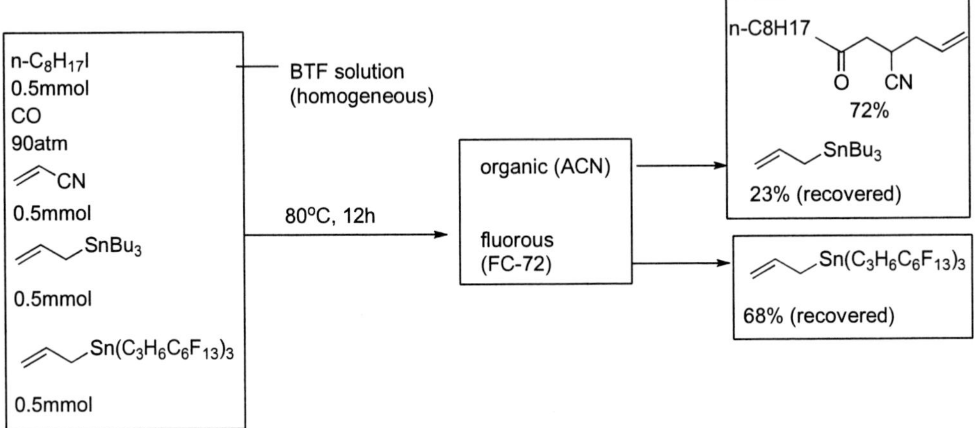

Scheme 17. Competition experiments between fluorousallyltin and allyltributyltin.

5.2.Fluorous Chemistry and Radicals Combined Efforts to the Rescue

Fluorous methods have emerged as new and powerful techniques that have greatly influenced the way in which preparative organic chemistry is currently performed.In cooperation with the Curran group, who first reported on fluorous tin hydride technology (Scheme 15) [34].

5.2.1.FluorousRadical Carbonylation Reactions: From Synthetic Approach to Practical Applications

Ryu developed "fluorous radical carbonylation" reactions [35]. The great advantage of fluorous radical carbonylation systems is the way in which they enable the easy separation of products and reagents by fluorous-organic liquid-liquid extraction or fluorous-solid phase extraction, to ensure the reuse of tin reagents (Scheme 16).

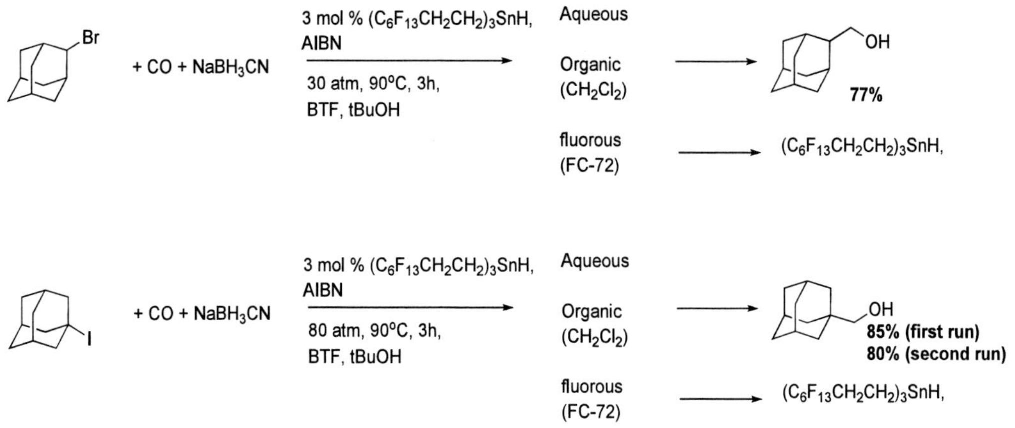

Br + CO + $NaBH_3CN$ → 3 mol % $(C_6F_{13}CH_2CH_2)_3SnH$, AIBN; 30 atm, 90°C, 3h, BTF, tBuOH

Aqueous

Organic (CH_2Cl_2) → OH 77%

fluorous (FC-72) → $(C_6F_{13}CH_2CH_2)_3SnH$,

I + CO + $NaBH_3CN$ → 3 mol % $(C_6F_{13}CH_2CH_2)_3SnH$, AIBN; 80 atm, 90°C, 3h, BTF, tBuOH

Aqueous

Organic (CH_2Cl_2) → OH 85% (first run) 80% (second run)

fluorous (FC-72) → $(C_6F_{13}CH_2CH_2)_3SnH$,

Scheme 18. Hydromethylation of RX using a catalytic amount of fluorous tin hydride.

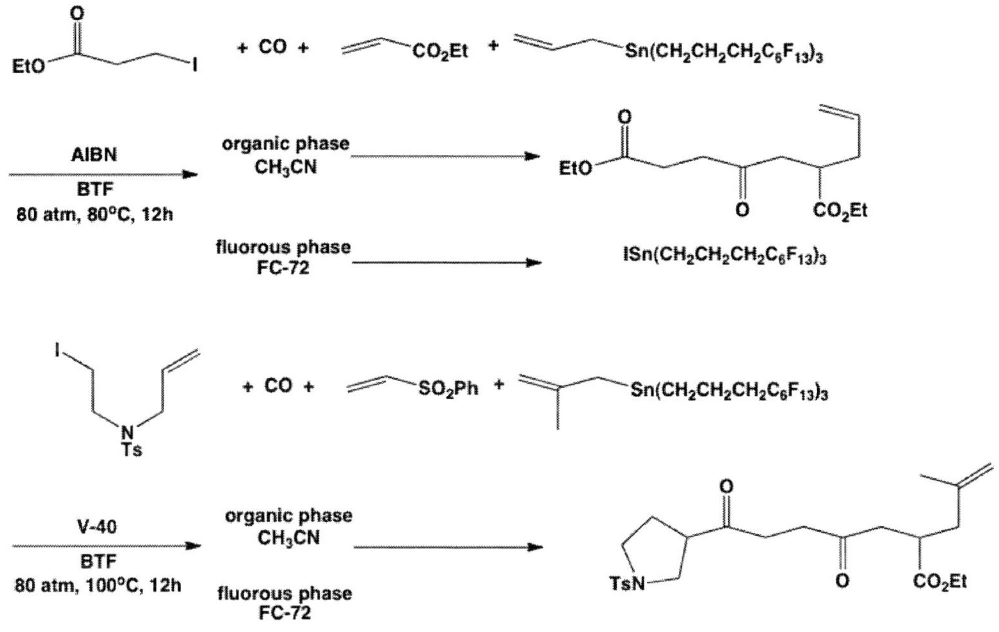

EtO, O, I + CO + CO_2Et + $Sn(CH_2CH_2CH_2C_6F_{13})_3$

AIBN, BTF, 80 atm, 80°C, 12h

organic phase CH_3CN → EtO, O, O, CO_2Et

fluorous phase FC-72 → $ISn(CH_2CH_2CH_2C_6F_{13})_3$

I, N, Ts + CO + SO_2Ph + $Sn(CH_2CH_2CH_2C_6F_{13})_3$

V-40, BTF, 80 atm, 100°C, 12h

organic phase CH_3CN → TsN, O, O, CO_2Et

fluorous phase FC-72

Scheme 19. Fluorousallyltin mediated four-component coupling reaction.

As illustrated in Scheme 16, Ryu obtained comparable results when compared conventional tin hydride with ethylene-spaced fluorous tin hydride.At an identical CO pressure, the use of tributyltin hydride afforded higher formylation / reduction ratios than ethylene-spaced fluorous tin hydride.This is consistent with the fact that the rate constant for primary alkyl radical trapping by fluorous tin hydride is about two fold that for tributyltin hydride at 20 °C.Thus, to obtain results identical to those for tributyltin hydride a higher CO pressure and/or a higher dilution are required for the fluorous analog. Fluoroushydroxymethylation of organic halides using a catalytic quantity of a fluorous tin hydride was also investigated and demonstrates a successful future direction of research and development in the area. Kahne and Gupta previously reported on a similar hydroxymethylation, using a catalytic amount of triphenylgermyl hydride and an excess amount of sodium cyanoborohydride [36].

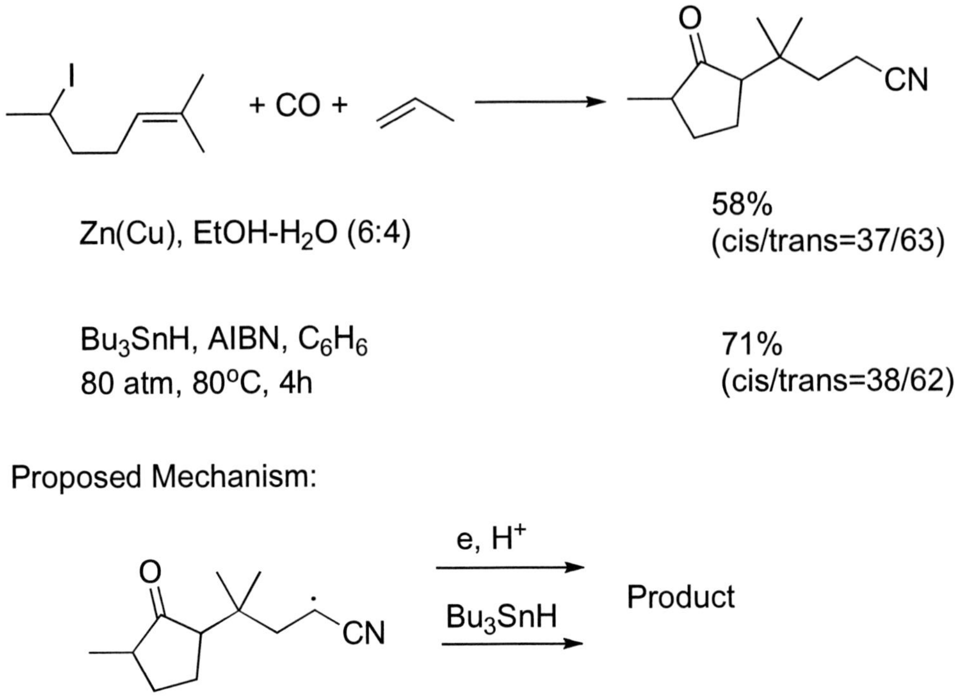

O

I

CN

+ CO +

Zn(Cu), EtOH-H_2O (6:4)

58%
(cis/trans=37/63)

Bu_3SnH, AIBN, C_6H_6
80 atm, 80°C, 4h

71%
(cis/trans=38/62)

Proposed Mechanism:

e, H^+

O

CN

Bu_3SnH

Product

Scheme 20.Three component coupling reaction under two different sets of conditions.

Ryu*et al* found that the reaction worked well even in the case of fluorous tin hydride (Scheme 17) [37]. Interestingly, this fluorous reagent, as is usually the case with the related fluorous radical reactions, permits simple purification through a three-phase (aqueous/organic/fluorous) extractive workup. After the workup, fluorous tin reagent is recovered from the perfluorohexane layer and reused. Very recently, Ryu reported on the use of F-626, 1H,1H,2H,2H-perfluorooctyl 1,3-dimethylbutyl ether, as a versatile alternative solvent for both fluorous and non-fluorous reactions [32].As illustrated in the second example of Scheme 18, the successful hydroxymethylation of 1-iodoadamantane was possible using this new solvent as a fluorous/organic amphiphilic solvent instead of BTF (benzotrifluoride).

Propylene-spaced fluorousallyltin and methallyltin proved particularly useful as mediators in the four-component coupling reactions, where alkyl halides, CO, alkenes, and allyltin are combined in a given sequence (Scheme 18) [39].Once the reaction was complete, the BTF (benzotrifluoride) was removed by vacuum-evaporation and the resulting oil was partitioned into acetonitrile and FC-72 (perfluorohexanes). Evaporation of the acetonitrile layer, followed by short column chromatography on silica gel provided with the desired product. This study also revealed that the FC-72-layer contained fluorous tin compounds such as fluorousallyltin and tin iodide and quantified reproduce fluorousallyltin reagents by treating the tin residue with an ethereal solution of allyl and methallyl magnesium bromides that were able to be reused without any appreciable loss in their activity.

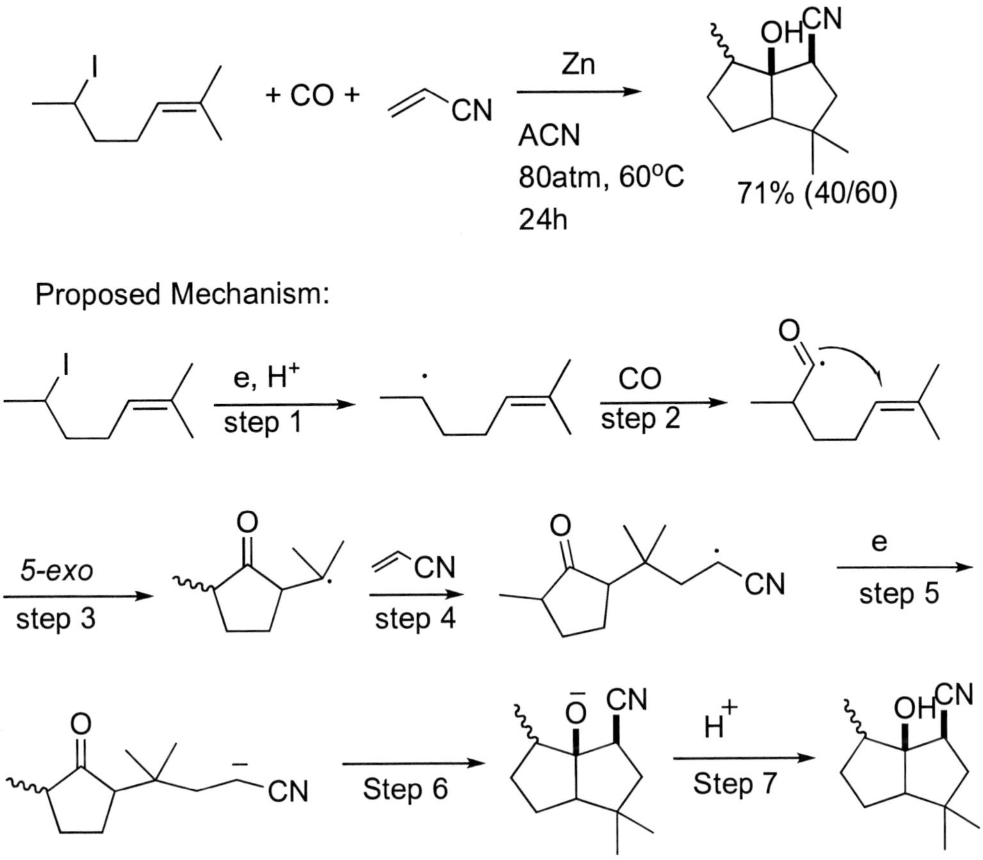

+ CO + CN

Zn

ACN

80atm, 60°C

24h

71% (40/60)

Proposed Mechanism:

e, H^+ step 1

CO step 2

5-exo step 3

CN step 4

e step 5

Step 6

H^+ Step 7

Scheme 21.Zn-induced dual annulation conducted in CAN.

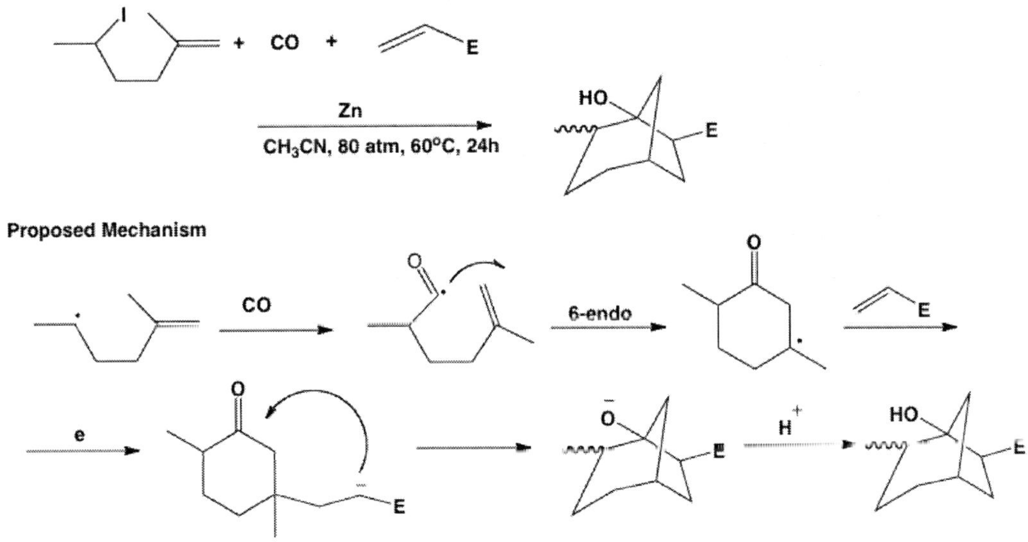

+ CO + E

Zn

CH_3CN, 80 atm, 60°C, 24h

Proposed Mechanism

CO

6-endo

e

H^+

Scheme 22. Dual [5+1]/[3+2] annulation leading to bicyclo[3.2.1]octanols.

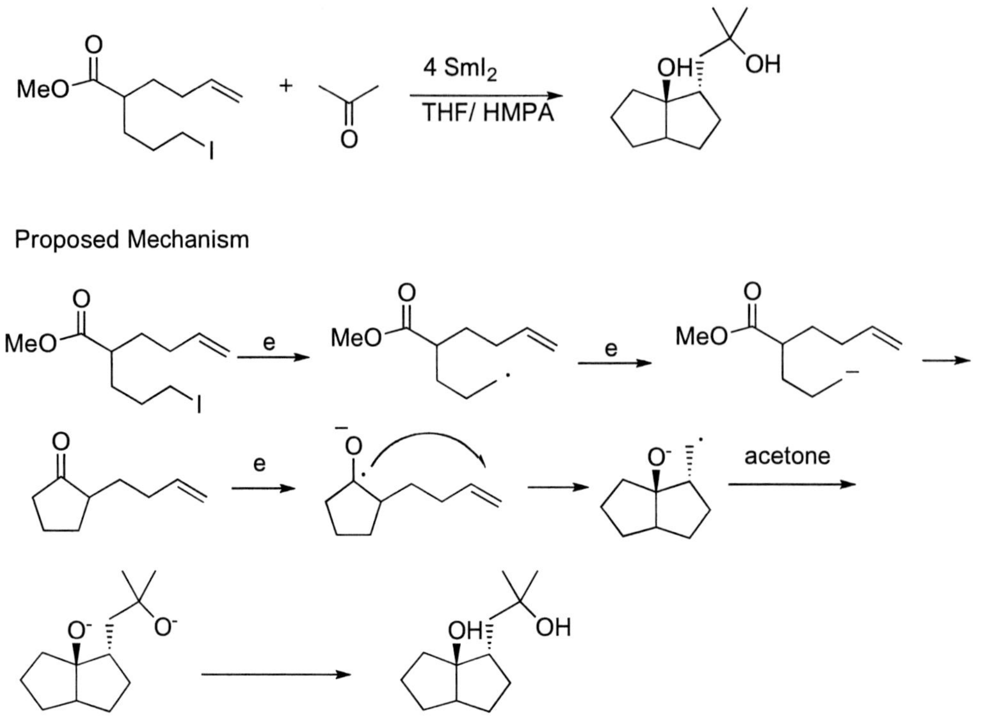

4 SmI_2

THF/ HMPA

Proposed Mechanism

acetone

Scheme 23.Molander's tandem processes by species hybridization.

Next, Ryu examined if acyl radical cyclization is being preferred over one-electron reduction using zinc to form an acyl anion, and, if so, would the resulting radical species consisting of a cyclic ketone moiety undergo a further radical addition reaction [40]. This idea was tested with the reaction of 4-alkenyl iodide, CO, and acrylonitrile as a radical trap, as a three-component coupling reaction took place to produce cyclopentanone derivatives (Scheme 20).

The observed stereochemical properties of the isolated products were identical to those obtained with the corresponding tin hydride mediated reaction and thus support the theory that free radical generation, radical carbonylation, and acyl radical cyclization take place simultaneously in a zinc-induced system. The overall transformation is equivalent to a tin hydride mediated system with the exception that the final step of the zinc/protic solvent system involves reduction to produce carbanions that are subsequently protonated (Scheme 21) [41].

The proposed mechanism that leads to the production of the bicyclic alcohol is as follows: (i) formation of an alkyl radical by one-electron reduction of the starting iodide; (ii) addition of the radical to CO; (iii) 5-exo-cyclization of the resulting acyl radical to give a tertiary alkyl radical; (iv) the addition of the tertiary radical to acrylonitrile; (v) one electron reduction of the resulting α-cyano radical leading to an α-cyano anion; (vi) addition of the anion to an internal carbonyl group; and (vii) proton quenching of the alkoxy anion to give bicyclo[3.3.0]octan-1-ol. Ryu applied the zinc-induced dual [4+1] and [3+2] annulation reaction to other alkenes, such as methyl acrylate and diethyl fumarate [32]. Also the construction of bicyclo[3.2.1]octanol skeletons was completed by using a zinc-induced cyclization. This system revealed that 6-endo cyclization is favored over 5-exo cyclization

(Scheme 22). The reaction of 5-iodo-2-methylhex-1ene with CO and acrylonitrile in the presence of zinc led to bicyclo[3.2.1]octanol at a 51% yield with a 57/43 exo/endo ratio. Similarly, trapping with methyl acrylate produced a 43% yield of the corresponding bicyclo[3.2.1]octanol. Authors have demonstrated that radical carbonylation, when combined with radical/anionic sequential reactions, provides a means of producing bicycle [3.3.0]octanols and bicyclo[3.2.1]octanols from readily accessible starting materials. Using this "one-pot" procedure, the authors were able to form four C-C bonds through three radical reactions and one anion reaction.

Ogawa et al. reported on the one-electron reduction of alkyl chlorides by SmI_2 [42]. When they coupled this system with photo-irradiation, they observed a highly efficient reaction.When they carried out this reaction under CO pressure, they obtained unsymmetrical ketones. Each product consisted of two molecules of alkyl chlorides and two molecules of carbon monoxide. Because their control experiments suggested that the dimerization of acyl anions is a likely key step, it is concluded that the one-electron reduction of an acyl radical to an acyl anion occurred quickly in this SmI_2 system. The rapid conversion to acyl anions is also supported by the absence of cyclized products, as in the case of 6-hexenyl chloride is used as a substrate (Scheme 23).

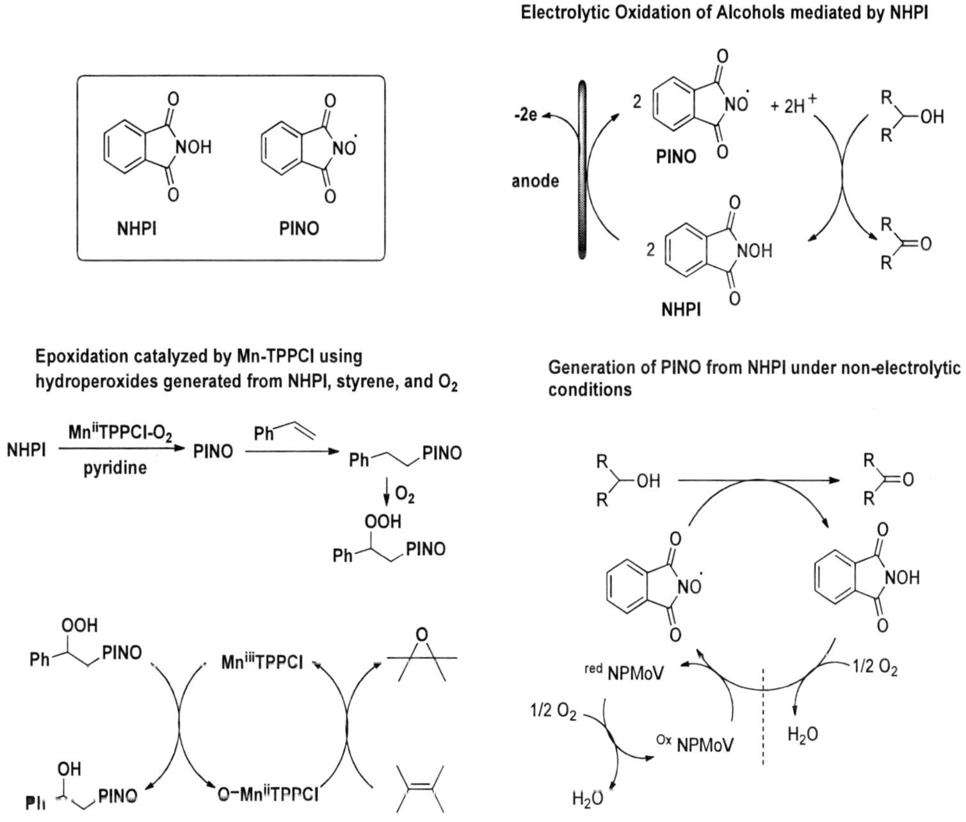

Scheme 24. Ishii oxidation in mechanistic detail: electrolytic oxidation, epoxidation catalyzedby Mn-TPPCl using hydroperoxides generated from NHPI, styrene, and O_2.

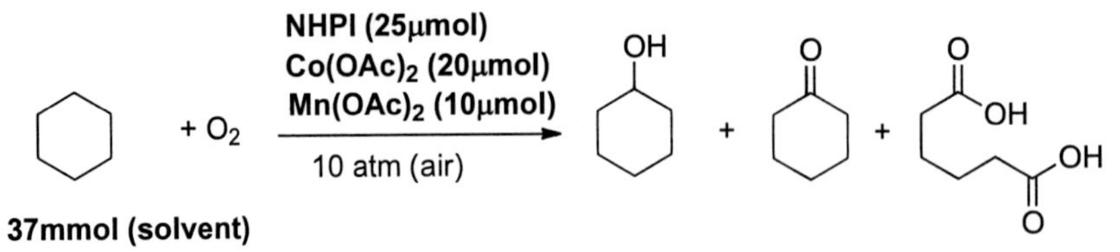

NHPI (25μmol)
$Co(OAc)_2$ (20μmol)
$Mn(OAc)_2$ (10μmol)
10 atm (air)

cyclohexane + O_2 → cyclohexanol (OH) + cyclohexanone (O) + adipic acid (OH, OH, O)

37mmol (solvent)

NHPI	6h 14h	622% 1060	468% 836	trace 5	TON = 10.9 19.0
F-NHPI	6h 14h	1278 2108	1485 2424	21 32	TON = 27.8 45.6

Scheme 25. Ishii's system for aerobic oxidation of cyclohexane.

5.3. Ishii Oxidation in Detail

An innovation of the aerobic oxidation of hydrocarbons through catalytic carbon radical generation under mild conditions was achieved by using N-hydroxyphthalimide (NHPI) as a key compound. Alkanes were successfully oxidized with O_2 or air to valuable oxygen containing compounds such as alcohols, ketones, and dicarboxylic acids by the combined catalytic system of Ishii and coworkers who have developed an innovative strategy for the catalytic carbon radical generation from hydrocarbons by a phthalimide N-oxyl (PINO) radical generated *in situ* from N-hydroxyphthalimide (NHPI) (Scheme 24) and molecular oxygen in the presence or absence of a cobalt ion under mild conditions. The carbon radical derived from a variety of hydrocarbons under the influcnce of molecular oxygen lead to oxygenated products like alcohols, ketones, and carboxylic acids in good yields (Scheme 24) [43].

The N-hydroxyphthalimide derivatives, F15- and F17-NHPI, bearing a long fluorinated alkyl chain, were prepared and their catalytic performances were compared with that of the parent compound, N-hydroxyphthalimide (NHPI). The oxidation of cyclohexane under 10 atm of air in the presence of fluorinated F15- or F17-NHPI, cobalt diacetate, and manganese diacetate without any solvent at 100 °C afforded a mixture of cyclohexanol and cyclohexanone (K/A oil) as major products along with a small amount of adipic acid. It was found that F15- and F17-NHPI exhibit higher catalytic activity than NHPI for the oxidation of cyclohexane without a solvent (Scheme 25). However, for the oxidation in acetic acid all of these catalysts afforded adipic acid as a major product in good yield and the catalytic activity of NHPI in acetic acid was almost the same as those of F15- and F17-NHPI. The oxidation by F15- and F17-NHPI catalysts in trifluorotoluene afforded K/A oil in high selectivity with little formation of adipic acid, while NHPI was a poor catalyst under these conditions, forming K/A oil as well as adipic acid in very low yields (Scheme 26). The oxidation in trifluorotoluene by F15- and F17-NHPI catalysts was considerably accelerated by the addition of a small amount of zirconium(IV) acetylacetonateto the present catalytic system to afford selectively K/A oil, but no such effect was observed in the NHPI-catalyzed oxidation in trifluorotoluene (Scheme 27) [44].

Cyclohexane + O_2 → cyclohexanol + cyclohexanone + adipic acid

37mmol (solvent)

NHPI (25μmol)
$Co(OAc)_2$ (20μmol)
$Mn(OAc)_2$ (10μmol)
10 atm (air)
100°C, 6h

F-NHPI	AcOH	2536%	408%	5120	TON = 82.6
F-NHPI	$CF_3C_6H_5$ (BTF)	1772	1504	36	33.1

Scheme 26.High selectivity in the oxidation of cyclohexane in BTF.

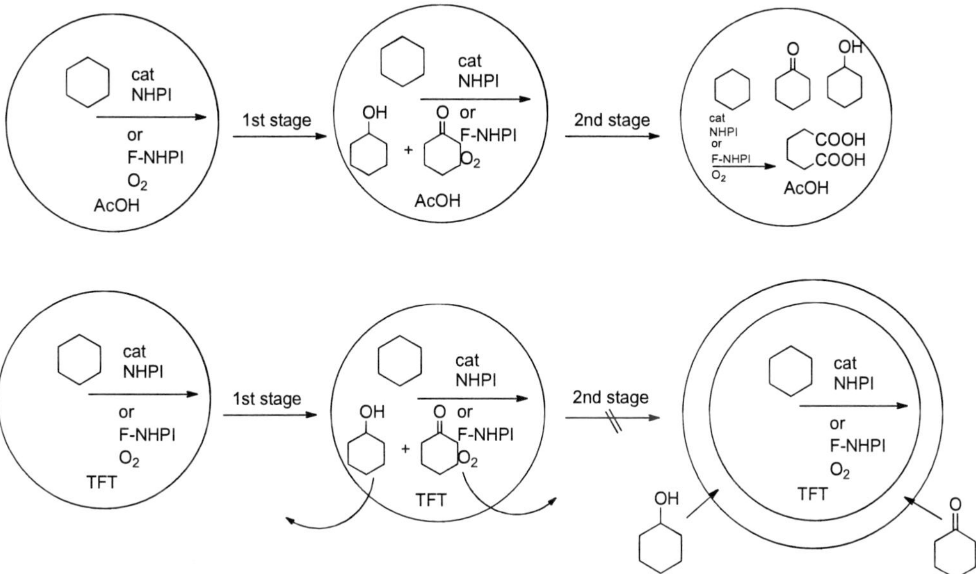

Scheme 27. Mechanistic aspects of oxidation and illustration of the selectivity observed for the aerobic oxidation of cyclohexane in AcOH and THF.

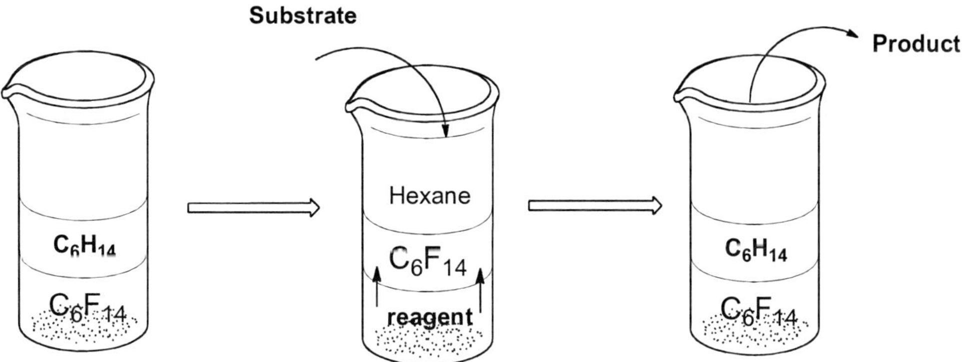

Scheme 28. General representation of phase-vanishing method.

5.4. From Phase-separation to Phase –vanishing Methods Based on Fluorous Phase Screen: A Simple Way for Efficient Execution of Organic Synthesis

Relying upon the unique properties of perfluorinated organic compounds, fluorous phase chemistry has opened fresh ground in the combination of organic reactions and separation processes. Perfluorinated compounds are generally immiscible with most organic solvents and are denser than typical organic compounds.Intrigued by these unique properties of fluorous solvents, Ryu and Curran [34,35] have developed a synthetically convenient triphasic system comprising of an organic phase containing substrates, fluorous phase, and a reagent phase (Scheme 28). In such a triphasic system, the fluorous layer functions as a liquid membrane, bringing the two separate layers.Thus, the reagent at the bottom layer diffuses slowly through the fluorous layer on the top layer. It encounters the substrate in the upper layer, then the reaction takes place.Eventually, the reagent phase disappears, leaving only two phases.

As a fluorous layer, typically FC-72 (perfluorohexane) is used for the triphasic phase-vanishing (PV) reaction. The more viscous and less volatile perfluorodecalin can also be used, but diffusion is slower. Recent studies reveal that less volatile and inexpensive perfluorinated polyether solvents, such as Galden HT-135, function as excellent fluorous phases in many cases. Inherent to the nature of a very slow reagent addition like a “micro” syringe pump, control of heat evolution is also possible with thermal diffusion. This allows for the test-tube reaction to be performed without any cooling equipment. In many cases of exothermic reactions, the disappearance of the bottom phase is a useful sign of the end of the addition. The change from the three layers to two layers is the origin of the wording a PV method, and therefore, the concept of the PV method is not restricted to the fluorous phase. However the superiority of fluorous media to water and some ionic liquids, which can also be used to constitute triphasic systems, has been demonstrated in many cases. Since the first report of Ryu and Curran in 2002, PV method has demonstrated its utility in the bromination of alkenes under various conditions and is summarized in the diagram below (Scheme 29).

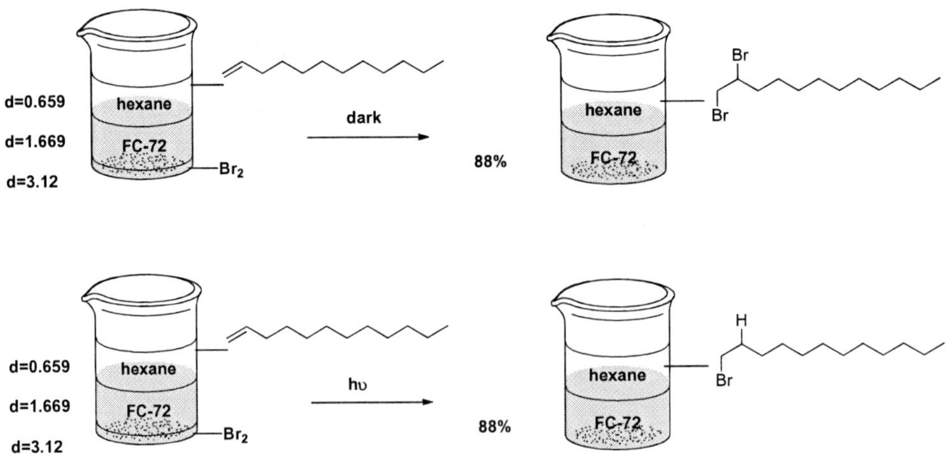

Scheme 29.Phase-vanishing bromination of alkenes.

5.5. Conclusions and Future Directions in Fluorous Chemistry

Over the past 15 years, fluorous chemistry has emerged as a real and viable environmentally attractive alternative to traditional reagents and catalytic systems. The section has attempted to highlight that these are no longer reasonable objections to rise: the costs of fluorous reagents and solvents are falling all the time and there are now specialist manufactures dealing solely with the production of fluorous reagents and compounds. Furthermore, large quantities of fluorous solvents are no longer required to effect reactions through the biphase system: there are light fluorous alternatives and, in addition to liquid-liquid extraction is a rapid method for the separation of both fluorous/organic and fluorous/fluorous compounds, without the need for fluorous solvents.

General Conclusion

The last 20 years has also seen the emergence of other so-called clean alternatives for the synthesis and catalysis, such as supercritical fluids and ionic liquids. It is becoming increasingly clear that no single system will, in its own right, ever be able to replace completely all conventional reagents and solvents as truly environmentally friendly alternative. There are drawbacks associated with all three systems, but each method does however, have its niche and role on the market and application. The great degree of crossover applications continuously emerging represents only a matter of time until fluorous technology, supercritical solvents and ionic liquids and aqueous media use in synthesis will be widely adopted bythe chemical community.

References

[1] Selected reviews: (a) Li, C.-J. Organic reactions in aqueous media with a focus on carbon-carbon bond formations: a Decade update, *Chem. Rev.*, 2005, *105*, 3095-3166; (b) Sheldon, R. A. Green solvents for sustainable organic synthesis: state of the art. *GreenChem.*, 2005, *7*, 267-278; (c) Li, C.-J.; Chen, L. Organic chemistry in water. *Chem. Soc. Rev.*, 2006, *35*, 68-82; (d) Perchyonok, V. T.; Lykakis, I. N.; Tuck, K. L. Recent advances in C–H bond formation in aqueous media: a mechanistic perspective. *Green Chem.*, 2008, *10*, 153-163; (e) Perchyonok, V. T.; Lykakis, I. N. Recent advances in free radical chemistry of c-c bond formation in aqueous media: from mechanistic origins to applications. *Mini Rev. Org. Chem.*,2008, *5*,19-32.

[2] Studer, A. Microwave chemistry and radicals review articles: Tin-free radical chemistry using the persistent radical effect: alkoxyamine isomerization, addition reactions and polymerizations. *Chem. Soc. Rev.*, 2004, *33*, 267-273 and references cited therein.

[3] Rogers, R. D.; Seddon, K. R. Ionic liquids - general reviews: Ionic liquids–solvents of the future? *Nature*, 2003, *302*(5646), 792 and references sited there in.

[4] (a) Zhang, S.; Chen, J.; Lykakis, I. N.; Perchyonok, V. T. Streamlining organic free radical synthesis through modern molecular technology: from polymer supported

synthesis to microreactors and beyond, *Current Org. Synth.*, 2010, *7*, 177-188. (b) Johnson, A. E.; Perchyonok, V. T. Recent advances in free radical chemistry in unconventional medium: ionic liquids, microwaves and solid state to the rescue, review article, *Current Org. Chem.*, 2009, *13*, 1705-1725.

[5] For an excellent general reference book in free-radical chemistrysee: B. Giese, Radicals in Organic Synthesis: Formation of Carbon–Carbon Bonds, Pergamon Press, 1986, and references cited therein.

[6] Perchyonok, V. T. *Radical reactions in aqueous media*, (RSC Green Chemistry Series), 1st edition, RSC publishing, 2009.

[7] Perchyonok, V. T.; Lykakis, I. N. Radical reactions in aqueous media: origins, reason and applications.*Curr.Org. Chem.*, 2009, *13*, 573-598.

[8] C. Chatgilialoglu, in Organosilanes in Radical Chemistry, JohnWiley & Sons, Ltd, West Sussex, UK, 2004, and references cited therein.

[9] Zard S. Z., Radical Reactions in Organic Synthesis, Eds. Compton, R. G., Davies, S. G., Evans, J., Oxford University Press.

[10] (a) Renaud, P.; Sibi, M. P. *Radicals in Organic Synthesis*, Vol. 1; Eds.; Wiley-VCH: Weinheim, 2001; (b) Postigo. Al.; Ferreri. C.; Navacchia. M. L.; Chatgilialoglu. C. The radical-based reduction with $(TMS)_3SiH$ 'on water'.*Synlett*, 2005, *18*, 2854-2856; (c) Postigo, Al.; Kopsov, S..; Ferreri, C.; Chatgilialoglu, C. Radical reactions in aqueous medium using $(Me_3Si)_3SiH$. *Org. Lett.*, 2007, *9*, 5159-5162; (d) Chatgilialoglu. C.; Guerra. M.; Mulazzani Q. G. Model studies of DNA C5' radicals. Selective generation and reactivity of 2'-deoxyadenosin-5'-yl radical. *J. Am. Chem. Soc.*, 2003, *125*, 3839-3848; (e) Jimenez. L. B.; Encinas. S., Miranda. M. A.; Navacchia. M. L.; Chatgilialoglu. C.The photochemistry of 8-bromo-2'-deoxyadenosine. A direct entry to cyclopurine lesions.*Photochem. Photobiol. Sci.* 2004, *3*, 1042-1051; (f) Chatgilialoglu. C.; Ferreri. C. Trans lipids: the free radical path. *Acc. Chem. Res.* 2005, *38*, 441-448; (g). Sugiyama. H.; Xu.Y. Photochemical Approach to Probing Different DNA Structures.*Angew. Chem. Int. Ed.* 2006, *45*, 1354-1362.

[11] Selected references on the use of organic hypohosphites as hydrogen donors in organic solvents: (a). Brigas, A. F.; Johnstone, R. A. W. Mechanisms in heterogeneous liquid-phase catalytic-transfer reduction: the importance of hydrogen-donor concentration. *J. Chem. Soc., Chem. Commun.*, 1991, 1041–1042; (b). Graham, S. R.; Murphy J. A.; Kennedy A. R., Hypophosphite mediated carbon–carbon bond formation: total synthesis of epialboatrin and structural revision of alboatrin. *J. Chem. Soc., Perkin Trans. 1*, 1999, 3071-3073; (c). Martin, C. G.; Murphy, J. A.; Smith, C. R., Replacing tin in radical chemistry: *N*-ethylpiperidine hypophosphite in cyclisation reactions of aryl radicals. *Tet. Lett.*, 2000, *41,* 1833–1836; (d). Perchyonok, V. T.; Tuck, K. L.; Langford, S. J.; Hearn, M. W., On the scope of radical reactions in aqueous media utilizing quaternary ammonium salts of phosphinic acids as chiral and achiral hydrogen donors. *Tet. Lett.* 2008, *49,* 4777-4779.

[12] Cho, D. H.; Jang, D. O. Enantioselective radical addition reactions to the C=N bond utilizing chiral quaternary ammonium salts of hypophosphorous acid in aqueous media. *Chem. Commun.*, 2006, *9*, 5045-5047.

[13] (a) Lindstrom U. M. Stereoselective Organic Reactions in Water. *Chem. Rev*, 2002, *102*, 2751-2772; (b) Jacobsen E. N.; Taylor. M.S. Asymmetric Catalysis by Chiral Hydrogen-Bond Donors. *Angew. Chem. Int. Ed.*, 2006, *45*, 1520-1543.

[14] Selected publications for enantioselectivity in free radical transformations in the organic solvents: (a) Dakternieks D.; Perchyonok. V. T.; SchiesserC. H. Single enantiomer free-radical chemistry-Lewis acid-mediated reductions of racemic halides using chiral non-racemic stannanes. *Tetrahedron: Asymmetry* 2003, *14*, 3057-3068; (b) Sibi. M. P.; Patil. K. Enantioselective H-Atom Transfer Reactions: A New Methodology for the Synthesis of β-Amino Acids. *Angew. Chem., Int. Ed.*, 2004, *43*, 1235-1238 and references sited there in.

[15] Selected publications on the application of radical cascade reactions in total synthesis: (a) Myers, A. G.; Condronski, K. R. Synthesis of (+,-)-7,8-Epoxy-4-basmen-6-one by a transannular cyclization strategy. *J. Am. Chem. Soc.*, 1993, *115*, 7926-7927; (b) Myers, A. G.; Condroski, K. R.Synthesis of (+,-)-7,8-Epoxy-4-basmen-6-one by a transannular cyclization strategy.*J. Am. Chem. Soc.*, 1995, *117*, 3057-3083.

[16] (a) Parson, A. F., An Introduction to Free Radical Chemistry, Blackwell Science; (b) Supercritical Carbon Dioxide: Separation and Processes: ACS symposium series 860, edited by Aravamudan S. Gopalan, Chien M. Wai, Hollie K. Jacobs and references sited therein; (c) Kishimoto, Y., Ikariya T., Supercritical carbon dioxide as a reaction medium for silane –mediated free-radical carbonylation of alkyl halides., *J. Org. Chem.*, 2000, *65*, 7656-7659; (d) Jessop, P. G.; Leitner, W. Chemical Synthesis Using Supercritical Fluids, Wiley-VCH, Weinheim, 1999, p1.

[17] Fletcher, B.; Suleman, N. K.; Tanko, J. M. Free radical chlorination of alkanes in supercritical carbon dioxide: The chlorine atom cage effect as a probe for enhanced cage effects in supercritical fluid solvents. *J. Am. Chem. Soc.,* 1998, *120*, 11839-11844.

[18] For an interesting example of use of supercritical CO_2, see: (a) Hadida, S.; Super, M. S.; Beckman, E. J.; Curran, D. P. Radical reactions with alkyl and fluoroalkyl (fluorous) tin hydride reagents in supercritical CO_2*J. Am. Chem. Soc.*, 1997, *119*, 7406-7407; (b) Tanko, J. M., "Free-Radical Chemistry in Supercritical Carbon Dioxide", in *Green Chemistry using Liquid and Supercritical Carbon Dioxide*, Joseph M. DeSimone and William Tumas, Ed.s; Oxford University Press: New York, 2003; chp. 4, p. 64.

[19] Kishimoto, Y.; Ikariya, T. Supercritical Carbon dioxide as a reaction medium for silane –mediated free-radical carbonylation of alkyl halides. *J. Org. Chem.,* 2000, *65*, 7656-7659.

[20] *Ionic Liquids in Synthesis*, Wasserscheid, P and Welton T.,Ed.s; Wiley-VCH: Weinheim, 2008; b) Pârvulesku, V. I.; Hardacre, C. Catalysis in ionic liquids. *Chem. Rev.*, 2007, *107*, 2615-2665; c) Harper, J. B.; Kobrak, M. N. Understanding organic processes in ionic liquids: achievements so far and challenges remaining *Mini-Rev. Org. Chem.*, 2006, *3,* 253-269; d) Chiappe, C.; Malvaldi, M.; Pomelli, C. S. Ionic liquids: solvation ability and polarity*Pure Appl. Chem.*, 2009, *81*, 767-776.

[21] Yorimitsu, H.; Nakamura, T.; Shinokubo, H.; Oshima, K., Omoto, K.; Fujimoto, H. Powerful solvent effect of water in radical reaction: triethylborane-induced atom-transfer radical cyclization in water *J. Am. Chem. Soc.*, 2000, *122*, 11041-11047.

[22] For some recent examples of 'green' radical reagents, see the references reported in: a) Walton, J. C. Linking borane with *N*-heterocyclic carbenes: effective hydrogen-atom donors for radical reactions, *Angew. Chem. Int. Ed.*, 2009, *48*, 1726-1728; b) Bencivenni, G.; Lanza, T.; Leardini, R.; Minozzi, M.; Nanni, D.; Spagnolo, P., Zanardi, G. Tin-free generation of alkyl radicals from alkyl 4-pentynyl sulfides via

homolytic substitution at the sulfur atom *Org. Lett.*, 2008, *10*, 1127-1130; c) Benati, L.; Bencivenni, G.; Leardini, R.; Minozzi, M.; Nanni, D.; Scialpi, R.; Spagnolo, P.; Zanardi, G. Reaction of azides with dichloroindium hydride: very mild production of amines and pyrrolidin-2-imines through possible indium-aminyl radicals *Org. Lett.*, 2006, *8*, 2499-2502; d) Healy, M. P.; Parsons, A. F.; Rawlinson, J. G. T. Consecutive approach to alkenes that combines radical addition of phosphorus hydrides with Horner−Wadsworth−Emmons-type reactions*Org. Lett.*, 2005, *7*, 1597-1600; e) Benati, L.; Leardini, R.; Minozzi, M.; Nanni, D.; Scialpi, R.; Spagnolo, P.; Strazzari, S.; Zanardi, G. A novel tin-free procedure for alkyl radical reactions *Angew. Chem. Int. Ed.*, 2004, *43*, 3598-3601; f) Studer, A.; Amrein, S. Tin hydride substitutes in reductive radical chain reactions *Synthesis*, 2002, 835. See also: Togo, H., *Advanced Free Radical Reactions for Organic Synthesis*, Elsevier: Amsterdam, 2004, chp. 12, p. 247.

[23] Pan, X-Q.; Zou, J-P.; Zhang,W. Manganese(III)-promoted reactions for formationof carbon–heteroatom bonds, *Mol Divers*, 2009, *13*, 421-438 and references cited there in.

[24] (a) Bar, G.; Parsons, A. F.; Thomas, C. B. A radical approach to araliopsine and related quinoline alkaloids using manganese(III) acetate. *Tetrahedron Lett.*, 2000, *41*, 7751–7755; (b) Bar, G.; Parsons, A. F.; Thomas, C.B. Manganese(III) acetate mediated radical reactions leading to araliopsine and related quinoline alkaloids. *Tetrahedron* 2001, *57*, 4719–4728; (c) Bar, G.; Parsons, A. F.; Thomas, C. B. Manganese(III) acetate mediated radical reactions in the presence of an ionic liquid. *Chem. Commun.* 2001, 1350–1351.

[25] (a) Nair, V.; Mathew, J.; Prabhakaran, J. Carbon-carbon bond forming reactions mediated by cerium(IV) reagents. *Chem. Soc. Rev.* 1997, 127–132; (b) Baciocchi, E.; Paolobelli, A. B.; Ruzziconi, R. Cerium(IV) ammonium nitrate promoted oxidative cyclization of dimethyl 4-pentenylmalonate. *Tetrahedron* 1992, *48*, 4617–4622; (c) D'Annibale, A.; Pesce, A.; Resta, S.; Trogolo, C. Ceric ammonium nitrate promoted free radical cyclization reactions leading to β-lactams. *Tetrahedron Lett.* 1997, *38*, 1829–1832.

[26] Bar, G.; Bini, F.; Parson, A. F. CAN-Mediated Oxidative Free Radical Reactions in an Ionic Liquid.*Synth. Commun.*,2003, *33*, 213–222.

[27] Ranu, B. C.; Adak, L.; Banerjee, S. Halogenation of Carbonyl Compounds by an Ionic Liquids, [AcMIm]X and Ceric Ammonium Nitrate (CAN). *Aust. J. Chem.*, 2007, *60*, 358-362.

[28] Riisagera, A., Fehrmanna, R., Haumannb, M., Wasserscheidb, P., Supported Ionic Liquids: Versatile Reactions and Separation Media, Topics in Catalysis, 2006, 40(1-4), 91-102

[29] Horvath, I. T.; Rabai, J. Facile catalyst separation without water: fluorousbiphasehydroformylation of olefins. *Science,* 1994, *266*, 72-75.

[30] Nostro, P. L. Phase separation properties of fluorocarbon, hydrocarbons and their copolymers. *Adv. Colloid Interface.*,1995, *56*, 245-287.

[31] Barthel-Rosa, L. P.; Gladysz, J. A. Chemistry in Fluorous media: a user's guide to practical consideration in the application of fluorous catalyst and reagents. *Coord. Chem. Rev.*, 1999, *190-192*, 587-605 and references sited therein.

[32] Clark, D.; Ali, M. A.; Clifford, A. A.; Parratt, A.; Rose, P.; Schwinn, D.; Bannwarth, W.; Rayner, C. M. "Reactions in Unusual Media", *Curr. Top. Med. Chem.*, 2004, *4*, 729-771.

[33] Kharasch, M. S.; Mayo, F. R. The peroxide effect in the addition of reagents to unsaturated compounds. The addition of hydrogen bromide to allyl bromide. *J. Am. Chem. Soc.,* 1933, *55*, 2468-2496.

[34] Curran, D. P.; Hadida, S. Tris(2-(perfluorohexyl)ethyl)tin Hydride: A new fluorous reagent for use in traditional organic synthesis and liquid phase combinatorial synthesis. *J. Am. Chem. Soc.*, 1996, *118*, 2531-2532.

[35] Ryu, I. New approaches in radical carbonylation chemistry: Fluorous applications and designed tandem processes by species-hybridization with anion and transition metal species. *Chem. Rec.*, 2002, *2*, 249-258.

[36] Gupta, V.; Kahne, D.; Direct introduction of CH_2OH by intermolecular trapping of CO. *Tetrahedron Lett.,* 1993, *34*, 591-594.

[37] Matsubara, H.; Yasuda, S.; Sugiyama, H.; Ryu, I.; Fujii, Y.; Kita, K. A new fluorous/organic amphiphilic ether solvent, F-626: execution of fluorous and high temperature classical reactions with convenient biphase workup to separate product from high boiling solvents. *Tetrahedron*, 2002, *58*, 4071-4076.

[38] Ryu, I.; Niguma, T.; Minakata, S.; Komatsu, M.; Luo, Z.; Curran, D. P. Radical carbonylation with fluorousallyltin reagents. *Tetrahedron Lett.,*1999, *40*, 2367-2370.

[39] Tsunoi, S.; Ryu, I.; Fukushima, H.; Tanaka, M.; Komatsu, M.; Sonoda, N. Free-Radical Carbonylation Using a Zn(Cu) Induced Reduction System. *Synlett,* 1995, 1249-1251.

[40] Tsunoi, S.; Ryu, I.; Yamasaki, S.; Tanaka, M.; Sonoda, N.; Komatsu, M. Tandem annulations: a one operation construction of bicyclo[3.3.0]octan-1-ol and bicyclo[3,2,1]octan-1-ol skeleton by a three-component coupling reaction of alk-4-enyl iodides with CO and alkenes in the presence of Zinc. *Chem. Commun.*, 1997, 1889-1890.

[41] Hope, E. G., Stuart, A. M., FluorousBiphase Catalyst, *Journal of Fluorine Chemistry*, 1999, *100,1-2*, 75-83.

[42] Ogawa, A.; Sumino, Y.; Nanke, T.; Ohya, S.; Sonoda, N.; Hirao, T.; *Journal of the American Chemical Society*, 1997, *119*, 2745-2746.

[43] Ishii, Y., and Yanagida, T., Single molecule detection in life science, *Single Mol.* 2000, *1*, 5-16.

[44] Guha, S. K.; Obora, Y.; Ishihara, D.; Matsubara, H.; Ryu, I.; Ishii, Y. Aerobic oxidation of cyclohexane using N-hydroxyphthalimide bearing fluoroalkyl chains. *Adv. Synth. Catal.,* 2008, *350*, 1323–1330.

In: Organic Radical Reactions in Water and Alternative Media ISBN 978-1-61209-648-3
Editor: J. Alberto Postigo, pp. 207-238

Chapter 5

ARTIFICIAL ENZYMES AND FREE RADICALS: THE CHEMISTS PERSPECTIVE

Ioannis N. Lykakis*, Stanislav A. Grabovskiy andV. Tamara Perchyonok#

[1]Department of Chemistry, University of Crete, Voutes 71003 Iraklion, Greece
[2]Institute of Organic Chemistry, Ufa Research Center, Russian Academy of Sciences, 71 prosp. Oktyabrya, 450054 Ufa, Russian Federation
[3]VTPCHEM PTY LTD, Melbourne, Australia

ABSTRACT

The field of artificial enzymes is rapidly evolving subject.As the barrier between chemistry and biology becomes less distinct a range of new methods, which combine expertise from both areas, are developing.In recognition of both the fact that the de novo design approach can be time consuming, and that a tiny miscalculation will have a detrimental effect, a trend in all these recent techniques is the use of "selection approaches".The natural processes of selection and amplification is after all, the way in which enzymes have evolved their sophisticated function and also the area is constantly evolving the sky is the limit in the desired crossing and eliminating the barrier between "traditional chemistry" and "traditional biological sciences" and developments and understanding of life science at the molecular level. Advances in the fields of molecular biology, biochemistry and more recently combinatorial and polymer chemistry have all furnished unique and often co-operative solutions to the synthesis of artificial enzymes, and it is the aim of this overview to discuss some of the more recent and diverse approaches taken by organic chemists towards the creation of effective enzyme mimics with particular focus on some free radical based mechanism enzymes in biocompatible media.The approaches to be discussed in detail are (a) "design approach" where a host molecule is designed with salient functionality (often also present in the natural enzyme counterpart), which is expected to be involved in catalysis of the chosen reaction. Catalytic cyclodextrins are one such example and will be discussed in some detail in this

*Tel: +30 2810545041, fax +30 2810 545001, e-mail: lykakis@chemistry.uoc.gr
#Te. +61414596304, fax +61395725223, email, tamaraperchyonok@gmail.com

chapter.(b) “transition state analogue-selection approach” where a library of hosts is generated in the presence of a transition state analogue (TSA) and the best host is then selected from the library with particular focus on the field of catalytic antibodies and has more recently inspired the process of `molecular imprinting' (vide infra) and (c) “catalytic activity-selection approach”. This chapter, written from the perspective of the organic chemist, will instead concentrate on less developed areas and will conclude with a discussion of some of the more recent developments in `selection approaches' towards artificial receptors.

1. INTRODUCTION

All living things in nature maintain their internal metabolic balances quite well when they are in healthy conditions. In other words, metabolic materials are in equilibrium in each living creature primarily in a collaboration of biological catalyses by numerous enzymes. Enzymes are sophisticated proteins having catalytic groups and often require specific cofactors or coenzymes for catalytic performance. If we look at enzymatic functions from physicochemical viewpoints rather than biological ones, catalytically active amino acid residues of enzyme proteins as well as coenzyme factors are buried in hydrophobic and water-lacking reaction sites furnished by enzyme proteins and well separated from the bulk aqueous phase needed to attain thermodynamic stabilities. In consideration of such physicochemical roles of enzyme proteins, we are allowed to use man-made materials for construction of artificial enzymes that are capable of simulating catalytic functions demonstrated by enzyme proteins [1, 2].

Two types of such artificial enzymes or apo-enzymes, macrocyclic compounds and molecular assemblies, are cited in this chapter as those which can provide specific microenvironments for substrate-binding and subsequent catalysis in aqueous media. Those micro-environmental properties are primarily due to hydrophobic internal cavity and the internal domain of molecular assemblies in aqueous media, and other non-covalent intermolecular interactions, such as electrostatic, charge-transfer, and hydrogen-bonding modes, between a substrate and an apoenzyme model are greatly enhanced in such microenvironments [3]. In this chapter, primarily we summarize recent studies on functional simulation of holoenzymes requiring coenzyme factors, such as vitamin B12, Artificial Methylmalonyl-CoA mutase and Glutamate Mutase as representative enzymes. Most of those coenzymes are soluble in aqueous media when separated from the corresponding apo-protein and, consequently, cannot be readily incorporated into hydrophobic microenvironments provided by the artificial systems. Modified proteins which are derived by mutagenic treatments of natural enzymes are often called artificial enzymes. It must be emphasized here that artificial enzymes described in this article are not directly related to protein structures but capable of carrying out functional simulation of enzymatic catalysis in the overall reaction schemes.

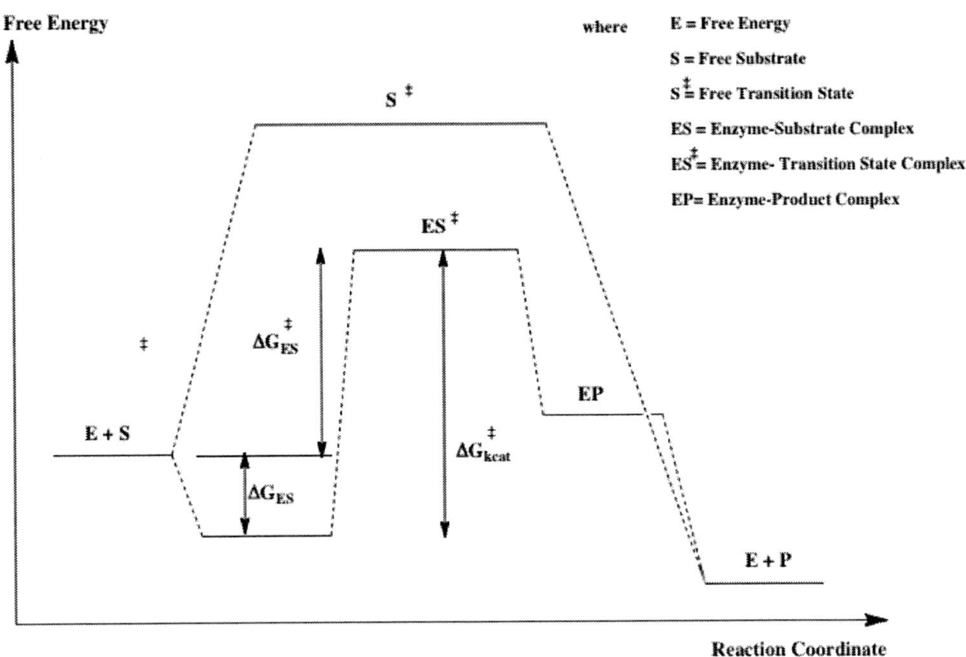

Figure 1. Energy diagram of an enzyme-catalysed reaction and the corresponding uncatalysed chemical reaction.

Enzymes have had billions of years to evolve into the sophisticated three dimensional structures of today. As chemists, we need to concentrate this period into a feasible timescale for research. In recognition of this fact, recent developments in the field of artificial enzymes have tended to move away from the rational design and multistep synthesis of complex molecules where the smallest flaw in conception can have catastrophic results. Instead, current strategies have tended to focus on selection approaches [3a, 2e].

Advances in the fields of molecular biology, biochemistry and more recently combinatorial and polymer chemistry have all furnished unique and often co-operative solutions to the synthesis of artificial enzymes, and it is the aim of this overview to discuss some of the more recent and diverse approaches taken by organic chemists towards the creation of effective enzyme mimics with particular focus on some free radical based mechanism enzymes.

In general these different approaches can be divided into three categories:

a) The "design approach". A host molecule is designed with salient functionality (often also present in the natural enzyme counterpart), which is expected to be involved in catalysis of the chosen reaction [4]. Catalytic cyclodextrins are one such example and will be discussed in some detail in this chapter.
b) The "transition state analogue-selection approach". A library of hosts is generated in the presence of a transition state analogue (TSA) and the best host is then selected from the library. This latter approach has been employed with considerable success in the field of catalytic antibodies and has more recently inspired the process of `molecular imprinting' (vide infra) [5].

c) The “catalytic activity-selection approach”. This takes advantage of the combinatorial chemistry revolution wherein a library of possible catalysts is generated and screened directly for enzyme-like activity. This area is justifiably deserving of review articles in their own right, and only selected examples will be discussed here. Equally, much research into enzyme mimetic systems has been carried out in the field of bio-inorganic and coordination chemistry and has been summarized elsewhere [6]. This chapter, written from the perspective of the organic chemist, will instead concentrate on less developed areas and will conclude with a discussion of some of the more recent developments in `selection approaches' towards artificial receptors.

2. Transition State TheoryBrief Introduction

The currently accepted view is that catalysis rests on the enzymes ability to stabilize the transition state of a reaction relative to that of the ground state. This principle is illustrated (Figure 1) for a uni-molecular example where the enzyme±substrate complex is stabilized relative to the free species in solution. The activation barrier to reaction is represented by the difference ΔG_{cat} and ΔG_{uncat} for the enzyme catalyzed and the uncatalyzed reactions respectively. It is clear from this picture that, for catalysis to work, the difference $\Delta G_{ETS\#}$ must be larger than ΔG_{ES}. In other words, the enzyme must stabilize the transition state of the reaction more than it stabilizes the ground state of the substrate. For true catalysis, a system also needs to exhibit turnover. If the product binds to the enzyme in a significantly stronger way than it binds to the substrate (E.S<E.P), then product inhibition of the reaction can result. In this case substrate binding can be beneficial. The most important consequence of this picture of enzyme action is that the design of an enzyme mimic must not only consider transition state binding relative to substrate binding, but also ensure that the active site is designed such that product release is a thermodynamically favorable process. The discussion above is a simplification of the real situation since enzyme catalysis of a transformation often involves an alternative reaction pathway from that taken in the non-catalyzed process, usually taking advantage of the enzymes ability to reduce the molecularity of multi-step sequences. The situation also becomes more complicated for bimolecular processes and reactions involving covalent enzyme bound intermediates, invoked in the initial step of the mechanism for amide bond cleavage by serine proteases. In these more complex systems, application of the above model of transition state stabilization is less straightforward [7].

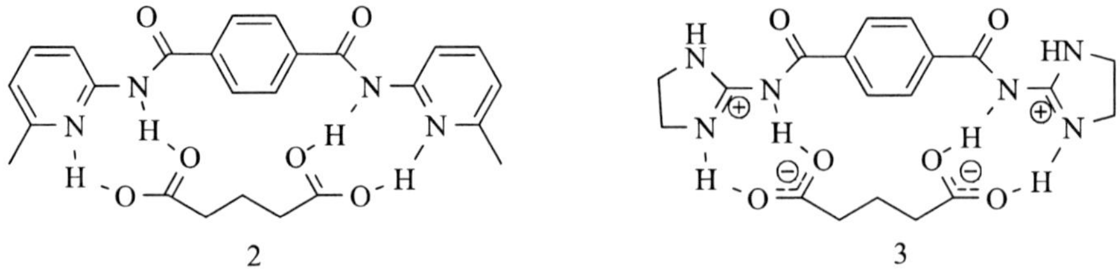

Figure 2. Difference between neutral and charged hydrogen bonding receptors demonstrated by comparison of the two model receptors 2 and 3 for glutaric acid.

However, the foregoing discussion is sufficient to appreciate the basic principle behind many of the various approaches to artificial enzymes.

Perhaps the biggest obstacle to the synthesis of enzyme mimics is that in order to design an artificial enzyme it is necessary to understand how enzymes achieve this selective binding of the transition state. Several studies have been carried out in attempts to quantify the contributions of the many weak intermolecular forces involved. In general the overall binding is a product of electrostatic forces, hydrogen bonding, and cumulative hydrophobic and Van der Waals influences. Since enzymes operate in water, desolvation effects and the resulting entropy changes are also very important factors [7b]. Hydrogen bonding and electrostatic interactions contribute significantly to the binding affinity between the substrate or transition state and the enzyme, although, since enzymes operate in water, the contribution of these effects is greatly moderated by solvation. In fact, in some cases, desolvation of both the polar group on the ligand and the complementary group in the enzyme may cost as much in enthalpy as is gained by bringing the two groups together. Detailed studies into the influence of hydrogen bonding in particular have been carried out, in an attempt to quantify the energy difference gained upon the formation of a ligand-host hydrogen bond in water. The value for a neutral±neutral hydrogen bond has been generally found to be in the order of 1.5 kcal mol [8], which represents a perhaps surprisingly modest energy gain [9]. As a result it has been suggested that such interactions `may play less of a role in enhancing association of the correct ligand than they do in creating a penalty for binding the wrong ligand, *i.e.*, in determining ligand specificity. By way of contrast, the contribution of charged hydrogen bonds to binding enthalpy is more significant.

This difference between neutral and charged hydrogen bonding is elegantly illustrated by comparison of the two model receptors 2 and 3 for glutaric acid (Figure 2). Both receptors create the same number of hydrogen bonds with glutaric acid. Both receptors create the same number of hydrogen bonds with glutaric acid [7b, 7c, 9-12].

However, although the neutral diamide 2 binds glutaric acid strongly in chloroform (K_{ass} =60000 M^{-1}), in DMSO binding is not observed [13], whilst the receptor 3 which incorporates two charged electrostatic hydrogen bond interactions is almost as good a receptor in DMSO (K_{ass}=50000 M^{-1}) containing 5% THF, moreover, binding is still measurable in the presence of 25% water [14]. In quantitative terms, the presence of charged hydrogen is distinguished between an active artificial enzyme and a synthetic receptor.Although the exact nature of the transition state is unknown the important feature is that the 'binding' of this transition state by the enzyme involves more than ordinary molecular recognition [11]. The partially formed covalent bonds at the reaction center represent 'dynamic' binding interactions, which have no conveniently modeled ground state counterpart. It can also be expected that these interactions must make a major contribution to transition state binding and stabilization, not least because they clearly represent difference between substrate and transition state recognition. All of the factors described above contribute to the overall binding of the transition state. Despite the many reviews available on the factors which influence molecular recognition and binding in enzyme systems, the practical application of these hypotheses remains the true test of our understanding. It is thus of relevance to investigate the design and synthesis of artificial enzymes since this will hopefully also lead to a better understanding of molecular recognition itself.

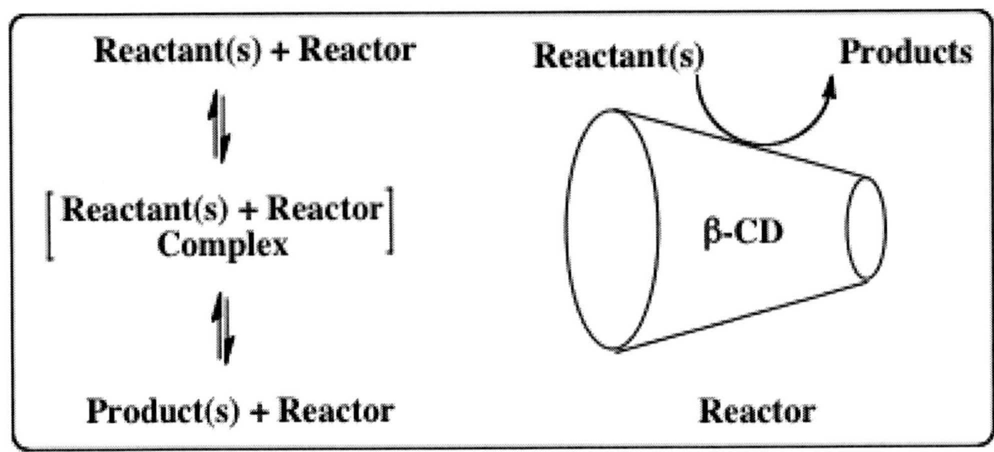

Scheme 1. CD-induced reaction of an included guest.

3. The "Design Approach"

3.1. Cyclodextrins as Enzyme Mimics

Cyclodextrins (CDs) have long attracted attention in catalysis and as enzyme mimics, due to the way in which they act as hosts to complex guest molecules and induce reactions of the complexes species (Scheme 1). The reactions exhibit kinetic characteristics, such as saturation, non-productive binding and competitive inhibition, that are typical of enzyme-catalyzed processes. In addition, the discrimination displayed by CDs in binding guests and promoting their reactions is analogous to the substrate selectivity displaced by enzymes [15].

With the natural CDs, hydroxy groups are the only functionality available to promote reactions of included guests. However, the introduction of a diverse range of new functional groups, through modifications of the natural CDs, results in catalysts, which mimic the entire range of enzyme behavior. The CD nucleus serves as a scaffold on which functional groups can be assembled. In some cases this has been accomplished with controlled alignment of both the functional groups and the CD annulus, to optimize the geometry for binding and reaction of a particular guest. This is an important factor in the catalytic activity of modified CDs, as it is with enzymes where the geometry at the active site is determined by the three dimensional structure of the protein. It is the catalytic activity of CDs which is the subject of this chapter.

3.2. Vitamin B12 Functions: Enzymatic Reactions

Vitamin B12 is a cobalt complex coordinated with a tetrapyrrole ring system, namely corrin, and linked to a 5,6-dimethylbenzimidazole moiety as a heterocyclic base (Figure 3). There are two B12 active forms: 5'- deoxyadenosylcobalamine and methylcobalamine Vitamin B12-dependent enzymes are known to catalyze two types of reactions: rearrangements as exemplified by methylmalonyl-CoA mutase and methylation by methionine synthetase. The rearrangement reactions involve the intramolecular exchange of a functional group (X) and a hydrogen atom between neighboring carbon atoms.

These reactions have attracted much attention because of their novel nature from the viewpoints of organic and organometallic chemistry. Carbon skeleton rearrangement reactions, mediated by methylmalonyl-CoA mutase, glutamate mutase, and *R*-methyleneglutaratemutase are shown by eqs 2-4.

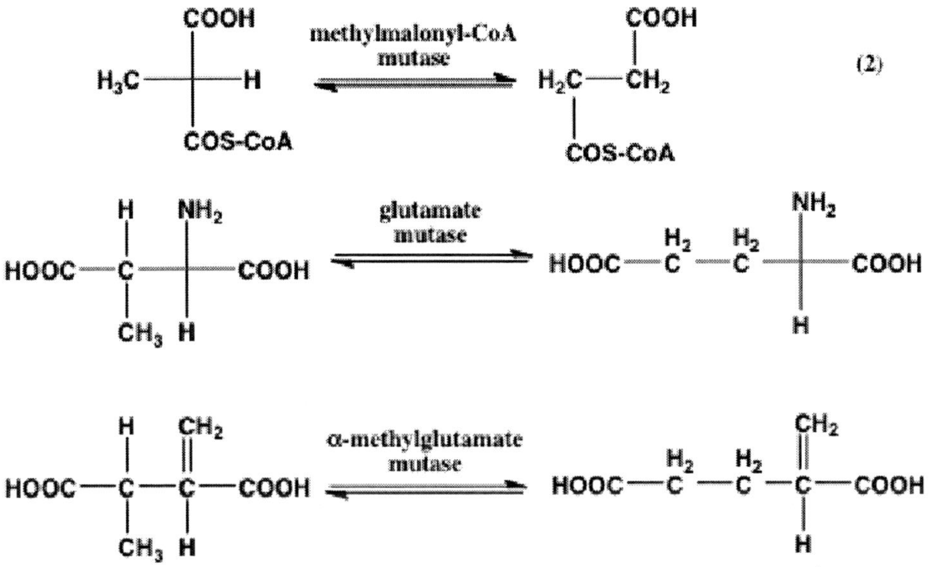

$$H_3C\text{-}CH(COOH)\text{-}COS\text{-}CoA \underset{}{\overset{\text{methylmalonyl-CoA mutase}}{\rightleftharpoons}} HOOC\text{-}CH_2\text{-}CH_2\text{-}COS\text{-}CoA \qquad (2)$$

$$HOOC\text{-}CH(CH_3)\text{-}CH(NH_2)\text{-}COOH \overset{\text{glutamate mutase}}{\rightleftharpoons} HOOC\text{-}CH_2\text{-}CH_2\text{-}CH(NH_2)\text{-}COOH \qquad (3)$$

$$HOOC\text{-}CH(CH_3)\text{-}C(=CH_2)\text{-}COOH \overset{\alpha\text{-methylglutamate mutase}}{\rightleftharpoons} HOOC\text{-}CH_2\text{-}CH_2\text{-}C(=CH_2)\text{-}COOH \qquad (4)$$

$$-\underset{X}{C_1}-\underset{H}{C_2}- \rightleftharpoons -\underset{H}{C_1}-\underset{X}{C_2}-$$

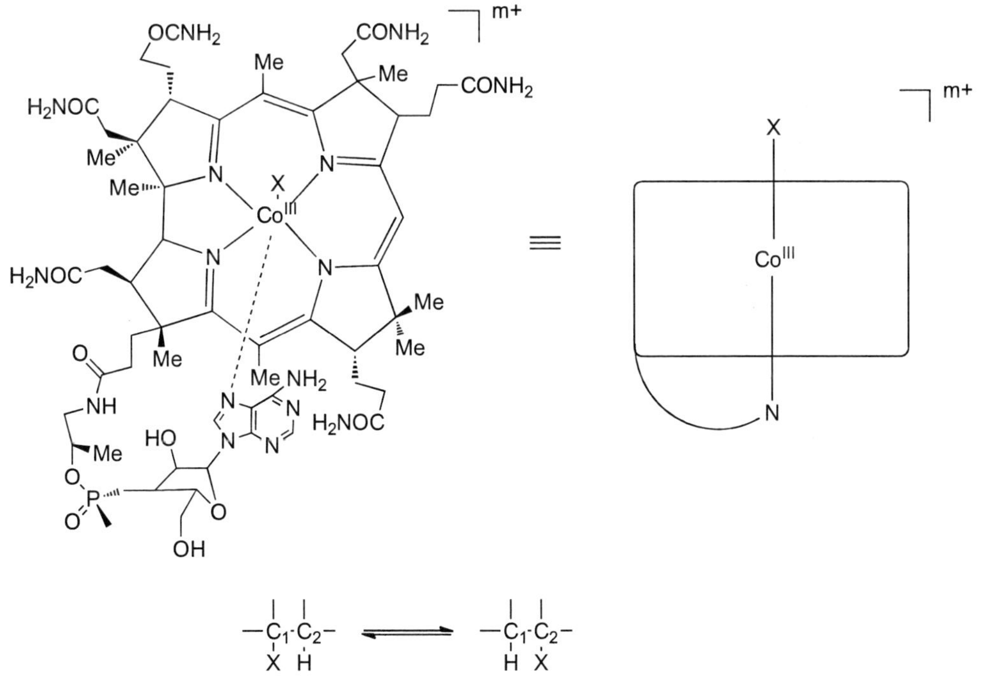

Figure 3. Vitamin B12-dependent enzymes catalyzes 1,2-Hydride shift in model system.

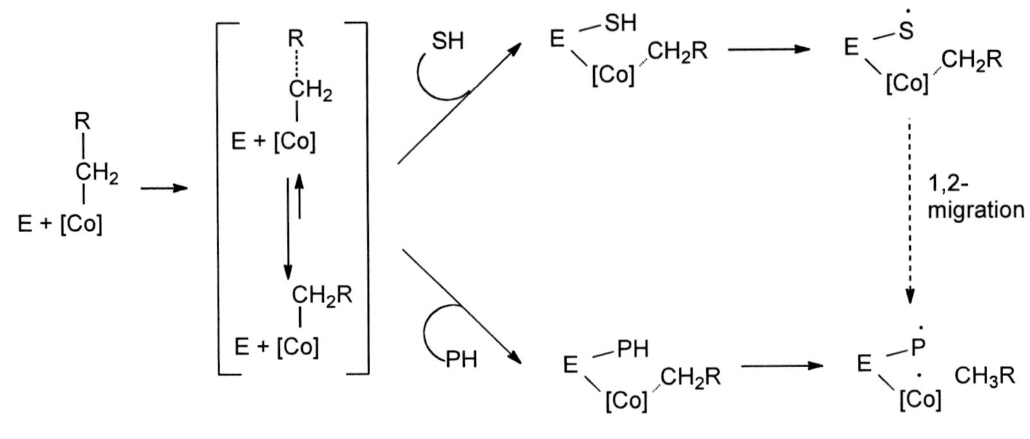

where R-CH2-, 5'-deoxyadenisyl; [Co], cobalamin

E= apoenzyme, SH= substrate, PH=product

Figure 4. A general feature of the radical mechanism for rearrangement reactions mediated by the 5'-deoxyadenosylcobalamin-dependent enzyme.

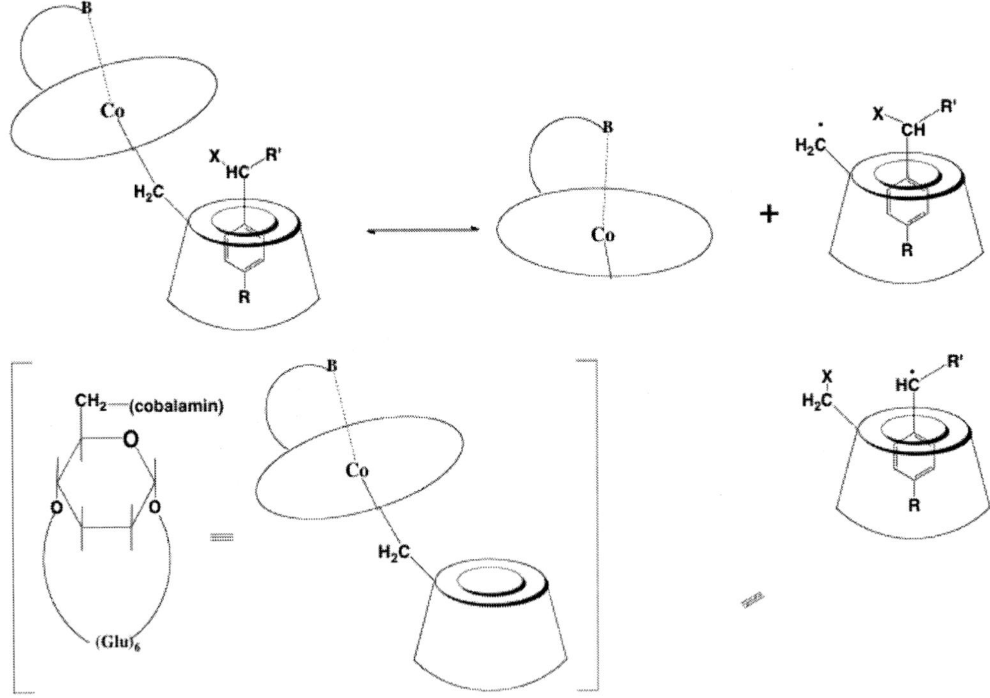

Scheme 2. Mode of action of a β-CD-B_{12} enzyme mimic.

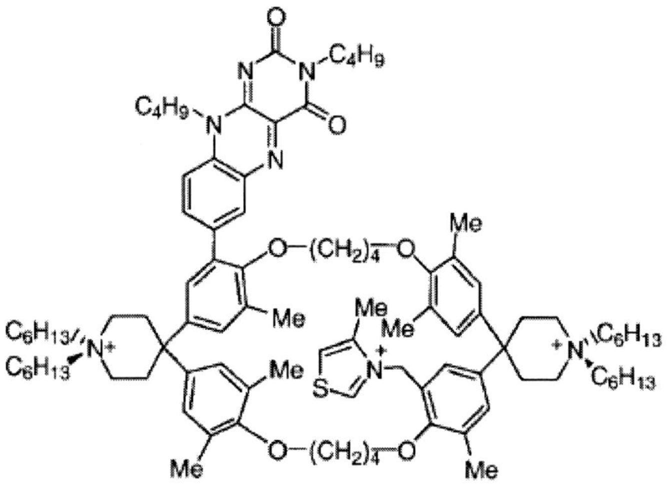

Figure 5. Model of Pyruvate Oxidase Mimic proposed by Diederich et.al.

Even though the real reaction mechanisms involved in the carbon-skeleton rearrangements have not been clarified up to the present time, radical mechanisms are considered to be the most plausible ones on the basis of ESR studies for methylmalonyl-CoA mutase, glutamate mutase, and *R*-methyleneglutaratemutase. A general feature of the radical mechanism is illustrated in Figure 4: the 5'-deoxyadenosyl moiety bound to cobalamin(vitamin B12) undergoes homolytic cleavage to give cobalamin in the Co(II) state and the 5'-deoxyadenosyl radical upon incorporation of a substrate into a specific microenvironment provided by the corresponding apoprotein; the 5'-deoxyadenosyl radical abstracts a hydrogen atom from the incorporated substrate to afford deoxyadenosine and the substrate radical in the active site of apoprotein; the substrate radical is eventually isomerized via 1,2-migration of a functional group to form the corresponding product radical; the product radical abstracts a hydrogen atom from deoxyadenosine placed in its vicinity; and the deoxyadenosyl radical is then bound to cobalamin to recover the original coenzyme state [16].

As for the role of the cobalt species, Halpern et al. proposed a reversible free radical carrier mechanism; coenzyme B12 is referred simply to a source of the 5'- deoxyadenosyl free radical that acts to generate a substrate radical by abstracting a hydrogen atom from the substrate to initiate the reaction, and behaves as a reversible free radical carrier.

3.3.Model Reactions with Apoenzyme Functions

It is obvious that an apoprotein plays an important role in such radical reactions as mediated by vitamin B12-dependent enzymes.The model reactions, which were designed in consideration of the role of apoenzymes, are as follows. Breslow et al. prepared a cyclodextrin-bound B12, in which cobalamin is directly linked to the primary carbon of â-cyclodextrin by a cobalt-carbon bond (Scheme 2) [17]. They expected that a hydrophobic substrate is incorporated into the cyclodextrin cavity in water, so that the cyclodextrinyl radical may undergo an intracomplex atom transfer to generate a substrate radical as shown in Scheme 2. Even though they did not mimic all steps of the B12- dependent rearrangement reaction, this is an interesting example showing that a substrate and B12 are bound together in a receptor site.

4. The "Transition State Analogue-Selection" Approach

4.1. The Transition State Analogue-selection Approach: General Introduction

The traditional approach to enzyme mimics is the design approach described above. Whilst this has furnished us with much information on the recognition processes involved in binding and the criteria required for successful catalysis, the realization of a project from original conception to experimental studies on an enzyme mimic can be a long and laborious process (Figure 5). A case in point is Diederich's pyruvate oxidase mimic which required an 18 step synthesis of the host.

In an attempt to move away from this linear approach, several techniques have been developed which make use of a selection strategy. This allows for the simultaneous screening of a wide range of possible candidates thus significantly reducing the time required and hopefully allowing for the detection of better hosts.

The earliest examples of a selection approach chose affinity for a transition state analogue (TSA) as the screening criteria. The logic behind this is that any macromolecule, which shows strong binding to a molecule resembling the transition state of a reaction, should also bind to and stabilize the real transition state. As this stabilization of the transition state is the basis behind enzyme catalysis, the hosts selected should behave as enzyme mimics for the chosen transformation.

More recently, it has been recognized that TSA binding alone may not be enough to obtain the rate accelerations needed to rival enzyme catalysis. Nowadays, the incorporation of catalytic groups in the host is often a designed aspect of the selection process and it is this, in combination with the TSA host selection, which has led to some of the most impressive advances described below [18].

4.2. Molecular Imprinted Polymers as a Method in the "Transition State Analogue-selection Approach

Molecular imprinting is a general method for synthesizing robust, network polymers with highly specific binding sites for small molecules. Molecular imprinting is a process by which polymeric materials are synthesized with highly specific binding sites for small molecules [19-25]. Molecularly imprinted polymers (MIPs) have been developed for a variety of applications including chromatography, enzymatic catalysis, solid-phase extraction, and sensor technology [26].Intermolecular forces that develop during polymerization between the template molecules (T), functional monomer (M) and developing polymer matrix are responsible for creating a polymer microenvironment for the template or imprint molecule [20b, 20c, 21a, 21b]. The resulting polymer network contains synthetic receptors that are complementary in size, shape and functional group orientation to the template molecule. The polymers typically employed in imprinting are complex thermo-sets, an insoluble, highly cross-linked network polymer. Because both the morphology of the bulk polymer and the chemical microenvironment of the binding site are critical to the overall performance, the number of experimental variables that influence these factors are large. Imprinted polymers are, therefore, ideal candidates for combinatorial synthesis and its screening technologies. Recently, combinatorial methods have been used to develop highly selective MIPs. Synthetic receptors produced by this technology, in turn, have been used in the screening of libraries of small molecules. This section covers both these emerging areas of MIP technology.

Optimization of MIP formulations many variables of the imprinting process influence the selectivity and capacity of a MIP. First, complementary interactions between the template and the functional and cross-linking monomers are necessary to create short-range molecular organization at the receptor site. These interactions include hydrogen bonding, electrostatic and/or Van der Waals forces. Second, the stoichiometry and concentration of the template and monomers influences both polymer morphology and MIP selectivity. Third, the solvent used in the polymerization process, also known as the 'porogen', plays a dual role. In addition to mediating the interactions between the functional groups and the template molecule, the porogen determines the timing of the phase separation during polymerization [20-25], which is an important determinant of polymer morphology, porosity and ultimately accessibility of the binding site. Finally, the temperature of polymerization influences the timing of phase separation. Also, the temperature dependence of the equilibrium between the functional monomers and template affects MIP selectivity and capacity.

A typical imprinting protocol involves thermally or photochemically induced, free-radical polymerization of a concentrated solution of monomers to produce bulk, monolithic insoluble polymers that are crushed, ground and sieved to micron size for analysis. Selectivity and binding are evaluated in the chromatographic mode by using the MIP as stationary phase for HPLC columns, or by batch rebinding studies. These methods can be tedious and time-consuming, and obtaining an MIP with optimal binding properties can take several days to weeks, especially if the variation in the formulation is made by trial-and-error. As a result of the number of variables that effect MIP performance, there has been an overuse of certain 'standard' formulations. The process of determining an optimal MIP formulation, therefore, is an ideal candidate for a combinatorial approach for screening various formulations.

This technique offers potential for developing tailor made catalysts, perhaps with catalytic functionalities not utilized in biology.Despite the inherent heterogeneity of the molecular recognition site produced, the increased stability of MIPs against heat, chemicals and solvents when compared to natural enzymes or artificial analogues means that the attainment of MIPs remains a highly sought-after aspiration.

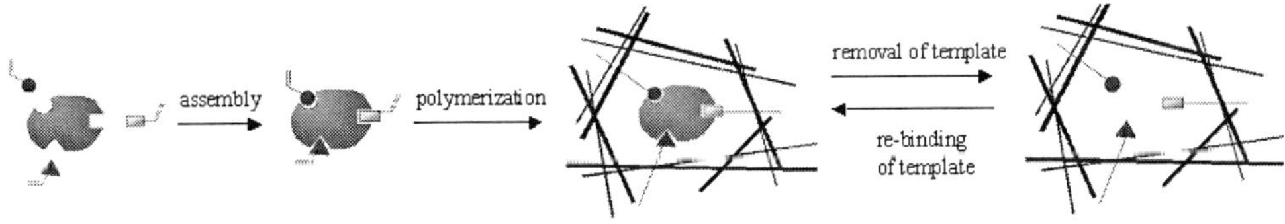

Figure 6. General scheme for preparation of molecular imprinted polymers.

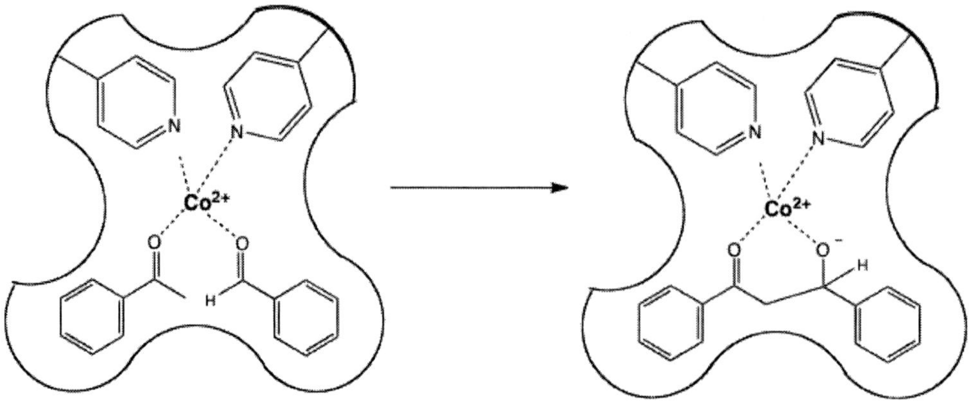

Scheme 3. MIP containing Co^{2+} ion used in aldol condensation reaction.

Mosbach reported the first true-enzyme-like catalysis of C-C bond formation using MIPs [19]. The molecular imprinting techniques was used in the development of a 4-vinylpyridine-styrene-divinylbenzene copolymers imprinted with aldol condensation intermediate analogues, dibenzoylmethane (DBM) and a Co^{2+}ion (Scheme 3). The imprinted polymer was able to catalyze the aldol condensation of acetophenone to benzaldehyde in manner analogues to Class II aldolases. In addition to metal coordination, the pyridinyl residues provided the base for generation of the enolate of acetophenone.

4.3. Imprinting an Artificial Proteinase

Another technique which utilizes the idea of 'imprinting' was reported by Suh for the creation of an artificial aspartic proteinase. The two aspartic carboxyl groups found within the natural enzymes are thought to act as key catalytic groups in-hydrolyzing peptide substrates [27]. In light of this fact, Suh synthesized an organic artificial protease which contained carboxyl groups in the active site (Scheme 4).

This involved complexation of three molecules of 5-bromoacetylsalicylateto an Fe(III) ion to give the resultant complex ($FeBAS_3$) which was cross-linked with poly(aminomethylstyrene-co-divinylbenzene) (PAD) to obtain ($FeSal_3$)-PAD. These were subsequently capped *via* acetylation to produce ($FeSal_3$)-Ac, and the Fe(III) ions removed under acidic conditions to give the active apo(Sal_3)PAD-Ac protease mimics. These were obtained as insoluble catalysts which reproduced the catalytic features of aspartic proteases. The activity of apo(Sal_3)PAD-Ac was tested in the hydrolysis of bovine serum albumin, in which it was revealed that albumin was cleaved into fragments smaller than 2 kDa. By looking at the pH profile for this reaction, it was found that it manifested optimum activity at pH 3, which is in agreement with conditions found within natural enzymes. Since the active site of apo(Sal_3)PAD-Ac contained both carboxyl and phenol groups, at pH 3, phenol was thought to be acting as a general acid since its activity is likely to be lower than that of the carboxyl groups. Therefore the activity of apo(Sal_3)PAD-Ac at pH 3 is attributable to cooperation of two or more carboxyl groups by a mechanism analogous to that found in natural enzymes. Moreover it has a $k$cat of over 0.17 h^{-1} at pH 3, indicating that it has a reasonably high catalytic activity. The idea of 'imprinting' has also been extended to include bio-molecules, mainly proteins, for use as efficient artificial enzymes.

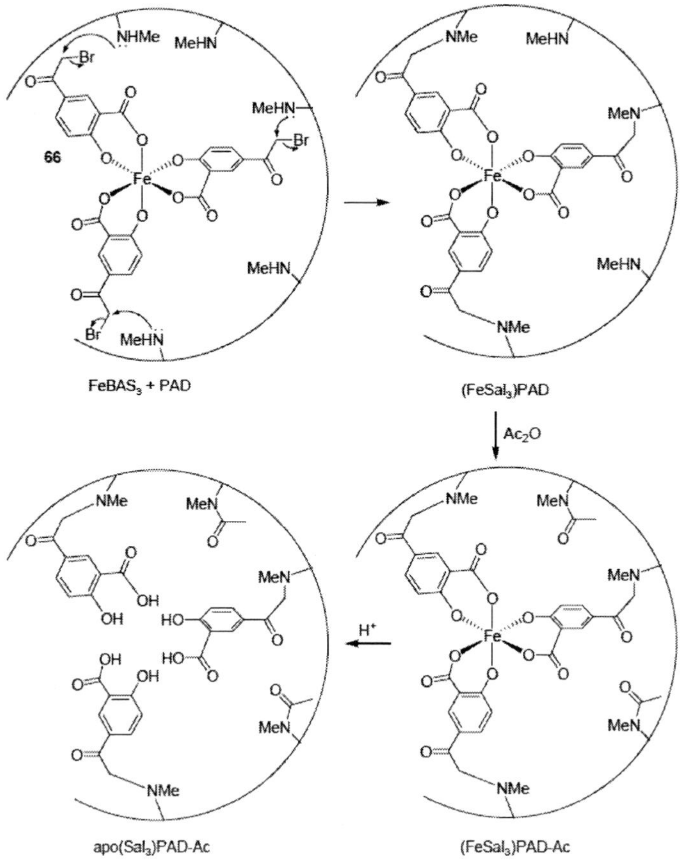

66

FeBAS$_3$ + PAD

(FeSal$_3$)PAD

Ac$_2$O

H$^+$

apo(Sal$_3$)PAD-Ac

(FeSal$_3$)PAD-Ac

Scheme 4. Imprinting process for the creation of an artificial aspartic proteinase.

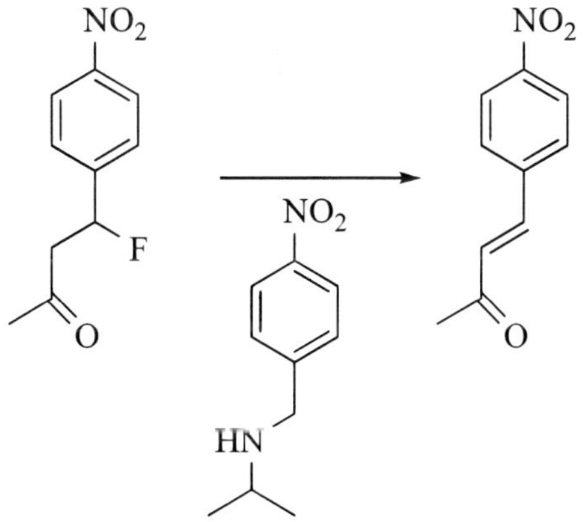

Scheme 5. β-elimination of 4-fluoro-4-(*p*-nitrophenyl) butan-2-one and the structure of TSA used for protein imprinting.

4.4.Bioimprinting

Biomolecular imprinting or bioimprinting refers to the induction of catalytic activity in proteins by lyophilisation (freeze drying) in the presence of a transition state analogue [28]. Slade has demonstrated that bioimprinting proteins in the presence of a TSA leads to a conformational change which either manifests itself in the form of a new catalytic site or as improvements of the pre-existing ones, which were then able to carry out catalysis.

This process was illustrated by bioimprinting β-lactoglobulin in the presence of TSA(Scheme 5). β-elimination of substratewas studied using this novel bioimprinted protein and compared with the results of the non-imprinted control.

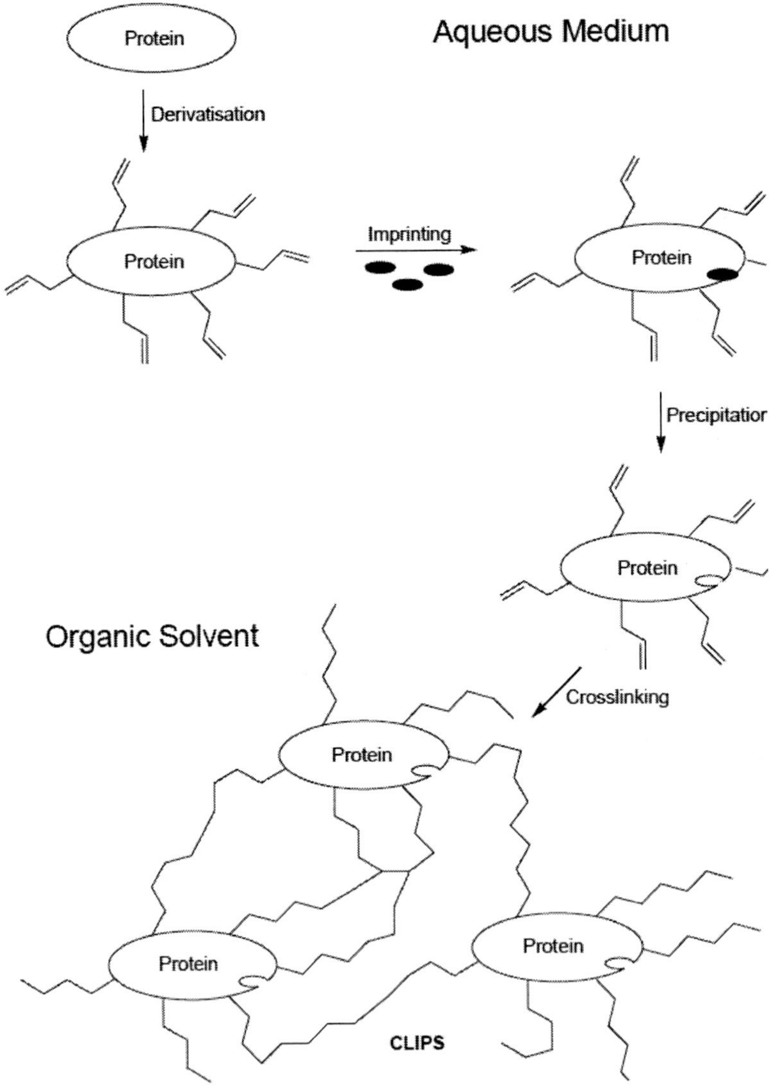

Scheme 6. Broadening the substrate selectivity using a combination of bio-imprinting and subsequent covalent immobilization technique.

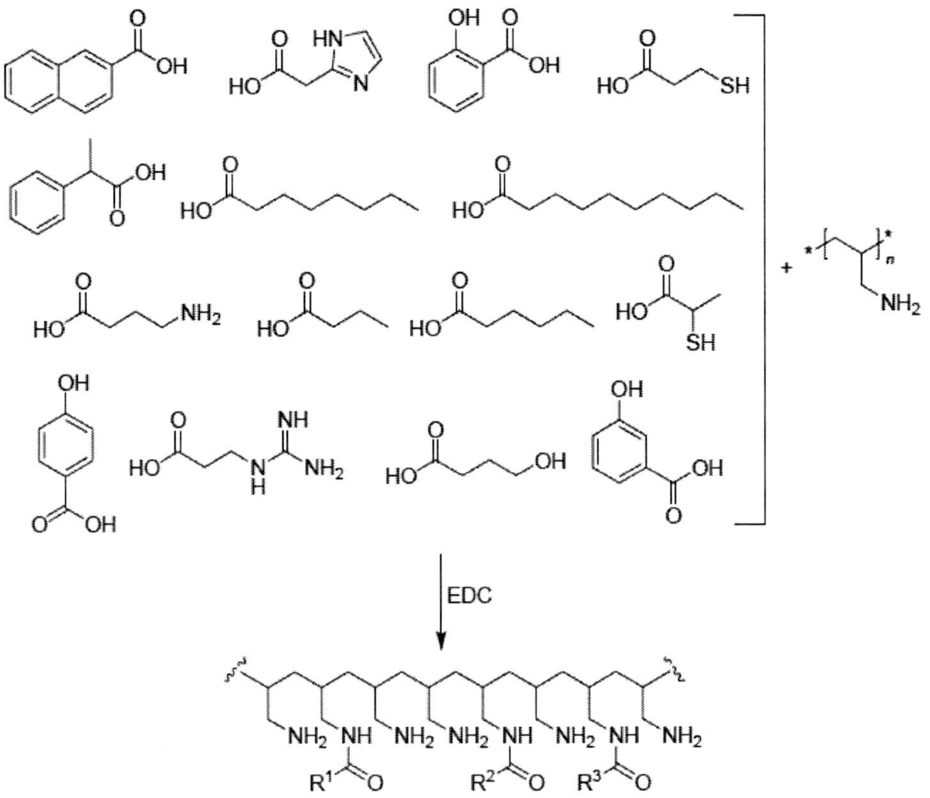

Scheme 7. Synthesis of functionalized polyallylamine using combinatorial chemistry.

The imprinted β-lactoglobulin showed catalytic activity three times that of the control reaction and almost four orders of magnitude higher than spontaneous β-elimination. Although this result may seem modest when compared to rate accelerations obtained using catalytic antibodies, it was found that the rate acceleration was almost identical to that observed for molecularly imprinted polymers.

A major drawback of this method of imprinting is that the enhanced properties of these proteins can only be sustained in nearly anhydrous environments, since hydration of these proteins causes re-naturation and therefore consequent loss of the imprinted binding sites. This problem however was solved to some degree by Peifβker and Fischer by combining the imprinting step with asubsequent immobilization method, resulting in the retention of the imprint by the enzymes, allowing their structure to be maintained in aqueous media (Scheme 6) [29].

This technique was used to stabilize the ligand induced acceptance for D-configured substrates by α-chymotrypsin or subtulisin Carlsberg. This involved the vinylation of the proteases by acylation with itaconic anhydride. Subsequent enzyme imprinting and crosslinking furnished the desired crosslinked imprinted proteins (CLIPs). Examples of the use of bioimprinting include Luo's glutathione peroxidase (GPX) mimic and those based on imprinting of myoglobin in the epoxidation of styrene [30].

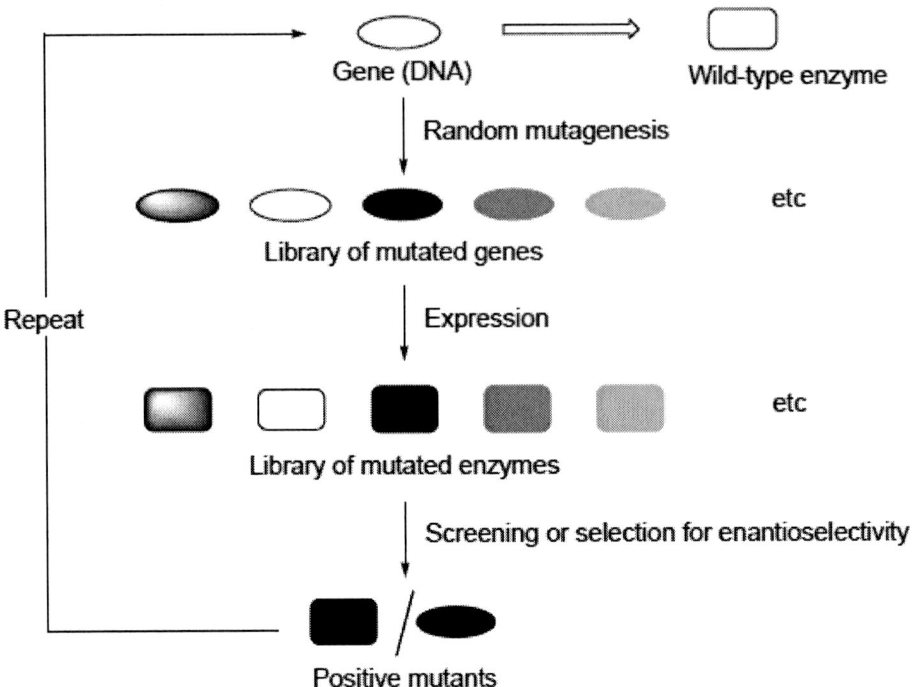

Scheme 8. Directed evolution of an enantioselective enzyme.

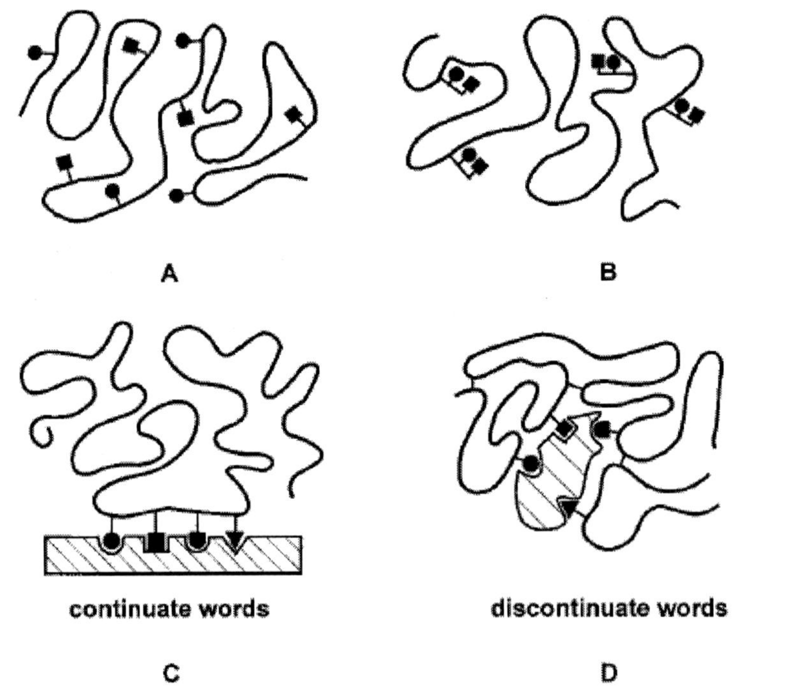

Scheme 9.Possible Arrangements of Functional Groups in Synthetic and Natural Polymers.

5. THE "CATALYTICACTIVITYSELECTIONAPPROACH": GENERAL INTRODUCTION

This approach utilizes the advances in combinatorial chemistry wherein a library of possible catalysts is generated and directly screened for enzyme-like activity. This not only provides a tool for synthesizing a large number of diverse compounds in a short amount of time but also allows for the discovery of effective catalysts, which exhibit potential activity when subjected to the relevant screening method.

5.1. Combinatorial Polymers as Enzyme Mimics

In the mid nineties, Menger introduced a highly original approach towards the creation of novel artificial enzymes. This involved the use of combinatorial chemistry to attach various combinations of three or four carboxylic acids onto poly(allylamine) (PAA) or poly(ethylenimine) (PEI) (Scheme 7). The figure below gives a general schematic for the synthesis of a functionalized polymeric library using poly(allylamine) [6].

Although no control was exercised over the attachment of the substituents, they were not necessarily randomly attached throughout the polymer. For example, a polymer with an octanoyl group at one site may be prone to receive another octanoyl group adjacent to it owing to hydrophobic interactions between the two functionalities. Although this combinatorial approach has yielded some impressive results, the major drawback is that each combinatorial polymer is a complex system, consisting of numerous polymeric variations. This not only makes the isolation of a pure component near impossible for sequencing and structural characterization but also provides very little detail to draw any significant mechanistic conclusions.

5.2. Directed Evolution of Enzymes

In quite a different manner to the previous catalytic selection approaches mentioned earlier, in 1995, Reetz began developing a high-throughput screening method for assaying the enantioselectivity of thousands of biocatalysts [31]. This involved exploitation of the tools of directed evolution in the creation of enantioselective enzymes for use in organic synthesis (Scheme 8).

The examples in this field include artificial esterases [32a] and artificial cytochrome P450 monooxygenases [32b,c].

5.3. Catalysis with Imprinted Silicasand Zeolites

5.3.1. "Footprint" Catalysis

The first attempts to obtain imprinted materials were achieved with silicas by Dickey [33]. He precipitated silica gel in the presence of dyes. After drying, the dyes were washed out, and the as prepared silica showed an increased affinity for the template compared to similar compounds. During this approach, the formation of the rigid silica gel matrix around the template facilitates shape selectivity, and additional silanol groups might be arranged such that interactions with the template can occur. No directed interactions between template and the growing silica chains were tried in these first experiments. One should expect a high stability of these arrangements in silica, but the selectivity disappeared readily [34]. Especially, racemic resolutions have been studied for some time with imprinted silicas [22, 35].

Much later, investigations of imprinted silicas started anew. The first to use this approach for the synthesis of catalytically active silicas was Morihara [36]. He used a quite unconventional approach to obtain catalytically active silicas by surface imprinting.

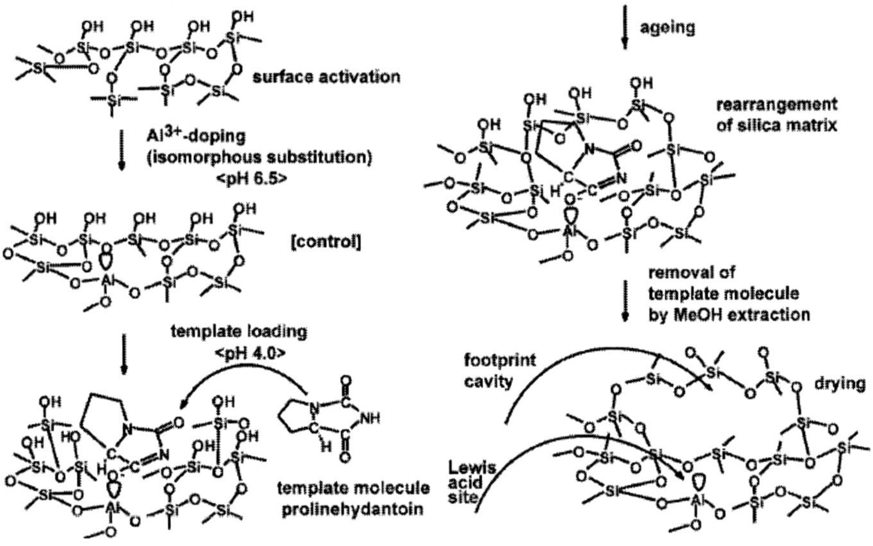

Scheme 10. Schematic representation of the preparation of "Footprints" on the surface of silica.

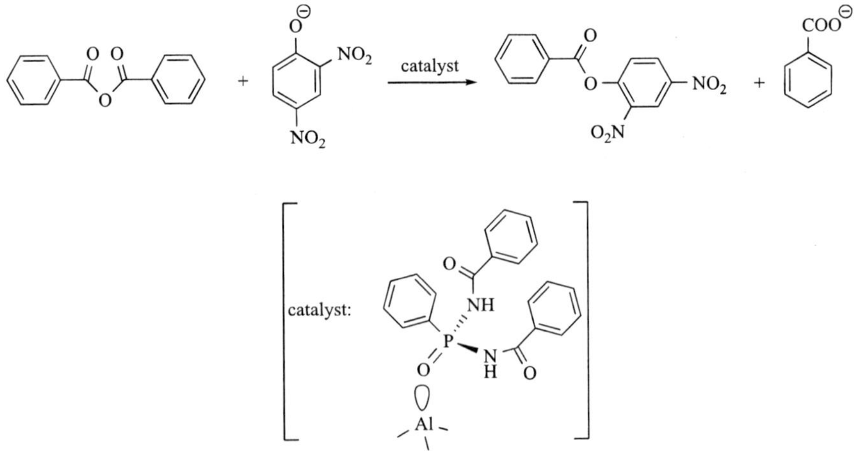

Scheme 11. Schematic representation of imprinting with a phosphonic diamide as a stable transition state analogue template.

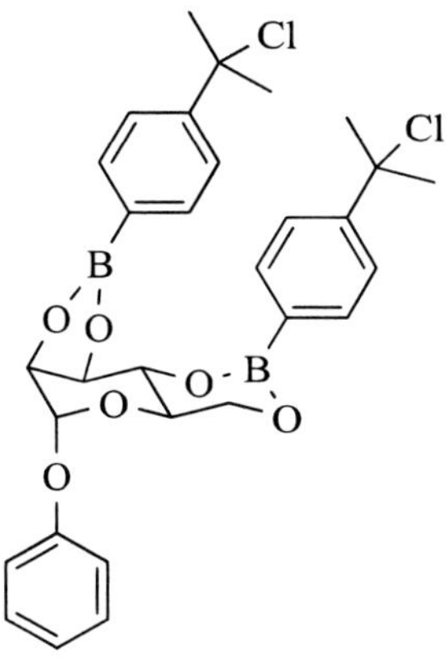

Figure 7. Template for a molecular imprinting.

This procedure was called "footprint catalysis". From 1988 onward many examples were published [37].In Morihara's procedure, commercial silica gel is treated at pH 6.5 with Al^{3+} ions [37d, 37i, 36]. The incorporation of Al^{3+} in the silica matrix by isomorphous substitution of silicate with aluminate causes the formation of surface Lewis acid groups (see Scheme 10). The surface is then loaded at pH 4.0 with template molecules that contains Lewis base groups. The silica is then aged and dried under highly controlled conditions. It is assumed that in this way the silicate matrix rearranges around the acid-base complexes by de-polymerization of the silicate matrix and re-polymerization under thermodynamic control. The complexes are stabilized by forming a maximum number of interactions between the acid-base complex and the surrounding silicate matrix. Afterward, the template is removed with methanol. Mostly acyl transfer reactions are investigated, since substrates and transition state analogues in this case are readily available. Scheme 11 depicts the surface imprinting with the transition state analogue phosphonic acid diamide (*N,N'*-dibenzoylbenzenepho-sphonediamide).

The substrate was an acid anhydride, and the nucleophile was 2,4-dinitrophenolate, the consumption of which was followed photometrically. In this case (one of the best in a large series), *k*cat increases by a factor of 10, and *K*m is improved by a factor of 3 with regard to non-imprinted materials [37d]. A comparison with other imprinting methods with regard to *k*cat and *K*m is difficult; since the exact number of active sites is not incorporated in the calculations, second-order rate constants are used for *k*cat and the reported *K*m values are indeed mostly *K*ass (M^{-1}) values. Therefore Morihara's*k*cat/ *K*m values do not refer to the usual Michaelis- Menten kinetics. In comparison with the own control silicas, though, the imprinted silicas show a remarkable catalytic activity.

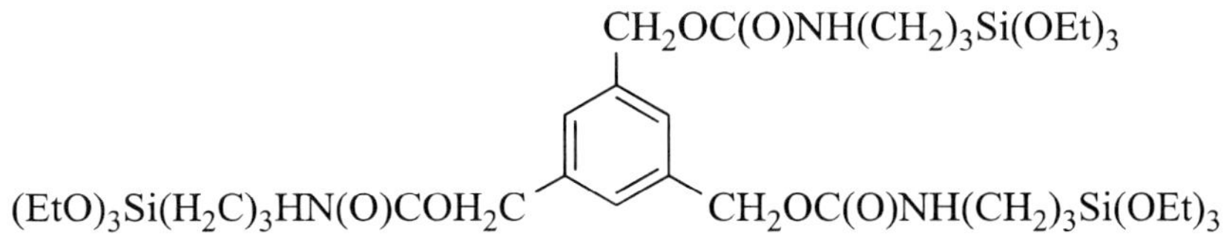

$CH_2OC(O)NH(CH_2)_3Si(OEt)_3$

$(EtO)_3Si(H_2C)_3HN(O)COH_2C$ $CH_2OC(O)NH(CH_2)_3Si(OEt)_3$

Figure 8. Functional silane as template for molecular imprinting.

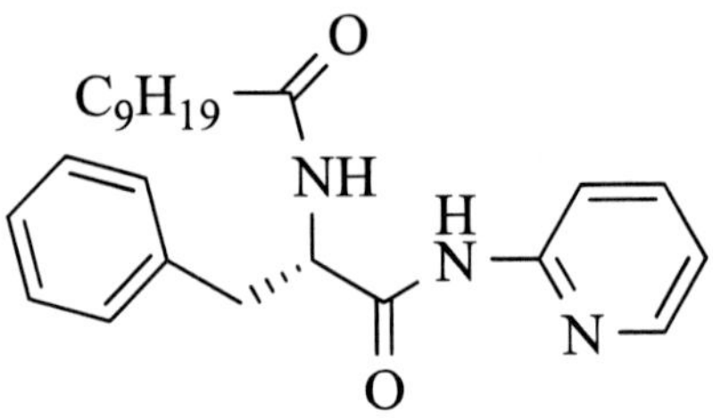

Figure 9. Chiral Template

In a number of examples, substrate specificity is shown. If an optically active template is used as a transition state analogue for imprinting, these catalysts cause enantiomers to react at different rates (kinetic racemic resolution). The substrate enantiomer corresponding to the template enantiomer reacts 2-4 times more rapidly than the other [37e,f,h,k,m]. Other reactions, such as crossed aldol condensation, enantioselective racemization, and asymmetric reductions, have also been investigated, though no details are available at present [36].

The footprint cavity approach has reached quite promising results. At the moment it is not easy to understand the mechanism completely. Apart from the Lewis acid-Lewis base interaction, no defined interactions between silica and template are known, and no investigations in this direction have been made. Only small shape selectivity should be possible in shallow cavities. Unfortunately, no other research groups have used this method until now. In view of the interesting results, more detailed knowledge of the method is desirable.

5.3.2.OtherExamplesfor Catalytically Active ImprintedSilicas

Imprinted silicas for catalysis were also prepared by Heilmann and Meier [38]. They used 1-triethoxysiloxy(phenyl)methanephosphonic cthyl hexyl ester as a transition state analogue for a trans-esterification reaction. After a sol-gel process with this compound and tetraethoxysilane, the template was removed by calcination at 523 K, leaving behind cavities with a shape of the transition state of the trans-esterification.

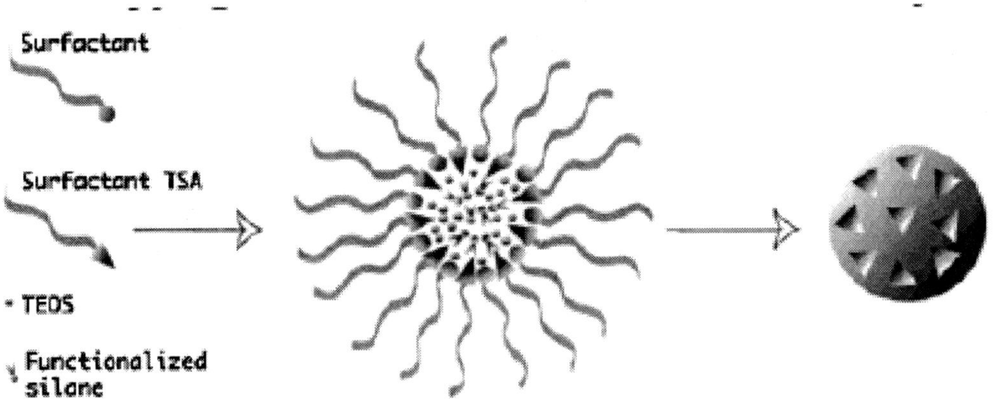

Scheme 12. Template directed molecular imprinting of silica particles.

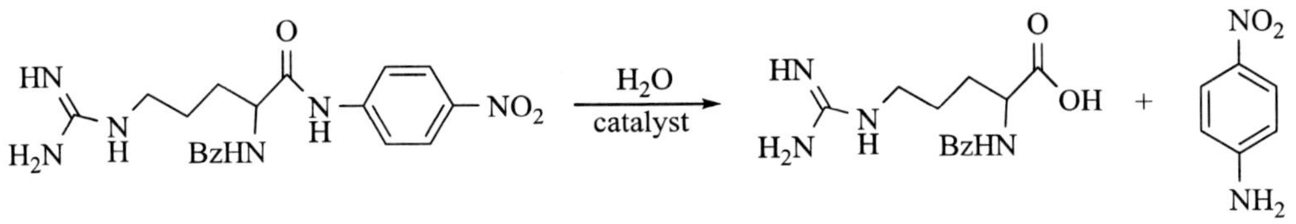

Scheme 13. Silica Imprinted Catalysis for Hydrolysis of an Arginine Derivative

An acceleration of the trans-esterification was indeed observed [38], but more detailed investigations showed that this effect was not due to an imprinting effect but to phosphoric acid left at the polymer and in solution [39]. More similar to the usual imprinting method in polymers, silicas with defined binding sites have been prepared that were introduced by poly-condensation of functionalized silanes together with tetraethoxysilanes. In this case the polymerizable double bond has been replaced by functionalized silanes (see, Figure 7). Selective recognition was also observed [22a, 23, 40]. This method has also been used to prepare catalytically active silicas. Via carbamate moieties to a template molecule, with one, two, or three amino groups (see Figure 8) are introduced in cavities during a sol-gel process. Initial results (without experimental details) indicate catalytic activity for a Knoevenagel condensation.

In several recent papers Markowitz et al. reported on a very original molecular surface imprinting method [41]. This method involves surface imprinting of silicas during the formation of the particles in a sol-gel process. A surfactant derivative of the imprint molecule is used as the template during the poly-condensation of tetraethoxysilane in the presence of added functionalized organosilanes. First, the surfactant imprint molecule, e.g., *N*-decanoyl-L-phenylalanine- *N*-pyridin-2-ylamide (Figure 9), is incorporated in a water-in-oil micro-emulsion with added nonionic surfactant. Tetraethoxysilane and a mixture of amine, dihydroimidazole, and carboxylate-terminated organosilanes were added to commence base-catalyzed surface imprinting and particle formation (see Scheme 12).

Since the silica particles are formed by a microemulsion process, the imprint molecule, which acts as the head-group of the surfactant, should be positioned at the surfactant-water interface of the reverse micelles within which the silica particles are formed. As a consequence, catalytic sites should only be formed at the surface and they should all have the same orientation with regard to the surface.

L-Phenylalanine-*N*-pyridin-2-ylamide can be regarded as a stable transition state analogue of the R-chymotrypsin- catalyzed amide-fission of peptides. Although the catalyst was imprinted with a phenylalanine derivative, the imprinted silica catalyzed best the hydrolysis of an arginine derivative (Scheme 13).

An enhancement in initial rate of 4.8-fold is observed compared to a nonimprinted control having the same functional groups. The most remarkable result is the enantioselectivity as silica imprinted with a L-phenylalanine derivative catalyzes the hydrolysis of the D-enantiomer of an arginine derivative 34 times faster than the L-enantiomer [41a]. This is by far the highest enantioselectivity for imprinted materials published to date. The reversed enantioselectivity (catalyst imprinted with L-enantiomer hydrolyses the D-substrate more actively) might be due to the surfactant moiety in the template. If the D-enantiomer is used as template instead of the L-enantiomer, the opposite enantioselectivity is observed. This result is to be expected, but it rules out possible experimental errors. Further development of this method might substantiate an important new way toward catalytic imprinted silicas and clarify some open problems.

5.3.3. Imprinting in Zeolites

Zeolites consist of a crystal lattice having defined pores and cavities throughout. Zeolites with very different cavity diameters are known. A very detailed review article deals with the numerous possibilities for using these compounds in catalysis [42]. Until now, no direct imprinting in zeolites has been published. There are reports on the synthesis of zeolites in the presence of certain organic molecules acting as "templates" in order to control the type of lattice being formed. This is no molecular imprinting in the original sense. Zeolites should be interesting candidates for imprinting inside the holes. This might reduce the polyclonality of the cavities. A problem can be foreseen with regard to the mass transfer inside the zeolites containing additional organic or inorganic polymers inside the pores and holes. Further possibilities of application of zeolites in the imprinting procedure are discussed by Davis et al [21a].

5.3.4. Future Prospects in Catalysis

What has been reached in the preparation of catalysts by molecular imprinting in polymers or silicas? Over the last 10 years a considerable progress in the preparation of efficient catalysts has been made.

Enzymes are in every case several orders of magnitude catalytically more efficient, but in a few cases imprinted polymers have reached the activity of catalytic antibodies, e.g., in the hydrolysis of carbamates [43]. This is surprising, since monoclonal antibodies are compared with "polyclonal imprinted" catalysts, and the imprinted materials are insoluble and rigid, whereas antibodies are soluble and more flexible. The binding site homogeneity in enzymes and monoclonal antibodies is high, whereas imprinted polymers, as discussed before, have a broad distribution of activity and there is no method available at the moment to really reduce this broadness.

A real advantage of imprinted catalysts is the ease of preparation and handling. They can be prepared in large quantities by suspension polymerization, and stable particles of uniform diameter can be easily obtained. Imprinted polymers can be applied directly in chemical processes. Such catalysts can also be prepared, in addition to beads or broken particles, in other very different forms, such as monoliths, microcapsules, membranes, surfaces. At the same time these materials are rather stable. Whereas enzymes and antibodies degrade under harsh conditions such as high temperature, chemically aggressive media, and high and low pH, imprinted polymers show better behavior in most cases. They have both good mechanical and thermal stability. Usually they can be used for a long time in a continuous process, or they can be reused many times. As a result of the insolubility of the materials, they can be easily filtered off after a reaction, or they can be placed in a flow reactor. All this brings a lot of advantages in the use of imprinted polymers or silicas. Though quite some progress has been made in the preparation of catalysts by molecular imprinting, for large application in industry and for broader application in research, further achievements have to be made. On one hand, the imprinting procedure has to be further improved and new approaches in the preparation of catalysts have to be used. At present, the following problems are in the forefront of investigations to improve the molecular imprinting procedure:

(a) molecular imprinting in microparticles during suspension or emulsion polymerization
(b) imprinting procedures in aqueous solutions
(c) imprinting with high molecular weight templates, biopolymers, or even bacteria by surface imprinting
(d) development of new and better binding sites in molecular imprinting
(e) improvement of the mass transfer in imprinted polymers
(f) reduction of the "polyclonality" of cavities
(g) increase of available active sites, especially with the usual noncovalent interaction
(h) development of extremely sensitive detection methods for use in chemosensors, and finally,
(i) development of further suitable groupings for catalysis.

5.4. Catalytic Antibodies and Few Examples of Radical Transformations

The most established applications of the above TSA selection strategy lie in the field of catalytic antibodies, pioneered by Lerner and Schultz in the mid eighties. The immune system generates a natural library of hosts, known as antibodies, in response to the introduction of a foreign molecule into the bloodstream. Advances in molecular biology techniques, notably the process of isolating monoclonal antibodies, allow the selection for a chosen antibody library members on the basis of function.

Traditionally, catalytic antibody technology focused on a purist TSA approach: the TSA was designed, synthesized and then used as a hapten in immunization (a hapten is a small molecule attached to a carrier protein which is used to stimulate the immune response). The desired monoclonal antibody was then selected from the polyclonal population on the basis of binding affinity to the TSA. Early efforts produced a range of successes in various synthetic transformations, affording artificial antibody catalysts for ester hydrolysis, the Diels-Alder reaction, cationic cyclizations, cyclopropanation, elimination reactions, the oxy-Cope rearrangement, and an allylicsulfoxide and sulfonate rearrangement, amongst others. However, rate accelerations have always fallen short of their enzyme catalyzed equivalents. Furthermore, detailed mechanistic investigations often revealed that a mechanism other than that originally assumed for the design of the TSA was involved. This has important consequences. Since the selection event is based on binding to the TSA and not on the basis of catalytic activity, the antibody selected may not be the best catalyst. The transition state is after all, not a discrete molecular entity and any TSA can only be expected to be an approximation of the true charge distribution required.

5.4.1. Antibody-catalyzedEnantioselectiveNorrish type II Cyclization

The most successful approach to limiting the broad variety of possible photoproducts has been based on solid-state reactions, in which the crystal-packing forces severely restrict the range of available conformations [44]. However, for syn-The Norrish type II photochemical reaction involves abstraction of a g hydrogen atom by an excited carbonyl oxygen atom (e.g. in 1) to produce a 1,4-diradical intermediate, such as A [45]. The latter can undergo three possible reactions: a) reverse hydrogen transfer to regenerate the ground state of 1; b) C-C bond cleavage to form an alkene 2 and an enol that tautomerizes to the carbonyl compound 3; or c) radical recombination [46] to produce the cyclobutanols 4 and 5 (Scheme 14). Usually, pathway b is the most common route, while c) represents a minor side reaction. Intense mechanistic studies over the past three decades have made the Norrish type II reaction one of the most well understood photochemical reactions [45]. Unfortunately, this important reaction has not yet enjoyed comparable status in synthetic chemistry, mainly as a result of a lack of control over product selectivity and, in particular, stereoselectivity [47].

5.4.2. Catalysis of the Photo-FriesReaction: Antibody-mediatedStabilization of High Energy States

A conformationally constrained hapten is presented that is capable of catalyzing the first antibody-mediated photo-Fries rearrangement have been synthesized [48]. In this reaction, absorption of light energy by a diphenyl ether substrate results in homolytic C-O bond cleavage followed by recombination to yield biphenyl-derived products (Scheme 15). The most proficient antibody studied converts 4-phenoxyanilineinto 2-hydroxy-5-aminobiphenylunder high-intensity irradiation at a rate of 8.6 *m*M/min. These results support a recent hypothesis stating that immunization with conformationally constrained haptens provides higher titers for the acquisition of simple binding antibodies; however, in this case, conformational constraint does not ensure the development of more efficient catalysts. Using the obtained antibodies, the presence of products resulting from escape of free radicals from the solvent cage can be suppressed, altering the excited state energy surface such that free radicals are funneled into the formation of the desired biphenyl product. However, studies also show the inactivation of the antibodies as a result of photo-decay of the biphenyl product. Using an isocyanate scavenging resin, the photo-decay product could be removed and the inactivation of the antibody drastically reduced. Furthermore, despite the observed photo-decay, turnover of the antibody was present; this represents the first case in which true turnover of a photochemical reaction using a catalytic antibody could be observed.

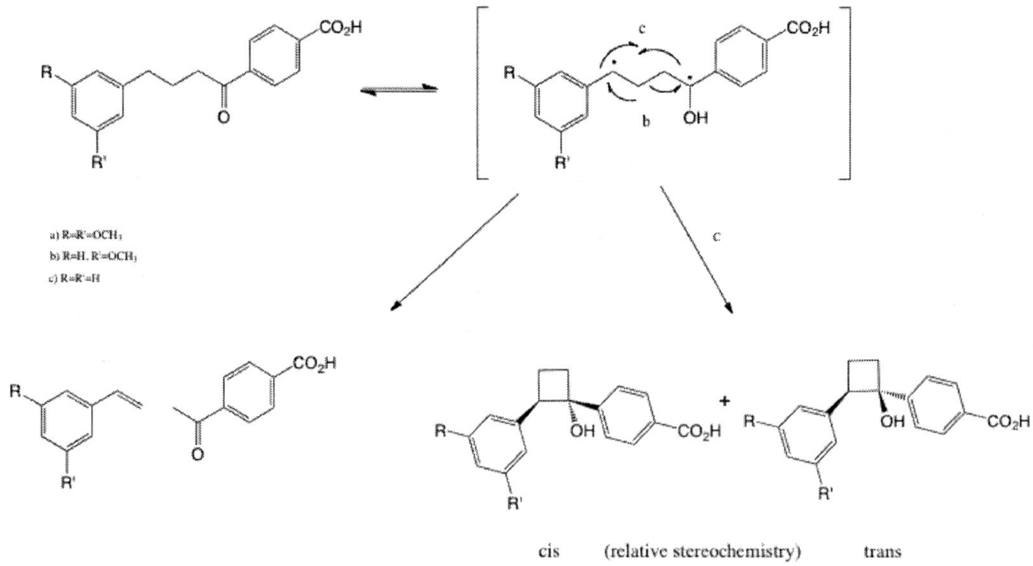

Scheme 14. Norish type II reaction with possible pathways of the 1,4-diradical intermediate.

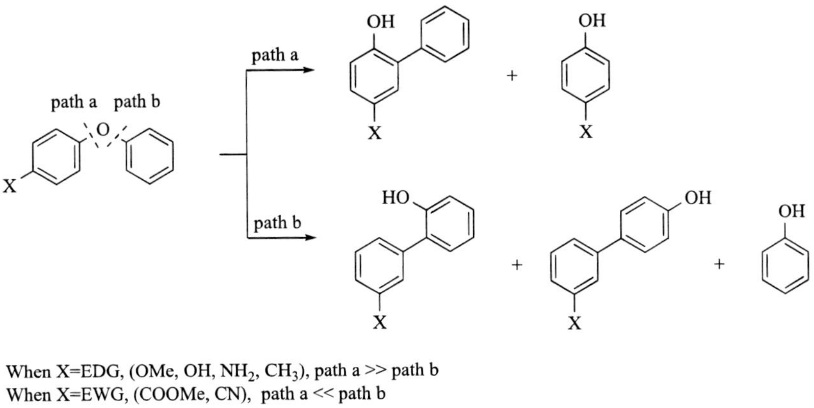
path a path b
O
X
path a
OH
X
+
OH
X
path b
HO
X
+
OH
X
+
OH
When X=EDG, (OMe, OH, NH2, CH3), path a >> path b
When X=EWG, (COOMe, CN), path a << path b

Scheme 15. Photo-Fries rearrangement of phenolic ethers.

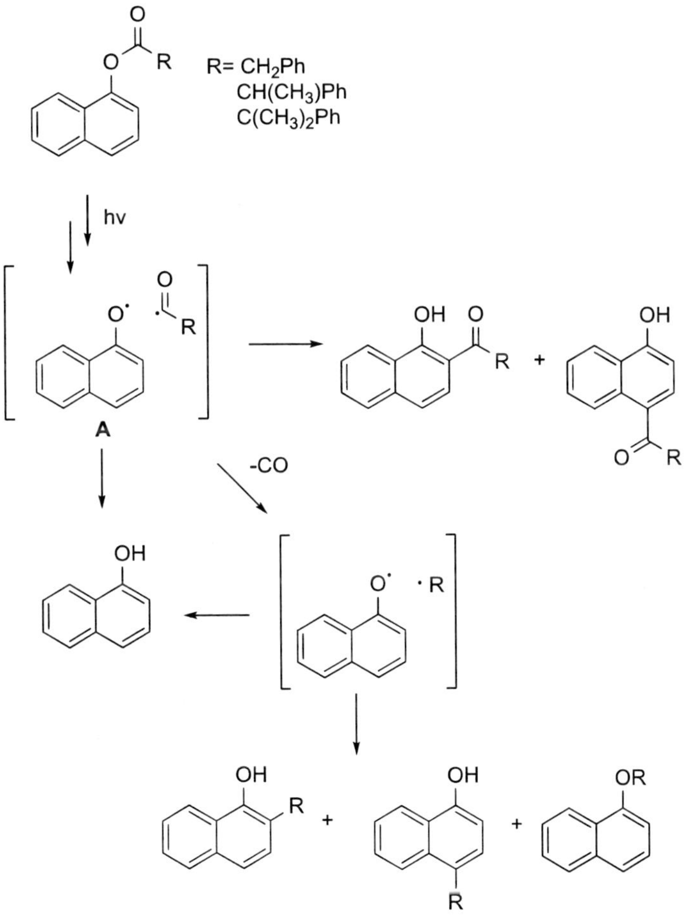
R= CH2Ph
CH(CH3)Ph
C(CH3)2Ph
hv
A
-CO
OH
OR
R

Scheme 16. Photochemistry of phenyl phenylacetates in zeolites.

5.4.3. Photochemistry of PhenylPhenylacetatesIncludedwithinZeolite and Nafion Membrane

The term microreactor refers to organized and constrained media where the chemical reactions occur. The substrate is usually a small molecule of dimensions of several angstroms, and the microreactors often have a size of tens of angstrom or larger. These, microreactors are also known as nanoreactors. Among the many classes of microreactors used in photochemical studies, molecular–sieve Zeolites, Nafion membranes, vesicles, and low-density polyethylene films are outstanding members. Molecular-sieve zeolite represents a unique class of materials [49]. The framework of the materials contains pores, channels, and cages of different dimensions and shapes. The pores and cages can accommodate, selectively according to size/shape, a variety of organic molecules of photochemical interest, and provide restrictions on the motions of the included guest molecules and reaction intermediates.

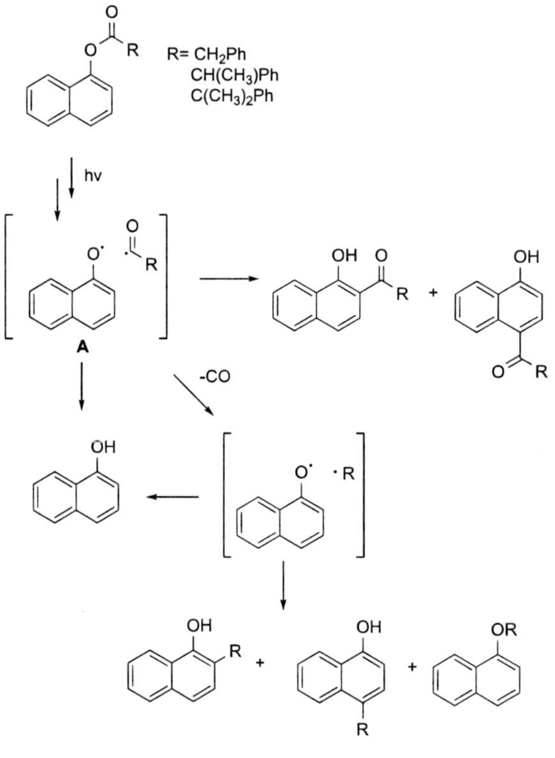

Scheme 17. Photo-Fries rearrangement of 1-naphtyl esters in zeolite compartment.

The photochemistry of the ester in Scheme 16, is expected to be analogues to that of phenyl acetate whose photochemistry in homogenous solutions has been well investigated and understood [50]. Scheme 16 gives the example of photochemical reactions of these esters. Upon photo-irradiation, esters undergo the C-O bond homolytic cleavage to give two paired radicals.These geminate radical pairs in cage recombinate to form the starting ester or *ortho-* and *para-*hydroxyphenones. The later reaction is known as photo-Fries rearrangement. The proposed mechanism is described in Scheme 16. The authors propose that product distribution of the reactions can be directly related with the shape, size and external surface of the zeolites involved, making the transformation unique and potentially *stereo* and *regio*-selective.

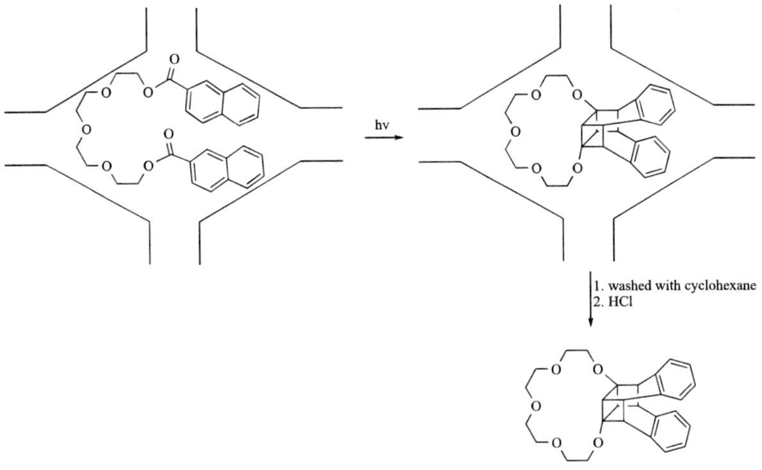

Scheme 18. Irradiation of N-P_n-N in organic solution inside of the Zeolite.

Using the photo-Fries rearrangements of three 1-naphthyl phenylacylates (Scheme 17), Ramamurthy *et. al.,* demonstrate that limiting the constraining space of a reaction cavity in an organized medium, such as zeolite, can be less important than wall-guest interactions in determining the selectivity of the reactions [51]. Reaction cavities of the media employed, cation-exchanged Y-zeolites, and a high density polyethylene film of 71% crystallinity possess very different properties. Irradiation of 1-naphthyl acetate or 1-naphthyl benzoate within alkali metal cation-exchanged X and Y-zeolites give a single photo-Fries photoproduct, 2-acyl-1-naphthol. In contrast, in hexane solution, the 2- and 4- isomers were formed in comparable yields. Selectivity in the zeolites was suggested to result from restrictions imposed on the naphthoxy and acyl radicals by cations along the cavity walls. Support of this observation comes from the work of Freiet. al., that shows that acetyl radicals (from 1-naphthyl acetate) live for $>10^{-5}$ s at room temperature within NaY zeolite [52]. The excellent yields of 2-acetyl-1-naphthol from 1-naphthyl acetate and the long radical lifetimes imply that the radical pairsA (R=methyl) are held tightly in place by cations before rejoining; they have the time and space, *but not the mobility,* to undergo other rearrangements (Scheme 17) [51].

5.4.4. Zeolites and LDPE Films as Hosts for Preparationof Large–ring Compounds: IntramolecularPhotocycloadditionof Diaryl Compounds

The construction of macrocyclic compounds continues to be an important topic of synthetic organic chemistry. A bifunctional molecule may undergo either intramolecular or intermolecular reactions. Intramolecular reactions give rise to macrocyclic ring-closure products, while intermolecular reactions result in dimmers, oligomers or polymers. The rates and therefore product distribution relies heavily on the concentration and the rate of addition of the reagent. Tung et. al., have reported a new approach towards synthesis large-ring compounds in high yields under high substrate concentration [53]. The approach involves microporous solids as templated and host for the cyclization reactions. The size of the micropore has been chosen to permit only one substrate molecule to fit within each. Thus intermolecular reactions are hindered, and cyclization can occur without competition under conditions of high loading with intramolecular ring closure photocyclomers being observed as major products (Scheme 18).

The authors have expanded the scope of the application of the zeolites, nafion membranes and vesicles towards photooxidation of alkenes and found that the photochemical reactions of organic compounds in microreactors usually show deviation of product distribution from their molecular photochemical reactions and, in some cases, result in the occurrence of reaction pathways that are not otherwise observed. The authors attributed this effect to: a) size and shape inclusion selectivities, b) restriction on rotational and translational motions of the included molecules and intermediates imposed by the microreactor, c) compartmentalization of the substrate molecules in the microreactor, and d) separation or close contact of the substrate with the sensitizer in photosensitization reaction.

Conclusion

The field of artificial enzymes is rapidly evolving subject.As the barrier between chemistry and biology becomes less distinct a range of new methods, which combine expertise from both areas, are developing.In recognition of both the fact that the de novo design approach can be time consuming, and that a tiny miscalculation will have a detrimental effect, a trend in all these recent techniques is the use of "selection approaches".The natural processes of selection and amplification is after all, the way in which enzymes have evolved their sophisticated function and also the area is constantly in the desired crossing and eliminating the barrier between "traditional chemistry" and "traditional biological sciences" and developments and understanding of life science at the molecular level.

References

[1] Voet, D.; Voet, J. G. *Biochemistry*; J. Wiley & Sons: Hoboken, NJ, 2004; Copeland, R. A. *Enzymes, a practical introduction to structure, mechanism, and data analysis*, Wiley-VCH: New York, 2000.

[2] (a) Pauling, L. *The nature of forces between large molecules of biological interest, Nature* 1948, *161*, 707; (b) Blow, D. M. *Structure and mechanism of chymotrypsin, Acc. Chem. Res.* 1976, *9*, 145; (c) Blow, D. M.; Birktoft, J. J.; Hartley, B. S. *Role of a buried acid group in the mechanism of action of chymotrypsin, Nature* 1969, *221*, 337; (d) Davis, A. M.; Teague, S. J. *Hydrogen bonding, hydrophobic interactions, and failure of the rigid receptor hypothesis*, *Angew. Chem., Int. Ed.* 1999, *38*, 737; (e) Motherwell, W. B.; Bingham, M. J.; Six, Y. *Recent progress in the design and synthesis of artificial enzymes*,*Tetrahedron* 2001, *57*, 4663.

[3] For leading references see: (a) Breslow, R.; Dong, S.D. *Biomimetic Reactions Catalyzed by Cyclodextrins and Their Derivatives, Chem. Rev.* 1998, *98*, 1997–2011; (b) Avalle, B.; Friboulet, A.; Thomas, D. *J.* Enzymes and abzymes relationships, *Mol. Catal. B: Enzymatic* 2000, *10* 39–45; (c) Wentworth, P.; Janda, K. D. *Catalytic*

antibodies,Curr. Opin. Chem. Biol. 1998, *2*, 138–144; (d) Smithrud, D. B.; Benkovic, S. J., *The state of antibody catalysis,Curr. Opin. Biotechn.* 1997, *8*, 459–466; (e). Kirby, A., *The potential of catalytic antibodies*, J *Acta Chem. Scand.* 1996, *50*, 203–210; (f) Schultz, P. G.; Lerner, R. A., *From molecular diversity to catalysis: lessons from the immune system, Science* 1995, *269*, 1835–1842.

[4] Breslow, R.; Schmuck, C. *Goodness of Fit in Complexes between Substrates and Ribonuclease Mimics: Effects on Binding, Catalytic Rate Constants, and Regio-chemistry,J. Am. Chem. Soc.* 1996, *118*, 6601–6605.

[5] (a) Tramontano, A.; Janda, K. D.; Lerner, R. A., *Catalytic antibodies,Science* 1986, *234*, 1566–1570; (b) Pollack, S. J.; Jacobs, J. W.; Schultz P. G., *Selective Chemical Catalysis by an Antibody, Science* 1986, *234*, 1570–1573.

[6] (a) Gordon, E. M.; Kerwin, J. F. *Combinatorial Chemistry and Molecular Diversity in Drug Design*, John Wiley & Sons (1998); (b) Wilson, S. R.; Czarnik, A. W. *Combinatorial Chemistry: Synthesis and Applications*, John Wiley & Sons (1997); (c) Menger, F. M.; West, C. A.; Ding J. *J. Chem. Soc., Chem. Commun.* 1997, 633–634; (d) Menger, F. M.; Eliseev, A. V.; Mingulin, V. A. *Phosphatase Catalysis Developed via. Combinatorial Organic Chemistry,J. Org. Chem.* 1995, *60*, 6666–6667.

[7] (a) Menger, F. M. *Analysis of ground-State and transition-state effects in enzyme catalysis, Biochemistry* 1992, *31*, 5368–5373; (b) Mader, M. M.; Bartlett, P. A. *Binding Energy and Catalysis:The Implications for Transition-State Analogs and Catalytic Antibodies, Chem. Rev.* 1997, *97*, 1281–1301.

[8] Fersht, A. *Enzyme Structure and Mechanism* (2nd ed.), W. H. Freeman and Co., New York (1985).

[9] Lemieux, R. U., *How water provides the impetus for molecular recognition in aqueous solution, Acc. Chem. Res.* 1996, *29*, 373–380.

[10] Dunitz, J. D., *The Entropic Cost of Bound Water in Crystals and Biomolecules,Science*1994, *264*, 670.

[11] Williams, D. H.; Searle, M. S.; Mackay, J. P.; Gerhard, U.; Maplestone, R. *Toward an estimation of binding constants in aqueous solution: Studies of associations of vancomycin group antibiotics, Proc. Natl. Acad. Sci. USA* 1993, *90*,.1172–1178.

[12] Fersht, A. R.; Shi, J. P. Knill-Jones, J.; Lowe, D. A.; Wilkinson, A. J.; Blow, D. M.; Brick, P.; Carter, P.; Waye, M. M. Y.; *Winter, G., Hydrogen bonding and biological specificity analysed by protein engineering,Nature 1985, 314, 235–238.*

[13] Garcia-Tellado, F.; Goswami, S.; Chang, S.-K.; Geib, S. J.; Hamilton, A.D. *Synthetic analogs of the ristocetin binding site: Neutral, multidentate receptors for carboxylate recognition , J. Am. Chem. Soc.* 1990, *112*, 7393–7394.

[14] Fan, A.; Van Arman, S.; Kincaid, S.; Hamilton, A. D. *J.Molecular recognition: hydrogen-bonding receptors that function in highly competitive solvents, Am. Chem. Soc.* 1993, *115*, 369–370.

[15] (a) Inoue, T.; Weber, C.; Fujishima, A.; Honda, K., *An Investigation of the Power Characteristics in Heterogeneous Electrochemical Photovoltaic Cells for Solar-energy Utilization,Bull. Chem. Soc. Jpn.* 1980, *53*, 334; (b) Komiyama, M.; Hirai, H. *General base catalyses by alpha-cyclodextrin in the hydrolyses of alkyl benzoates,Chem. Lett.* 1980, 1251; (c) Shokat, K.; Uno, T.; Schultz, P. G., *Mechanistic Studies of an Antibody-Catalyzed Elimination Reaction,J. Am. Chem. Soc.* 1994, *116*, 2261; (d) Breslow, R.; Nesnas, N. *Tetrahedron Lett.* 1999, *40*, 3335: (e) Tastan, P.; Akkaya, E.

U., *A novel cyclodextrinhomodimer with dual-mode substrate binding and esterase activity, J. Mol. Catal. A: Chem.* 2000, *157*, 261; (f) Yu, J. X.; Zhao, Y. Z.; Holterman, M. J.; Venton, D. L., *Combinatorial search of substituted β-cyclodextrins for phosphatase-like activity, Bioorg. Med.Chem. Lett.* 1999, *9*, 2705.

[16] Chen, H-L.; Zha, B. Cyclodextrin in artificial enzyme model, rotaxane, and nano-material fabrication, *J. Inclusion Phenom. Macrocyclic Chem.* 2006, *56*, 17–21.

[17] Breslow, R.; Duggen, P. J.; Light, J. P. *Cyclodextrin-B12, a potential enzyme-coenzyme mimic, J. Am. Chem. Soc.* 1992, *114*, 3982-3983.

[18] Jencks, W. P. *Catalysis in chemistry and enzymology;* McGraw-Hill: New York, 1969.

[19] Matsui, J.; Nicholls, I. A.; Karube, I.; Mosbach, K., *Carbon−Carbon Bond Formation Using Substrate Selective Catalytic Polymers Prepared by Molecular Imprinting:An Artificial Class II Aldolase , J. Org. Chem.* 1996, *61*, 5414.

[20] (a) Wulff, G.; Sarhan, A., *On the use of enzyme-analogue-built polymers for racemic resolution, Angew. Chem. Int. Ed. Engl.* 1972, *11*, 341; (b) Wulff, G.; Sarhan, A. German Patent, Offenlegungsschrift DE-A2242796, 1974. *Chem. Abstr.* 1975, *83*, P 60300*w*; US Patent,continuation in part US-A 4127730, 1978.(c) Wulff, G.; Vesper, W.; Grobe-Einsler, R.; Sarhan, A., Enzyme-analogue built polymers, 4. On the synthesis of polymers containing chiral cavities and their use for the resolution of racemates,*Makromol. Chem.* 1977, *178*, 2799-2816;.

[21] (a)Davis, M. E.; Katz, A.; Ahmad, W. R. *Rational Catalyst Design via Imprinted Nanostructured Materials,Chem. Mater.* 1996, *8*,1820-1839; (b) Davis, M. E. *CATTECH* 1997, 19-26.

[22] (a) Wulff, G., *Molecular Imprinting in Cross-Linked Materials with the Aid of Molecular Templates,Angew. Chem., Int. Ed. Engl.* 1995, *34*, 1812-1832; (b) Wulff, G. *CHEMTECH* 1998, *28*, 19-26; (c) Wulff, G. In *Polymeric Reagents and Catalysts*; Ford, W. T., Ed.; ACS Symposium Series; American Chemical Society: Washington, DC, 1986; Vol. 308, pp 186-231; (d) Wulff, G. In*Templated Organic Synthesis*; Diederich, F., Stang, P. J., Eds.; Wiley-VCH: Weinheim, 1999; pp 39-73.

[23] Mosbach, K., *Molecular imprinting,Trends Biochem. Sci.* 1994, *19*, 9-14.

[24] Haupt, K.; Mosbach, K. *Molecularly Imprinted Polymers and Their Use in Biomimetic Sensors, Chem. Rev.* 2000, *100*, 2495-2504.

[25] Mosbach, K.; Ramstrom, O., *The emerging technique of molecular imprinting and its future impact on biotechnology,Biotechnology* 1996, *14*, 163-170.

[26] (a) Tonge, S. R.; Tighe, B. J. Responsive hydrophobically associating polymers: a review of structure and properties, *Adv. Drug Deliv. Rev.* 2001, *53*, 109–122; (b) Alvarez-Lorenzo, C.; Concheiro, A. Molecularly imprinted polymers for drug delivery, *J. Chromatogr., B* 2004, *804*, 231– 245; (c) Langer, R.; Peppas, N. A. Advances in biomaterials, drug delivery, and bionanotechnology, *AIChE J.* 2003, *49*, 2990– 3006; (d) Cui, D. X.; Gao, H. J. Advances and prospects of bionanomaterials, *Biotechnol. Prog.* 2003, *19*, 683– 692.

[27] Suh, J.; Hah, S. S. Organic artificial proteinase with active site comprising three salicylate residues, *J. Am. Chem. Soc.*, 1998, *120*, 10088–10093.

[28] Slade, C. J.; Vulfson, E. N., *Induction of catalytic activity in proteins by lyophilization in the presence of a transition state analogue,Biotechnol. Bioeng.* 1998, *57*, 211.

[29] Peissker, F.; Fischer, L., *Crosslinking of imprinted proteases to maintain a tailor-made substrate selectivity in aqueous solution,Bioorg. Med. Chem.* 1999, *7*, 2231.

[30] Ozawa, S.; Klibanov, A. M., *Myoglobin-catalyzed epoxidation of styrene in organic solvents accelerated by bioimprinting,Biotechnol. Lett.* 2000, *22*, 1269.

[31] Reetz, M. T., *Combinatorial and evolution-based methods in the creation of enantioselective catalysts, Angew. Chem., Int. Ed.* 2001, *40*, 284.

[32] (a) Bornscheuer, U. T.; Altenbuchner, J.; Meyer, H. H., *Directed evolution of an esterase: screening of enzyme libraries based on pH-indicators and a growth assay,Bioorg. Med. Chem.* 1999, 7, 2169; (b) Joo, H.; Lin, Z.; Arnold, F. H., *Laboratory evolution of peroxide-mediated cytochrome P450 hydroxylation, Nature*1999, *399*, 670; (c) Joo, H.; Arisawa, A.; Lin, Z. L.; Arnold, F. H., *Directed evolution of glucose oxidase from Aspergillusniger for ferrocenemethanol-mediated electron transfer, Chem. Biol.*1999, *208*, 699.

[33] Dickey, F. H.,*The Preparation of Specific Adsorbents,Proc. Natl. Acad. Sci. U.S.A.* 1949, *35*, 227-229. Dickey, F. H., *Molecularly imprinted polymers: a new approach to the preparation of functional materials, J. Phys. Chem.* 1955, *59*, 695-707.

[34] Bernhard, S. B., *The preparation of specific adsorbants, J. Am. Chem. Soc.* 1952, *74*, 4946-4947.

[35] Nicholls, I. A.; Andersson, H. S. In *Molecularly Imprinted PolymersMan-Made Mimics of Antibodies and Their Application in Analytical Chemistry*; Sellergren, B., Ed.; Elsevier:Amsterdam, 2001; pp 1-19.

[36] Morihara, K. In *Molecular and Ionic Recognition with Imprinted Polymers*; Bartsch, R. A., Maeda, M., Eds.; ACS SymposiumSeries; American Chemical Society: Washington, DC, 1998; Vol.703, pp 300-313.

[37] (a) Morihara, K.; Kurihara, S.; Suzuki, S. *Bull. Chem. Soc. Jpn.* 1988, *61*, 3991-3998; (b) Morihara, K.; Nishihata, E.; Kojima, M.; Miyazaki, S. *Bull. Chem. Soc. Jpn.* 1988, *61*, 3999-4003; (c) Morihara, K.; Tanaka, E.; Takeuchi, Y.; Miyazaki, K.; Yamamoto,N.; Sagawa, Y.; Kawamoto, E.; Shimada, T. *Bull. Chem. Soc. Jpn.* 1989, *62*, 499-505; (d) Shimada, T.; Nakanishi, K.; Morihara, K. *Bull. Chem. Soc. Jpn.* 1992, *65*, 954-958; (e) Morihara, K.; Kurokawa, M.; Kamata, Y.; Shimada, T. *J. Chem. Soc., Chem. Commun.* 1992, 358-360; (f) Matsuishi, T.; Shimada, T.; Morihara, K. *Chem. Lett.* 1992,1921-1924; (g)Shimada, T.; Kurazono, R.; Morihara, K. *Bull. Chem. Soc. Jpn.* 1993, *66*, 836-840; (h) Morihara, K.; Kawasaki, S.; Kofuji, M.; Shimada, T. *Bull. Chem. Soc. Jpn.* 1993, *66*, 906-913; (i) Morihara, K.; Doi, S.; Takiguchi, M.; Shimada, T. *Bull. Chem. Soc. Jpn.* 1993, *66*, 2977-2982; (j) Morihara, K.; Iijima, T.; Usui, H.; Shimada, T. *Bull. Chem. Soc.Jpn.* 1993, *66*, 3047-3052; (k) Matsuishi, T.; Shimada, T.; Morihara, K. *Bull. Chem. Soc. Jpn.*1994, *67*, 748-756; (l) Shimada, T.; Hirose, R.; Morihara, K. *Bull. Chem. Soc. Jpn.*1994, *67*, 227-235; (m) Morihara, K.; Takiguchi, M.; Shimada, T. *Bull. Chem. Soc. Jpn,* 1994, *67*, 1078-1084.

[38] Heilmann, J.; Maier, W. F. *Angew. Chem., Int. Ed.* 1994, *33*, 471-473.

[39] (a) Ahmad, W. R.; Davis, M. E. *Catal. Lett.* 1996, *40*, 109-114; (b) Maier, W. F.; Mustapha, W. B. *Catal. Lett.* 1997, *46*, 137-140.

[40] Wulff, G.; Heide, B.; Helfmeier, G. *J. Am. Chem. Soc.* 1986, *108*, 1089-1091.

[41] (a) Markowitz, M. A.; Kust, P. R.; Deng, G.; Schoen, P. E.; Dordick, J. S.; Clark, D. S.; Gaber, B. P. *Langmuir* 2000, *16*, 1759-1765; (b) Markowitz, M. A.; Deng, G.; Gaber, B. *Langmuir* 2000, *16*, 6148-6155; (c) Markowitz, M. A.; Kust, P. R.; Klaehn, J.; Deng, G.; Gaber, B. P. *Anal. Chim. Acta*2001, *435*, 177-185.

[42] Dyer, A. *An Introduction to Zeolite Molecular Sieves*, Wiley, New York, 1988.

[43] Strikovsky, A. G.; Kasper, D.; Gru¨n, M.; Green, B. S.; Hradil, J.; Wulff, G. *J. Am. Chem. Soc.* 2000, *122*, 6295-6296

[44] Saphier, S.; Sinha, S. C.; Keinan, E. Antibody-catalyzed enantioselective Norrish type II cyclization. *Angew. Chem. Int. Ed.*, 2003, *42*, 12.

[45] Wagner, P. J. in *CRC Handbook of Organic Photochemistryand Photobiology*, CRC, Boca Raton, FL, 1995, p. 449.

[46] Yang, N. C.; Yang, D. H. *J. Am. Chem. Soc.* 1958, *80*, 2913.

[47] Griesbeck, A. G.; Heckroth, H. *Res. Chem. Intermed.* 1999, *25*, 599.

[48] Dickerson, T. J.; Tremblay, M. R.; Hoffman, T. Z.; Ruiz, D. I.; Janda, K. D. Catalysis of the Photo-Fries reaction: Antibody-mediated stabilization of high energy states, *J. Am. Chem. Soc.*, 2003, *125* , 15395–15401.

[49] (a) Breck, D. W., *Zeolite Molecular Sieves: Structure, Chemistry and Use*, Wiley, New York, 1974; (b) Dyer, A., *An Introduction to Zeolite Molecular Sieves*, Wiley, New York, 1988; (c) Van Bekkum, H.; Flanigen, E. M.; Jansen. J. C., *Introduction to Zeolite Science and Practice*, Elsevier, Amsterdam, 1991.

[50] (a) Tung, C. -H.; Ying. Y. M., Photochemistry of phenyl phenylacetates adsorbed on pentasil and faujasite zeolites, *J. Chem. Soc., Perkin Trans. 2*, 1997, 1319-1322; (b) Tung, C. -H.; Xu, X. H., Selectivity in the photo-Fries reaction of phenyl phenylacetates included in a Nafion membrane, *Tetrahedron Lett*, 1999, *40*, 127-130.

[51] (a) Pitchumani, K.; Warrier, M.; Ramamurthy, V., Remarkable Product selectivity during Photo-Fries and Photo-Claisen rearrangements within zeolites, *J. Am. Chem. Soc*, 1996, *118*, 9428-9429; (b) Gu, W.; Warrier, M.; Ramamurthy, V.; Weiss, R. G., Photo-Fries reactions of 1-naphthyl esters in cation-exchanged zeolite Y and polyethylene media, *J. Am. Chem. Soc*, 1999, *121*, 9467-9468.

[52] (a) Vasenkov, S.; Frei, H., Observation of acetyl radical in a zeolite by time-resolved FT-IR spectroscopy, *J. Am. Chem. Soc*, 1998, *120*, 4031-4032; (b) Vasenkov, S.; Frei, H., Time-resolved study of acetyl radical in zeolite NaY by step-scan FT-IR spectroscopy, *J. Phys. Chem. A*, 2000, *104*, 4327-4332.

[53] (a) Tung, C.-H.; Wu, L. Z.; Yuan, Z.Y.; Su, N., Remarkable product selectivity in photosensitized oxidation of alkenes within Nafion membranes, *J. Am. Chem. Soc*, 1998, *120*, 11874-11879; (b) Tung, C.-H.; Wu, L. Z.; Yuan, Z.Y.; Su, N., Zeolites as templates for preparation of large-ring compounds: Intramolecularphotocycloaddition of diaryl compounds, *J. Am. Chem. Soc*, 1998, *120*, 11594-11602; (c) Tung, C.-H.; Wu, L. Z.; Li, H. R., Reactions of singlet Oxygen with olefins and sterically hindered amine in mixed surfactant vesicles, *J. Am. Chem. Soc*, 2000, *121*, 2446-2451; (d) Tung, C.-H.; Wu, L. Z.; Li, H. R., Vesicle controlled selectivity in photosensitized oxidation of olefins, *Chem. Commun*, 2000, *12*, 1085-1086.

In: Organic Radical Reactions in Water and Alternative Media ISBN 978-1-61209-648-3
Editor: J. Alberto Postigo, pp. 239

Chapter 6

CONCLUSION

This book is intended for the practical synthetic organic chemist whose interest is devoted to studying radical reactions of synthetic utility in water and non-conventional media.

The chapter dedicated to Carbon and Sulfur Centered Radicals in water, Chapter I, intends to cover recent literature, up to 2010, on these types of radicals, with a synthetic goal, thus helping the Radical Synthetic Chemist to grab the fundamental aspects devoted to the synthesis of a wide array of organic compounds through radical methods in water. Some aspects of the effects of water on Radical Chemistry are discussed.

Chapter II is concerned with the chemistry of silicon-centered radicals in water. The attention devoted to silicon reactions in water is not casual, given the need to replace the known reactivity of tin radicals in water as fast reductive agents. Thus, a series of interesting transformations employing silyl radicals were recently uncovered.

As for radical synthetic reactions carried out in water with the aid of metal-centered radicals (Chapter III), emphasis is again made on the types of organic transformations achieved by these metallic radicals. The radical transformations employing metals encompass Reformatzky –type reactions, alkylation and allylations of both carbonyl compounds and imine derivatives, radical conjugate addition reactions, metal-mediated Mannich-type reactions and cyclization reactions, pinacol coupling reactions, metal-mediated reduction and oxidation reactions performed in water, and miscellaneous reactions. The array of metal-centered radicals employed for these latter transformations encompasses main-group elements (Groups XIII, XIV), and several transition metals. This Chapter is presented in terms of the organic transformations that can be accomplished through radical methods in water, employing metals, rather than focusing on the radical transformation accomplished by a certain metallic species. This manner of presentation, useful for the practical synthetic chemist, to the best of our knowledge, has never been described before in a review Chapter.

In Chapter IV, radical reactions performed in alternative media are discussed. These alternative media, supercritical carbon dioxide fluids and radical reactions in ionic liquids broaden the scope of radical reactions for the synthetic chemist.

Chapter V deals with combinatorial polymers and enzyme mimics and other types of catalysis. These state of the art approaches shows a new facet of radical chemistry and assets to the general scope achieved by these intermediates, both for laboratory organic chemists and for biochemists as well.

INDEX

A

B

C

D

E

F

G

H

M

N

O

P

Q

R

S

T

U

V

W

Y

Z